Mathematical Modelling of Immune Response in Infectious Diseases

Mathematics and Its Applications

Volume 395

Mathematical Modelling of Immune Response in Infectious Diseases

by

Guri I. Marchuk

Institute of Numerical Mathematics,
Russian Academy of Sciences,
Moscow, Russia

Springer-Science+Business Media, B.V.

A C.I.P. Catalogue record for this book is available from the Library of Congress.

ISBN 978-90-481-4843-1 ISBN 978-94-015-8798-3 (eBook)
DOI 10.1007/978-94-015-8798-3

The manuscript was translated from Russian by
Dr Guennadi Kontarev and Dr Igor Sidorov.

Printed on acid-free paper

Table of Contents

Author's Preface to the English Edition

Beginning his work on the monograph to be published in English, this author tried to present more or less general notions of the possibilities of mathematics in the new and rapidly developing science of infectious immunology, describing the processes of an organism's defence against antigen invasions.

The results presented in this monograph are based on the construction and application of closed models of immune response to infections which makes it possible to approach problems of optimizing the treatment of chronic and hypertoxic forms of diseases.

The author, being a mathematician, had creative long-lasting contacts with immunologists, geneticist, biologists, and clinicians. As far back as 1976 it resulted in the organization of a special seminar in the Computing Center of Siberian Branch of the USSR Academy of Sciences on mathematical models in immunology. The seminar attracted the attention of a wide circle of leading specialists in various fields of science. All these made it possible to approach, from a more or less united stand point, the construction of models of immune response, the mathematical description of the models, and interpretation of results.

The formation of the author's methodology in the questions of immunology and related theory of mathematical processing of clinical data was influenced essentially by his contacts and discussions with R.V. Petrov, I.B. Pogozhev, Yu.M. Lopukhin, L.D. Sidorova, I.I. Zubikova, E.P. Berbentsova, and many others. The author also had fruitful discussions with R. Mohler, C. Burton, C. Bruny, G. Bell, R. De-Boer, A. Perelson, J-P. Oben, L. Segel, G. Weisbuch, G. Koch, and other colleagues. The author is deeply gratefull to all of them.

My colleagues, who work in the Institute for Numerical Mathematics, Russian Academy of Sciences, were of great help in the research studies on mathematical modelling in immunology and medicine and while preparing this monograph. I would like to indicate the sections, where they made essential contribution and, in point of fact, are coauthors of the corresponding sections of the book: §2.1–2.3, 10.1–10.3 – G. Bocharov; §3.3, 3.3, 3.5 – L. Belykh; §3.4 – A. Asachenkov; §5.1, 7.1, 7.2, 8.1 – G. Bocharov, A. Romanyukha; §5.2 – S. Zuev; §5.3 – I. Pogozhev, R. Usmanov, S. Zuev; §6.1, 6.2 – G. Bocharov, A. Romanyukha; §7.3 – A. Romanyukha, I. Sidorov; §8.2 – A. Karpov, A. Romanyukha; §9.1, 9.2 – D. Kalyaev, S. Zuev.

The author thanks Dr. G.A. Bocharov for scientific editing of the monograph, and Dr. G.R. Kontarev and Dr. I.A. Sidorov for the translation.

G. Marchuk

Introduction

Aspects of an organism's defence against viral and bacterial infections and the reaction of immune system to infection are the main problems in practical immunology. In addition to antiviral and antibacterial defence, the immune system plays a decisive role in tissue incompatibility reactions, antitumour immunity, autoimmune diseases, and allergies.

Understanding of regularities in immune response provides the researchers and clinicians new powerful tools for the stimulation of the immune system in order to increase its efficiency in the struggle against antigen invasion. Such general regularities are revealed, as a rule, on the basis of the analysis of all the main components of an organism's vital activities along with the system of immune defence. In this connection the construction of models of immune response to an antigen irritant seems to be the only right tactics in the cognition of the above regularities; that is why the monograph is dedicated to the analysis of all the facts accumulated in immunology as a united system on the basis of logical concepts and mathematical models.

The task of this author was made easier with the appearance recently in immunology of a series of first-rank books presenting both fundamental facts and general logical conclusions made on the basis of modelling. First of all, these are the books by Benacerraf and Unanue, Burnet, Nossal, Paul (ed.), Roit, Weissman et all; they introduce the reader to modern ideas and conceptions.

Many elements in models are introduced in the form of hypotheses, where the immunology does not yet have any final conceptions. However, all these hypotheses are based, as a rule, on balance relationships, and may, therefore, be used in mathematical experiments. Our experience has shown that many of our initial assumptions proved to be correct with some refinements which are not essential for models.

A few words about the methodology. The reader will notice, that the models discussed in the monograph are, as a rule, multiparametric. The number of parameters grows sharply with the complication of a model. A question arises: is there a sense to discuss complex models while it is impossible to determine many parameters for an individual patient at the contemporary state of medicine?

We believe, that there is a sense by two reasons. On the first hand, such models allow one to penetrate ever more deeply the dynamics of an organism's defence reactions against antigens and reveal the general regularities in the dynamics of disease. On the other hand, complex models state the problems of the identification of their parameters and thus stimulate both the mathematicians and clinicians to search for optimal systems estimating the model's parameters for an individual patient. After all, future medicine is the optimal therapy for an individual patient based on the monitoring of immune, endocrine, and vascular features of a patient with regard to the chronic locuses of various aetiology that are continuously acquired with age.

In order to widen the circle of potential readership, the author begins with the discussion of models of the immune response at logical level without mathematical formalism. The second part deals with the quantitative description of the mechanisms of viral and bacterial infections in a human organism, and with the parameter identification methods for mathematical models. Some conclusions made in the first part are exemplified in the second part by the results of mathematical modelling. This construction of the monograph, as the author believes, will help to assimilate its methodology actively both by specialists-medical men and mathematicians interested in the modeling of complex biological processes.

The diversity of mathematical models of the immune processes reflects complex organization of the immune system. The models of low degree of detailing and large-scale models supplement obviously each other while analysing the regularities of immune reactions. As a rule, the first ones are used for the analysis of *qualitative*, fundamental laws of immune reactions, realized over long time intervals. Models of the second type are constructed, as a rule, for the quantitative organization of conceptions and data and *quantitative* prediction over a concrete time interval.

It is necessary to notice that theoretical studies in immunology on the basis of mathematical approaches are not restricted by the modeling of the dynamics of immune response and immune network. These approaches are also applied to a complex of fundamental interaction processes on cellular and molecular levels investigated with the methods of chemical kinetics and to homeostasis in immune system. There are new fields in immunology where mathematics now intrudes: estimation of B- and T-cells, of the size and structure of receptors, principles of coding of antibody molecules, maturation of the affinity, recognition of self–nonself, and problems of autoimmune diseases.

In its turn the solution of problems in theoretical and experimental immunology stimulates new statements of problems in the divisions of applied mathematics such as functional equations, nonlinear dynamical systems, control and adaptation theory, vitality theory.

On the whole, the studies in theoretical immunology on the basis of mathematical models are considered nowadays as a priority direction in the investigations of complex systems in biological sciences which is supported by the European Science Foundation [11] and the European Society of Mathematical and Theoretical Biology, and implies intensive exchange of ideas between various schools.

PART I

Fundamental Problems in Mathematical Modeling of Infectious Diseases

CHAPTER 1

General Knowledge, Hypotheses, and Problems

This chapter deals with general knowledge in immunology, whose rapid development has inspired new ideas in biology and medicine. The general scheme of an approach is presented for the analysis of mechanisms of infectious diseases that helps understand the major components of processes and is the basis for further investigations.

Discovery of cellular and humoral immunities have led to a new understanding and conceptions of an organism's vital functions as a whole and its defence systems against foreign cellular neoplasmas, as well as against bacteria and viruses that can affect this or another organ of a human being. It was an important discovery of the last decades that the immune reaction of an organism to antigenes does not depend practically on their specificity as related to pathological processes they initiate in the organism. This means that the immune response is connected with the universal character of an organism's defence against bacterial and viral invasions and against the intoxication by products of viral and bacterial vital functions or poisoning by foreign agents of biological nature. Therefore the cognition of the immune response mechanisms provides a key to understanding disease processes and methods of effective medical treatment.

The development of immunology nowadays proceeds so fast that general and special knowledge, being the basis of this science, are changing rapidly, bringing into the conception arsenal still new facts and hypotheses that refine or change radically separate elements of theory. It should be noted, however, that most general characteristics of the immune system have been well studied due to investigations by Burnet [49], Feldmann [99], Doherty [84, 85], Good [119], Nossal [241], Benacerraf and Unanue [28], Petrov [273], Zinkernagel [351–353],

Paul [255], and others; they form a solid base for the construction of models simulating the basic features of the immune processes.

We begin by describing the system of an immune response with the simplest scheme that is rather of principal character since it neglects many (sometimes very important) details. At the same time, this system is open for the inclusion of the most complicated mechanisms at phenomenological level, which requires, of course, separate investigations.

The next chapters contain more detailed schemes of the formation and development of immune response, which will supplement our simple considerations with thorough analysis. There is the hope of building finally a bridge between the phenomenology of processes and their molecular-cellular realization.

1.1. Basic Components of Immune Response

We assume that precursors of immunocompetent cells (lymphocytes and leukocytes) as well as precursors of blood cells are produced in the bone marrow. Let us denote these cells as S cells. A part of these cells getting into the thymus initiates the cloning of T-lymphocytes that proliferate and differentiate into helper T_H-lymphocytes and effector (killer) T_E-lymphocytes (Fig. 1). We do not consider here the suppressor T_S-lymphocytes since the existence of a separate suppressor cell population is questioned, although, as a rule, it is mentioned in the textbooks on immunology.

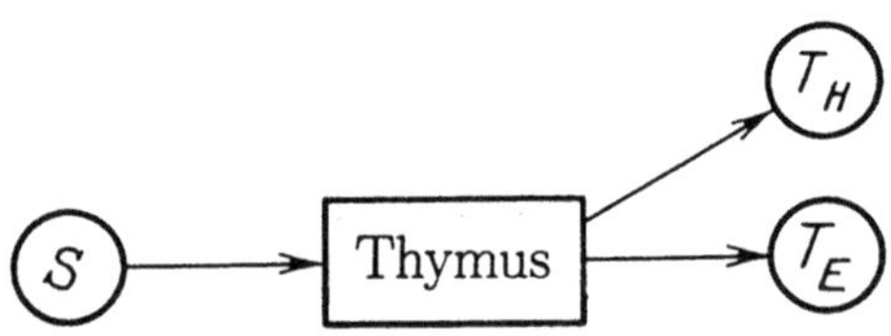

Fig. 1. Differentiation of a stem cell S into T-lymphocytes: T_H-helper-lymphocytes, T_E-effector-lymphocytes.

Another part of S-cells generates precursors of B-lymphocytes, which are transformed into B-lymphocytes in the specific organ called bursa in birds (bursa of Fabricius) and in the bone marrow in mammals (Fig. 2). Part of stem cells remaining in the bone marrow transforms into mature macrophages M (Fig. 3) and other types of peripheral blood leukocytes.

Each of the T–lymphocyte populations performs its own function in immune process. So T_H-helpers interact with specific antigens and enhance the transformation of B-cells into plasma cells. The effector T_E-lymphocytes on the whole are responsible for genetic "purity" of their host organism's cells. They destroy degenerated cells whose genetic structure was changed by mutations or affected by an antigen. B-lymphocytes and macrophages M seem to differ in their various functions. But for the sake of simplicity we shall regard them in our models as homogeneous populations.

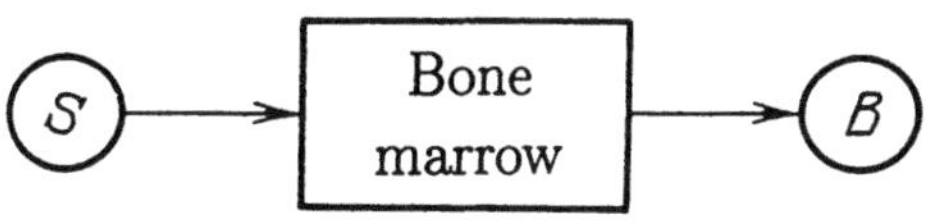

Fig. 2. Differentiation of a stem cell S into B-lymphocytes.

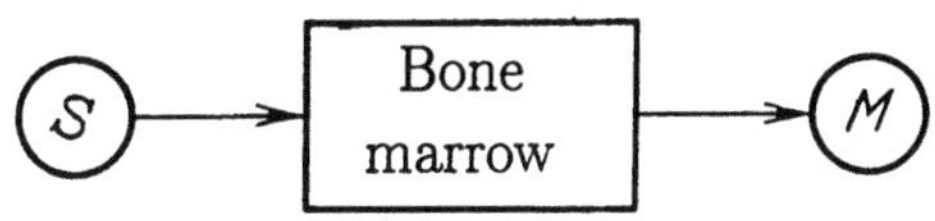

Fig. 3. Differentiation of a stem cell S into macrophages M and other types of peripheral blood leukocytes.

The plasma cells are formed continuously in a healthy organism and produce immunoglobulins IgM, IgG, IgA, IgD, and IgE (antibodies) capable of binding and neutralizing the antigens.

Consider the simplest mechanism of immune reaction to an antigen. Bacterial or viral disease first of all has a latent period of the disease course when the antigens (bacteria, viruses) that have penetrated into an organism proliferate meeting no sufficiently pronounced reaction of immune system. In this period the immune system adjusts to the reaction of neutralization of a specific antigen. The principal scheme of this adjustment was suggested in terms of humoral immune response in the works by Feldmann [99], Miller and Mitchel [220], Petrov [272, 273]. Here is the essence of this adjustment.

The process of generation of antibodies is initiated by joint work of three types of cells: macrophage, B-lymphocyte, and helper T-lymphocyte. It is now established that the macrophage is the major cell that presents an antigen to lymphocytes.

Antigens V, encountering macrophages M, are trapped by them and processed, and then the macrophages form a cluster of antigenic determinants on their surface. This situation is shown schematically in Fig. 4.

Fig. 4. Formation of antigenic determinant cluster on the surface of a macrophage.

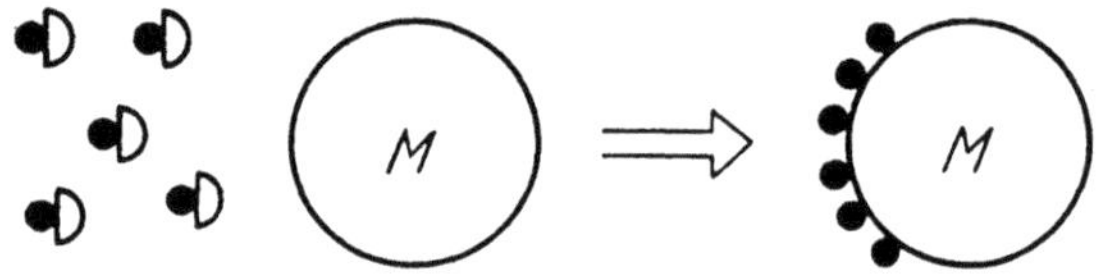

The macrophages M present this cluster to B-lymphocytes which results in binding the antigen determinants V with the surface immunoglobulin receptors of B-lymphocytes (IgMs, IgGs) (Fig. 5).

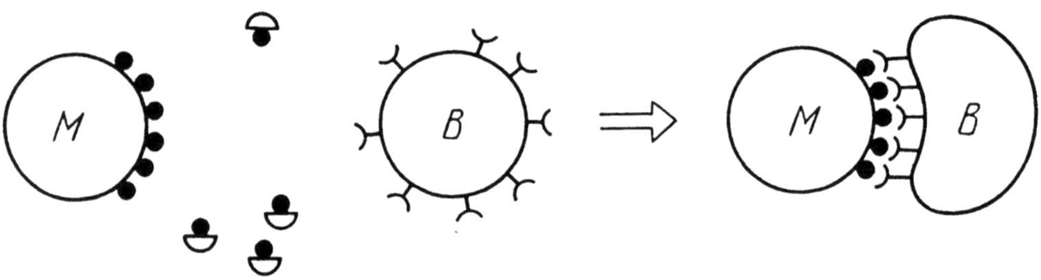

Fig. 5. Macrophages present the determinants to B-lymphocyte.

The presence of T_H−helper cells activated by antigens presented by macrophages in the context of MHC (Major Histocompatibility Complex molecules) stimulates B-lymphocytes which begins to divide and differentiate toward plasma cells (Fig. 6). Such cascade process of formation of plasma cells clones lasts from several hours to several days [241] (Fig. 7).

Thus, in agreement with the simplified scheme presented here, a complex of signals is needed to include B-lymphocytes into the process of antibodies production that is, binding the surface receptors with an antigen with the participation of a T_H-helper. Let us consider this complex in the simplest model of immune response as a single generalized initiating signal determined by the quantity of complexes of an antigen with Ig-receptors of B-lymphocyte, and let us call it VF-complex.

In the case when two or more types of antigens penetrate into human organism, the process of plasma cells cascade formation becomes more complex. Indeed, if we have two types of antigens V_1 and V_2, then two different types of antigen are presented by a macrophage. Then the stimulation of B-lymphocyte seems to occur specific to prevailing antigenes. Usually this stimulation happens in the presence of T_H-helper cell.

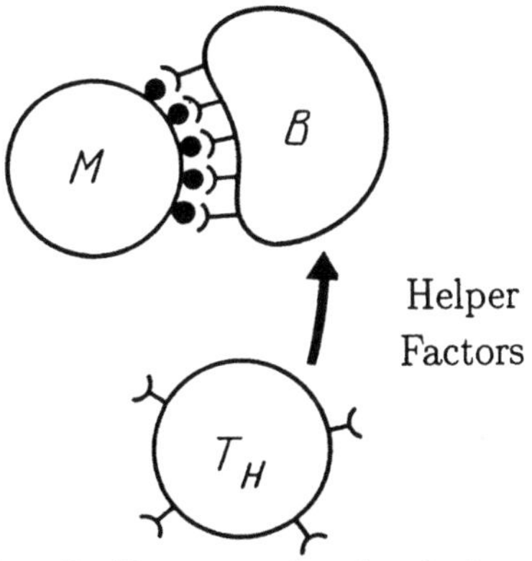

Fig. 6. Process of stimulation of B-lymphocyte.

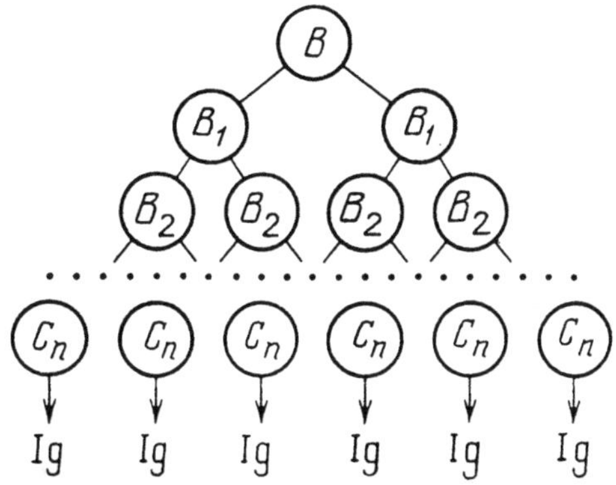

Fig. 7. Cascade process of consecutive divisions and differentiation of B-cells into a clone of plasma cells.

At present there are two basic hypotheses about the functioning of antibody forming cells. The first one suggests that the same plasma cell produces initially IgM and then switches over to IgG production, and so on. Another hypothesis assumes that every cascade produces a strictly determined type of antibodies IgM, IgG, or IgA, etc. This accounts for two different models we may consider. However, both ways can exist simultaneously, as the WHO Commission maintains.

We will assume that if an antigen is "new" for an organism, then plasma cells producing IgM are formed first. Such immunoglobulins have multideterminant basis, and it is "advantageous" for an organism to use them during the initial period, when the immune system is adjusting to struggle with specific antigens. When the optimization of the process is over (it is not yet clear how this happens at the molecular level), intensive production of IgG starts, which are lighter and more mobile than IgM and, therefore, the probability of their collision with a specific antigen is much higher than that for IgM. At this moment the most active immune reaction of the organism to the antigen occurs. When the antigen population is almost suppressed, the synthesis of IgG antibodies switches to the production of IgA antibodies.

Information about the dynamics of immunoglobulins D and E is rather insufficient, so their role in the dynamics of immune response has not yet been defined completely. It is only clear that an increase in IgE level is connected with the appearance of hypersensitivity of immediate type.

We have considered the immune process connected with the formation of antibodies. Such a type of immunity was called humoral; it plays an exceptional role in the organism's vital functions. However, cellular immunity is important too, which is realized by T_E-lymphocytes (effectors) that destroy degenerated cells infected by a virus, or cells foreign to the organism.

Mathematical models that are discussed in the chapters that follow, allow us classify the immune response to foreign antigens according to the following scheme: subclinical form, acute form with recovery, acute form with lethal outcome, chronic form.

1.2. Subclinical Form of Disease

This form of infectious disease is usually latent and is not connected with physiological disorder of an organism. It is usual contact of an organism with a familiar antigen, and the organism has the resources sufficient to suppress the antigen: specific immunoglobulins, effector T_E-lymphocytes, interferon, macrophages, and other components of the immune system. In this case the proliferating population of viruses or bacteria is suppressed by available resources and the antigen is destroyed before it reaches the concentration level that provokes noticeable immune and physiological reactions of the organism. The pathogen population dynamics in this case is shown schematically in Fig. 8.

Since an organism contacts numerous antigen irritants every day, the immune process of struggling against them occurs in the subclinical form. It is one of most remarkable features of the immune system.

Subclinical forms of disease are accompanied by some increase in production of IgG and IgA with no considerable increase in IgM. Such a situation happens usually in the case of recurring contacts with the antigen that is already "familiar" to the memory T- and B-cells.

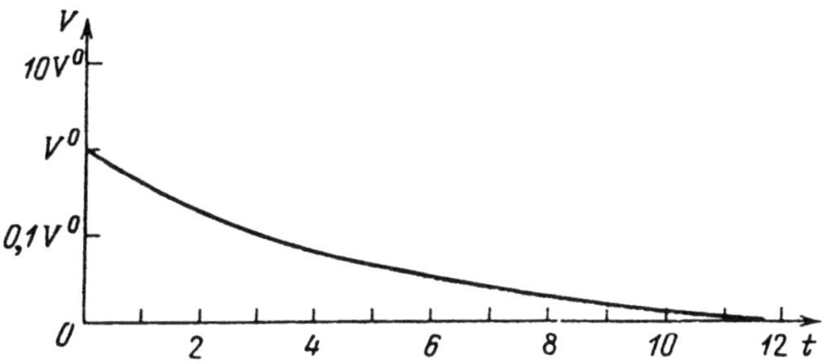

Fig. 8. Antigen concentration dynamics in case of subclinical form of disease.

1.3. Acute Form of Disease

If an antigen that has penetrated into an organism is found to be "unfamiliar", then the increase in its concentration due to proliferation is accompanied by the process of recognition and formation of plasma cells producing IgM. This process usually proves to be slow (it lasts for several days), and during this time the antigen concentration attains the levels that exceed the level of appreciable physiological and pathological changes. This is the case of the normal acute form of a disease's shown schematically in Fig. 9. It is a classical form of disease course with a rise in temperature, intoxication of the organism, with considerable or medium pathological changes in the affected organ.

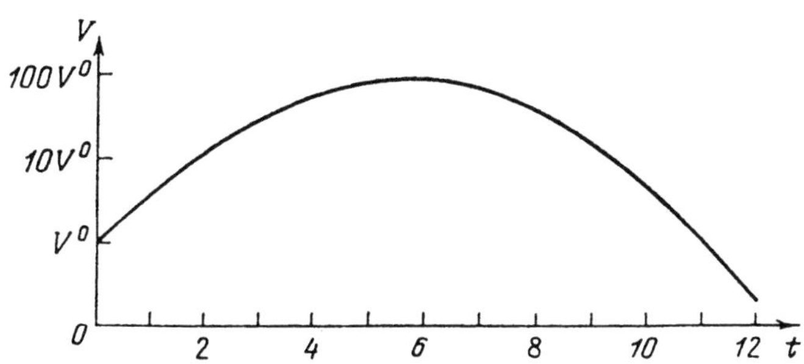

Fig. 9. Antigen concentration dynamics in the case of the acute form of disease.

Finally, the acute form of a disease with lethal outcome is possible. Intoxication of an organism by toxic products of viral or bacterial activity is possible, but there is one significant case when the outcome can be unfavorable for an organism. This happens when immune rĕsponse has been delayed for some reason so that considerable pathological changes had occurred in an organ under the influence of viruses or bacteria, and the affected organ can no longer provide normal efficiency of the organs responsible for the formation of immuno-valued components (T- and B-lymphocytes, leukocytes, interferon, complement, etc.). A serious form of disease with possible lethal outcome occurs in this case.

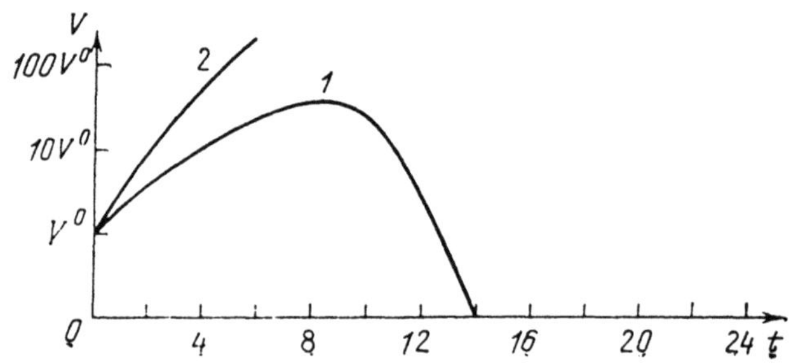

Fig. 10. Dynamics of antigen concentration: 1, case of acute form of disease with recovery; 2, case of lethal outcome.

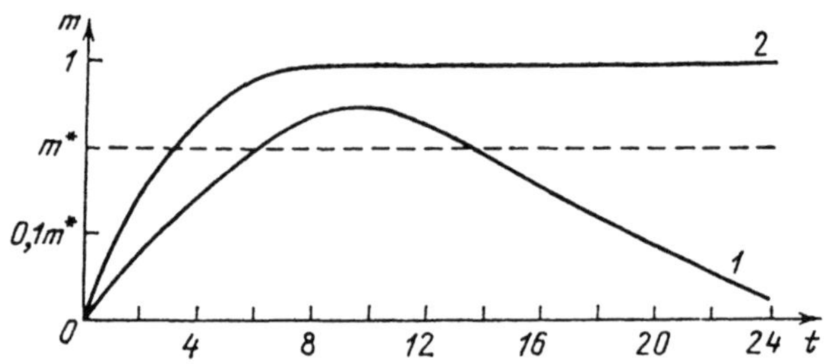

Fig. 11. Changes of relative quantity of damaged part of a target organ in case of acute form of disease: 1, recovery; 2, lethal outcome.

Curve 1 in Fig. 10 characterizes the acute form of a disease with recovery, curve 2 characterizes a lethal outcome. Fig. 11 shows graphically the changes of relative characteristic of damage of an organ corresponding to cases 1 and 2. Here, m^* is a threshold value of relative damage of an organ. As soon as this value is reached, normal activity of this organ is impossible.

1.4. Chronic Form of Disease

Let us proceed now with the immunological interpretation of chronic diseases. It is generally known that chronic diseases are the most serious and exhausting forms, which often last for many years. Medicine in many cases is still unable to cure these diseases, whereas their range, unfortunately, tends to grow. Consider the principal scheme of the origin of chronic disease.

Mathematical models reveal that chronic diseases represent the stable form of an immune process and have either almost cyclic or time-independent dynamics (Fig. 12). In the case of chronic disease the antigen concentration tends to a certain value $V^* > 0$ rather than to zero. During the chronic process the IgA (possibly, IgD) play special role. As for IgM and IgG, they usually remain within the limits of their normal level.

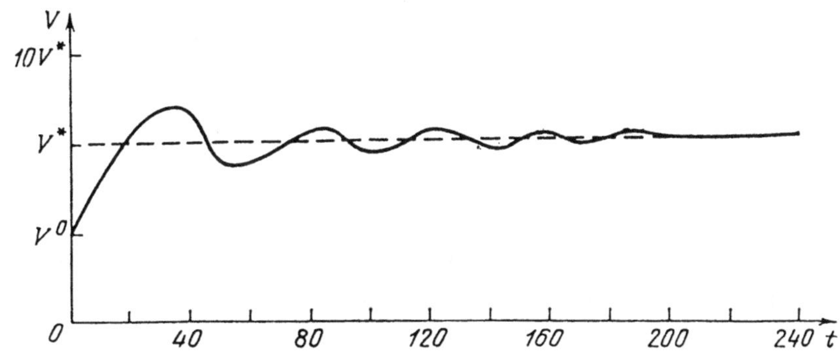

Fig. 12. Antigen concentration dynamics during chronic disease.

Stationary asymptotic level of antigens V^* and antibodies (for example, IgA) is achieved due to equilibrium in the processes of neutralization and generation of viruses or bacteria occurring every second.[1] A certain number of antibodies continuously produced by an organism are spent to destroy the antigens. Thus an equilibrium between the antigens and all components participating in the immune process is established in an organism. A very stable state of disease classified as the chronic form of the disease thus appears; the only way of escape

[1]Apart from specific neutralization of antigens by antibodies, nonspecific systems and processes (interferon, proteolytic enzymes, nonspecific phagocytosis, etc.) destroying viruses, bacteria, and antibodies affect their concentration.

from this state is complete suppression of the entire antigen population, either with the help of some medical drugs, or with the help of the immune system itself with subsequent self-healing of organism.

Consider the first way – passive treatment with drugs injected into a patient's organism. This is the conventional approach to the treatment of chronic disease. Medicine today has a large arsenal of antiviral and antibacterial drugs that can be injected into an affected organ directly; they allow us in many cases to get rid of populations of pathogenic viruses and bacteria. However, when viruses or bacteria of a certain kind have been removed from the organism completely, after some remission the relapse of the same disease can occur again.

Indeed, having cleared the organism of the pathogenic viruses and bacteria, we have not rehabilitated its immune resistance. The exacerbation of chronic process is usually caused by new disease, when the defence systems of chronic loci are weakened. Furthermore, intensive antibiotic therapy usually leads to suppression of the immune system, which is weak enough as it is. This makes it easier for the process to relapse. Therefore, the repeated antiviral or antibacterial therapy does not usually guarantee final remission from chronic disease, although it may be successful in some cases.

It seems that effective therapy of chronic disease must combine methods that activate the immune system, with antiviral and antibacterial therapy (Ermol'eva, [92]).

Consider the problem of immunodeficiency acquired as a result of chronic disease. It seems that one may present the following assumption. For many years of chronic disease the immune system of an organism is directing a part of its resources (antibodies) to the neutralization of antigen population. One of these resources is the production of antibodies of class IgA carrying out the immunologic surveillance over the mucosal systems during the quiet course of chronic disease. Since some part of IgA is constantly used in the struggle with antigens, their concentration level in blood is lower than normal one. This new level becomes a new norm of IgA in the organism and this norm remains unchanged after the elimination of pathogenic viruses or bacteria (the source of the chronic disease) from the organism. Since this level happens to be lower than the normal one, the organism's resistance against these infestants is lowered and, as a result, the organism can be exposed to a relapse of the chronic disease. This means that after complete elimination of pathogenic sources of chronic disease from the organism the immune components must be stimulated up to the level corresponding to the norm. In this case one can consider the chronic disease as completely cured.

1.5. Estimation of Disease Severity

A series of methodologies is justified in literature on the processing of medical data for diagnosis. We assume here that a diagnosis has been made correctly and the task is to estimate the disease's severity and control its dynamics.

This is important in the case of infectious diseases, since during epidemics they demand simultaneous routine work with a large number of patients.

A quantitative estimate is also necessary while establishing the severity of chronic diseases, especially in the case when new infectious diseases happen against the background of other infectious diseases, for example, respiratory virus infection and pneumonia aggravating the severity of chronic bronchitis.

All this stimulates the development of methods for the objective estimate of the disease's severity based on generalized expert data of qualified physicians and results of laboratory analyses. We had decided, when beginning this work, to have two values: clinical and laboratory indices of severity.

Let us start with examining the general approach to the construction of a severity index. Let it be a clinical index denoted by Πc and calculated on the basis of estimations for the manifestation of symptoms. It is necessary, in order to construct the clinical severity index, to define a basis of indicators (symptoms) whose manifestation is taken into account by a physician while estimating the disease's severity. Let us designate the manifestation of each separate symptom by P_i and select m $(i = 1, 2, \ldots, m)$ such indicators. Introduce one more index j

in order to mark that the indicators belong to a certain patient; if a sampling of patients with the same clinical disease severity contains n patients, then $j = 1, 2, \ldots, n$. The manifestation of a symptom P_i for the patient numbered j can be written as $P_i(j)$. Find the average value for each of the symptoms of a given sampling to obtain

$$P_i = \frac{1}{n} \sum_{j=1}^{n} P_i(j). \tag{1.5.1}$$

The averaged severity index Πc for the whole sampling of m patients is representable in the form

$$\Pi c = \sum_{i=1}^{m} \alpha_i P_i. \tag{1.5.2}$$

We shall use this index in order to estimate the severity of a disease.

While using formula (1.5.2) it is still necessary to find the unknown coefficients α_i. They can be found from the condition of the best coincidence for values $\Pi c(j)$ and $\sum_{i=1}^{m} \alpha_i P_i(j)$, that is,

$$\sum_{j=1}^{n} \left[\Pi c(j) - \sum_{i=1}^{m} \alpha_i P_i(j) \right]^2 = \min. \tag{1.5.3}$$

Here $\Pi c(j)$, the disease severity of patient number j, is determined by qualified physicians and $P_i(j)$ is, as was noted above, the manifestation of symptoms for this patient.

The problem (1.5.3) of finding α_i is solved with well known methods. As a result, all unknown coefficients α_i are found and we can use the generalized index of clinical severity (1.5.2) for the estimation of disease severity for each patient from the whole sampling.

Now take this formula and notice one of its peculiarities. This formula yields not only integer numbers 0, 1, 2, 3 that correspond respectively to a healthy person ($\Pi c = 0$), a light form of disease ($\Pi c = 1$), middle ($\Pi c = 2$), severe ($\Pi c = 3$), and severe with unpredictable outcome ($\Pi c = 4$ and greater). Consequently, this formula gives a more detailed estimate of disease severity compared to the conventional integer scale used in clinical practice. This is very important while studying and observing the dynamics of disease.

They use formula (1.5.1) in practice as follows. All symptoms P_i ($i = 1, 2, \ldots, m$) are estimated for a particular (j)th patient in a mark system. As a result an estimate is obtained in differential form. If the obtained value Πc is rounded off to an integer we get marks for disease severity. If rounding off is performed to the closest integer or to a half integer, we get not only four major forms of severity, but transitional ones too. The values of $\Pi c(j)$ with no rounding off are the most interesting. It is efficient in this case to use a graphical representation of the function $\Pi c(j)$ with respect to time.

Conditionally we have divided all indicators into clinical and laboratory ones. For example, the temperature reaction of an organism to disease is of explicitly quantitative character, though, as a matter of fact, the temperature is practically always regarded as the general clinical indicator. Therefore the construction of correct and independent sets of clinical and laboratory indicators is creative in character, which demands the participation of experienced physicians.

For most of acute infectious viral and bacterial diseases general indicators are the most informative, such as temperature, shivering, headache, etc. These indicators usually reflect the general reaction of an organism to antigens and intoxication by the metabolism products of bacteria and cells infected by viruses. Specific (or local) symptoms are also essential in inflammation processes; they determine, in fact, the severity estimation. Obviously a set of symptoms for one acute disease differs from a set of symptoms for another acute disease, therefore it is important to establish an optimal set of symptoms.

Chronic diseases have those or other sets of specific symptoms. In many cases they are crucial in the estimation of disease severity. At the same time, for chronic diseases, when an organism and its immune system are adjusted to cope with a permanent antigen population, general indicators may be not informative any more. It is possible, for a given disease, to form a listing and estimation in marks of clinical symptoms connected with certain viral or bacterial disease.

The following general gradation was used for the manifestation of clinical symptoms: 0 for no symptom; 1 fora symptom expressed weakly; 2 for a moderate symptom; 3 for sharply expressed symptom.

Let us introduce a laboratory index Πl. To this end, use formula (1.5.1) again and select, in the case of acute pneumonia, the following laboratory indicators: concentration of fibrinogen in blood, seromucoid, ESR, leukocytes, neutrophiles, lymphocytes, C-reactive protein, and pulse rate. Coefficients for Πl were obtained as a result of processing the individual data with the method of least squares. Here is the formula for the quantitative estimation of Πl [197]:

$$\begin{aligned}\Pi l \; = \; & \frac{F-2.50}{4.50}+\frac{C-0.20}{0.30}+\frac{S-16}{30}+\frac{L-5.0\cdot 10^{9}\mathrm{l}^{-1}}{20.0\cdot 10^{9}\mathrm{l}^{-1}} \\ & +\frac{Fr-70}{90}+\frac{N-1}{40}+\frac{30-l}{50}+\frac{Cr}{20}.\end{aligned}$$

The variables mean here: F, fibrinogen concentration in blood (in g/l); C, seromucoid concentration in blood (in g/l); Cr, concentration of C-reactive protein (quantity of "crosses" by the method of sedimentation in capillaries); S, precipitation rate for erythrocytes (mm/h); L, number of leukocytes in blood (10^9/l); l, percentage content of lymphocytes in blood; Fr, pulse rate per minute; N, neutrophiles (in %).

Calculations of Πl for 600 patients with various forms of acute pneumonia have revealed that $1 < \Pi l < 3.5$ in acute site pneumonia connected with various bacterial flora and viruses. For patients with complicated pneumonia Πl was always higher than 4.5 and could reach 12 marks.

These investigations have shown that Πl is universal in character and does not depend on the aetiology of disease; it characterizes the general activity of the inflammation process in all infection sites in an organism. It means that if acute pneumonia is the only bacterial disease for a patient, then the inflammation index coincides with Πl of acute pneumonia. If acute pneumonia develops in the background of another active bacterial process, then Πl expresses general severity of all inflammation bacterial processes.

Fig. 13 shows an example of Пc and Пl dynamics during the treatment of chronic bronchitis.

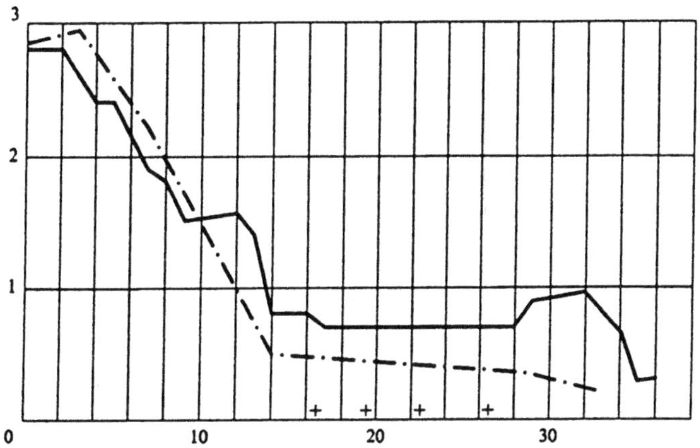

Fig. 13. Clinical observation [198]. Patient C., 38. Dynamics of clinical (Пc, solid line) and laboratory (Пl) severity indices.

The results of processing the clinical data for several of infectious diseases (hepatitis, chronic bronchitis, acute pneumonia, dysentery) made it possible to construct a quantitative representation of the typical course for infectious diseases of various aetiology, severity, and outcome; to estimate quantitatively the influence of treatment, initial state of a patient, and other factors on the course of disease. These results were used while constructing mathematical models discussed below.

1.6. General Scheme of Infectious Disease

Fig. 14 summarizes the general diagram of typical forms of infectious disease. Curves 1 characterize the diseases in latent form. An antigen is removed from an organism due to high normal level of antibodies specific to this antigen. Curves 2 characterize normal acute process of disease; curves 3 characterize the disease with lethal outcome; curve 4 characterizes chronic form of disease.

In some cases the antigens are blocked in an organism so "closely", for various reasons, that the process of disease is extremely slowed down and does not manifest itself in the indices characterizing functional or pathological changes in the organism. This state may last for months or years. A person seems to be absolutely healthy. However, if a patient is infected with viruses or bacteria of another kind during this period which results in the acute form of the viral or bacterial disease, then after some time new, more complicated form of disease arises, which can be even more dangerous. This situation occurs, for instance, during an influenza epidemic. Schematically, a complicated disease process in this case can be described as follows.

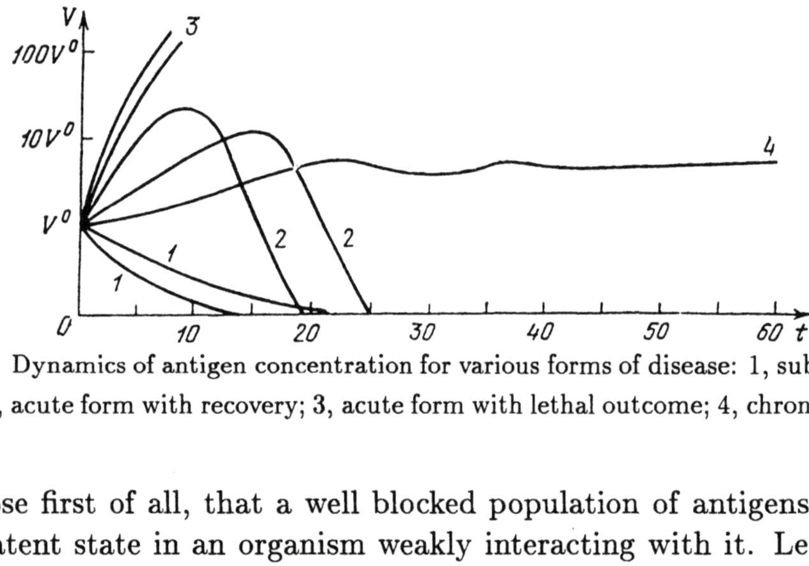

Fig. 14. Dynamics of antigen concentration for various forms of disease: 1, subclinical form; 2, acute form with recovery; 3, acute form with lethal outcome; 4, chronic form.

Suppose first of all, that a well blocked population of antigens exists in a latent state in an organism weakly interacting with it. Let it be staphylococci, for example. The organism produces constantly specific antibodies IgG and IgA and other necessary components of defence in order to block the staphylococci activity. Influenza viruses that have penetrated into respiratory tracts and have overcome the intracellular defence mechanisms, exhaust and affect a target organ using cellular genetic apparatus for the reproduction of their own population. When the process becomes acute the immune system of the organism is being switched on; it forms the mechanisms of defence, such as specific immunoglobulins and T_E-effector-lymphocytes destroying the infected cells. As a result the immune system adjusts to the struggle with the new infestant and immune surveillance over the latent staphylococcus is weakened. Thus, favourable conditions are created for intensive propagation of staphylococcus.

However, being involved in the struggle with the main enemy (influenza viruses) the organism does not notice the increase in staphylococci concentration for a long time, since the concentration of staphylococcus antigen is small in comparison with concentration of antigens of influenza viruses (see Section 1.1). Finally, the staphylococci that are left without appropriate immune control start propagating vigorously and affect the lungs. After that the organism seems to switch on some of its resources and later, after having destroyed the influenza viruses, turns all resources to developing the defence means against the staphylococci. This enemy latent till now becomes the main one for the organism. But this is a beginning of a complication, which, as we see, has been programmed a priori by the very logic of events. A clear boundary between latent and chronic forms of disease does not exist, therefore the above considerations are equally applicable to a quiet period of chronic disease.

One more possibility for complications of disease to appear must be noticed. Gavrilov, Zhdanov, and Semenov emphasize in [111] that a viral infection sometimes provokes an immunodepression related to the antigen introduced into the organism after the viral infection has occurred. This immunodepression is likely to be one of the reasons for disease complications.

1.7. Immunological Models of Infectious Diseases

When we speak about a model of immune response we have in mind the defence mechanism of an organism against antigen invasion with various degree of detailing. As for mathematical models, they are an adequate mapping of immunological models based on theoretical and experimental conceptions on the defence system of organism. For example, while considering the simplest model of immune response we investigate the interaction of only four components of the system: antigen (virus or bacterium), antibody, plasma cell producing antibodies, and quantitative characteristics of affected organ. A system of these indices allows us in principle to investigate theoretically only general features of the human defence reaction, and an antibody in this system is just a generalized element of the immune defence, no matter whether it is a T_E-effector-lymphocyte or immunoglobulin. This system becomes a model only in the case when the connections between all four components are established on the basis of immunological conceptions: between antigen, antibody, plasma cell, and scale of damage to an organ. These connections are established by the immunology.

Indeed, in designing the simplest model of immune defence we have used the following conceptions of immunology. First, an antibody binds an antigen forming antibody–antigen complexes. In proportion to the quantity of these complexes, plasma cells are formed in an organism in a time τ which carry out the mass production of antibodies. The quantity of plasma cells forming in response to antigenic stimulation depends on the viability of the affected organ: the more severe is the damage to this organ the less is the quantity of plasma cells because of the deficiency arising that affects the immune defence activity. It is seen clearly that many details are missing in this model; however, all the essential components of the immune defence mechanism are taken into account.

Detailed investigation of the defence mechanism on the bases of immunology conceptions requires the introduction of more differentiated components, such as T- and B-lymphocytes, immunoglobulins IgM, IgG, and IgA, macrophages, system of plasma cells producing the immunoglobulins of all three types, mechanism of T- and B-lymphocyte proliferation, etc. And again, this system becomes a model only when all essential quantitative bonds between all parts of the system are determined. We have in mind these very models while investigating the defence mechanism of an organism against the antigens.

The simplest immunological model of viral or bacterial disease described above includes just most general laws of reaction of an organism's immune system to antigen invasion. In real conditions, of course, this reaction is more complex.

It should be noticed that the system of cellular and humoral immunity is in fact more diverse. For instance, B-lymphocytes binding with a complement become killers of cells infected by viruses. Notice that macrophages are also able to kill the cells infected by viruses. Therefore it is convenient while investigating the cellular defence mechanism to consider just killers and macrophages. Their main task is to destruct the cells infected by viruses and excrete the decomposition products. The ability to destruct the self cells infected by viruses which helps to stop the virus multiplication process is a really surprising property of an organism's system of defence.

All that occurs against the background of a continuous process of recognition of self cells based on the major hystocompatibility complex system (Zinkernagel [351], Benacerraf and Unanue [28]). This system uses the fact that all organisms' cells, including macrophages, lymphocytes, etc., are equipped with receptors that serve to distinguish a self cell from an alien one. It is a splendid invention of nature. Furthermore, the immune system performs so called double recognition during the interaction of T_H-helper-lymphocytes with macrophages, which helps to avoid mistakes while recognizing a foreign antigen – the main enemy of an organism. Just after such, so to say insurance, the immune system starts the proliferation of necessary cells and humoral antibodies.

1.7.1. Immunological model of viral infection. Let us begin with a viral antigen as the most insidious for an organism. Having got into an organism, a virus finds the cells of sensitive tissue and infects them, overcoming the cellular defence mechanism consisting of intracellular interferon and other means of counteraction (Burnet [49]). As soon as all "obstacles" are overcome the virus starts to reproduce itself using the genetic resources of the cell. As a result the cell becomes exhausted and dies, and viruses move out of the cell into extracellular space or blood, finding new target cells where they continue to develop their population.

Thus an organism has two viral enemies: the virus in a cell, and the virus outside the cells in plasma. In order to destroy the viruses in cells the immune system uses T-effectors and intracellular interferon; the T-effector also kills the organism's infected cells. It is a very powerful weapon, since hundreds and thousands of viruses capable of infecting new cells come out into extracellular space or plasma from each infected cell. That is why it is better for an organism to destroy viruses within the cell even if it is its own cell that must be destroyed. But if some of the viruses get into extracellular space or plasma, they are neutralized by antibodies which cannot get into a cell but defend their organism outside the cells. Thus here is another reliable defence against viruses. But this barrier proves to be reliable just in the case when an antigen is recognized fast enough and the immune system has no primary or secondary immunodeficiencies. Keeping in mind these two mechanisms of immunity, we shall try to analyse the forms of disease gravity and disease outcomes.

We will not analyse the subclinical form of a viral disease, since in this case the system of cellular and humoral immunity manages to cope with penetrating antigens cloning no new plasma cells and using available reserve of specific components of immune system. Let us consider more serious cases of acute viral diseases, when pathological changes occur in organs which are infected by this type of viruses. Three forms of pathological process severity are possible: mild, average, and serious one. In immunological aspects these three forms are connected by a scale of viral damage of a target organ.

When the immune system functions normally, the mild form is the usual form of disease. The damage to an organism's cells is usually minimal, since it takes about one day to tune the immune system for the formation of clone of plasma cells producing specific antibodies; in two or three days there are enough antibodies to suppress the whole virus population in plasma. The viruses which are reproduced in cells are destroyed by the killer system. Thus in case of the mild form of the disease the virus population is destroyed in two or three days, and then affected cells of an organ regenerate gradually. That takes another few days. As a result the patients with a mild form of influenza or another viral infection recover completely in about one week.

Now consider an average form of disease. In this case, due to various reasons, plasma cells cloning in an organism is delayed, and viruses, meeting no proper resistance from the immune system, infect a considerable part of the target organ. The redoubling time of a virus population when there is no immune defence is about several dozens of minutes, and it is natural that the delay in the formation of specific clones even for several hours leads to a situation when the viruses affect a considerable part of an organ. This means that the destruction of viruses in the organism is delayed, but the delay is insignificant: about one or two days. However, it takes about two or three weeks for the affected organ to recover, which depends on the virulence of the viral infection and the scale of damage to the target organ.

A serious form of a disease is usually characteristic of patients with primary or secondary immunodeficiencies of cellular or humoral types. In both cases the delay in synthesis of antibodies occurs against the background of initially low levels of immunocompetent cells and antibodies. This creates conditions for considerable damage to target organ cells and for acute reaction of destruction by killers of own cells infected by viruses. As a result of the immune system's prolonged struggle against viruses, the destruction of viruses takes place, followed by a long period of recovery. Active process of elimination may last for one or two weeks, whereas a month or more is required for seriously damaged organs to recover. In certain cases, however, a chronic form of the disease develops, when homeostasis changes and the patient becomes a virus carrier.

1.7.2. Hypertoxic form of viral disease. Together with the above forms of disease, there is one more hypertoxic or malignant form of disease with unpredictable outcome which is thought to be the most dangerous. This particular form takes the lives of a considerable number of patients during influenza epidemics. This case of the serious form of sudden disease is connected with extensive viral damage of cells caused, as a rule, by too slow a formation of clones of immunocompetent cells. If, for instance, the target organ for viruses is a lung, the infection of hypertoxic form leads to general lung oedema and oxygen starvation. As a result, the oxygen supply of lymphatic tissue drops sharply, and consequently the production of T-lymphocytes and antibodies necessary for the organism's defence drops too. A lethal outcome is almost inevitable, unless special measures are taken.

Since the treatment of the hypertoxic form of viral disease is extremely important, we will discuss this problem in detail, emphasizing the main principles of the therapy that are used here. These methods were worked out by Marchuk, Berbentsova, and Astakhova in 1980.

We had the following considerations to begin with. In the case of the hypertoxic form the viruses infect many cells of a target organ very rapidly due to the delay of formation of plasma cells' clones. Destroying the cells, viruses penetrate into intracellular space and blood and infect new cells. Having no proper opposition on the part of humoral immunity because of the delay in antibodies production, viruses affect ever larger numbers of cells of vitally important organs which results in a dangerous deficiency in the activity of organism.

First of all one should, in this critical situation, weaken the virus' aggression for the time needed for the clones' formation to be completed and for the mass of antibodies to be produced. This usually takes one or two days. It has been noticed before that the rise in temperature of a body depresses the pathological activity of viruses. But in the case of this form of the infectious process the temperature, as a rule, is already critically high, and it should not be decreased by febrifuges excepting for emergency. This means that this factor of virus depression is switched on by the organism itself. One more remedy still remains: the glucocorticosteroid hormones depressing the virus reproduction.

Although the glucocorticosteroid hormones depress the immune system too, during such a short period (one or two days) this influence does not become dangerous for the plasma cells' cloning. At the same time it is necessary to find some temporary substitution for antibodies. The best means, from our viewpoint, is passive immunization by antiviral drugs with directed or polyvalent action.

Usually these are the drugs made on the basis of globulins; getting into plasma they actively bind the viruses and then these complexes are eliminated by macrophages.

Finally, it is necessary to increase the membrane's permeability which activates the effector T-lymphocytes and macrophages as related to infected cells, i.e., improves the chemotaxis. The most effective means in this case is calcium chloride. This complex of therapy controlled by objective estimation of the disease's severity allows one usually to break the course of the disease and prevent serious complications.

1.7.3. Immunological model of bacterial infection. It is generally known that in the case of bacterial attack bacteria do not penetrate into the cells of affected organ, but circulate in the liquid medium (blood, lymph, and so on) and become localized in the epithelium of various organs or tissues. Bacteria damage and destroy the cells by their toxins using the cell's material to develop their population. Thus the killer activity against the organism's own cells, observed in the case of viral infection, is simply absent in a bacterial process. This means that an organism struggles against the bacteria in affected loci using its immune and macrophage system, and the main role belongs to killers and macrophages destroying the bacteria with enzymes. The antibodies are of great importance in the immune defence of an organism: they create favourable conditions for the phagocytosis of bacteria by macrophages. Three forms of a disease's severity are possible here, too: mild, average, and serious forms. But all of them, unlike viral infection, depend on the degree of the inflammatory process, which is the main characteristic of pathological activity of bacteria. It was found that the heavier is bacterial damage of an organ, the higher is the degree of the inflammatory process.

As for the mechanisms of bacterial diseases, it must be noted that there are many data supporting the suggestion that a viral infection starts a bacterial one: the viral infection, damaging the cells of a target organ, paves the way for the bacterial one. It seems quite possible that not only the pathogenic property, characteristic of the acute form of a virus attack, but distraction of all immune components for the struggle against the virus, that is the most dangerous enemy of an organism, is important here: the redoubling time for a virus population, as was noticed above, is several dozens of minutes, whereas the redoubling time for a bacterium, as a form of higher order of organization, is about 12 hours. During the redoubling time for bacteria a quantity of viruses in an organism can increase by 20–60 times, even with regard to intracellular means of defence and first of all to interferon. It is a really fantastic rate! That is why an organism directs all its means of defence to its struggle against the virus and weaken the surveillance over the bacteria which are not pathogenic for the present. Then the bacteria become active and pathogenic and, while there is no appropriate immunologic surveillance, form the loci which are known to be the seats of bacterial disease. Usually this is the way the bacterial process starts. Meanwhile, having coped with the viral infection, the organism switches its immune resources for the struggle with new enemy, i.e., bacterial infection. The formation of T-lymphocytes and plasma cell clones specific to bacteria of a given type starts, and the immune defence against bacteria is created. The recovery process after bacterial infection has begun.

This model demonstrates, how important is the early diagnosis of a viral infection, since the latter creates favorable conditions for a bacterial disease to develop. In this case the task of clinicians is to identify the types of viral and bacterial infections and to begin a course of antiviral treatment by means of passive immunization by trying to suppress the viral population with simultaneous antibacterial therapy with necessary antibiotics.

Sometimes, in clinical practice, in the case when a viral disease precedes the bacterial process, the antibiotics are prescribed for the purpose of preventing the bacterial complication. However, this therapy, due to immunodepressive properties of most of antibiotics and sulphamides, leads sometimes to the opposite result, i.e., to the activation of viral infection. The disease takes a more grave form and often transforms into an acute viral-bacterial form. In order to avoid this situation it is necessary to use antiviral drugs in combination with such antibiotics, and at such concentrations that do not depress the immune system or depress it slightly. But in all cases one must keep in mind that the viral beginning of an infectious process is the most dangerous, and the main task of therapy is to interrupt it as early as possible. After that one can apply an active antibacterial therapy maintaining strict control of the disease's dynamics.

1.7.4. Immunological model of viral-bacterial infection. We have already shown that bacterial and viral infection can be interlocked in the case of bacterial disease, when a virus "paves the way" for bacterial infection by its pathological activity. Let us consider now a typical case of the formation of acute viral and bacterial infection against the background of chronic bacterial infection. It is usually formed in the case of the lingering course of the bacterial process due to immunodeficiency of cellular and humoral components. Long exposure of an organism under conditions of continuous and lingering bacterial process forms in an organ a new level of antibodies of given specificity against the pathogenic bacteria. If the homeostasis has prevented before the presence of pathogenic bacteria, now it not only lets them be, but even fixes a certain quantity of them. The immune system now tries to conserve this quantity of bacteria in a locus instead of destroying them. It appears that if the quantity of bacteria in a locus becomes greater than has been supported by the homeostasis, additional antibodies production starts and these antibodies restore the normal quantity of bacteria. If the quantity of bacteria in a locus decreases, the immune system weakens its surveillance in order to restore the stable level of bacteria concentration. This state is extremely stable, which accounts for the viewpoint familiar to physicians: it is impossible to recover from chronic disease, whereas it is possible to transfer it into more or less protracted remission. It should be noted that in the remission state the level of antibodies concentration which is regulated by the homeostasis is attained. Thus pathogenic bacteria are present constantly in the organism of a patient with chronic bacterial disease and, therefore, there is the constant threat of the exacerbation of the chronic process. Consider this in detail.

Viral infection is usually a "trigger device" for the exacerbation of the chronic bacterial process. Indeed, let a patient with chronic bacterial infection be in remission. Let us suppose that we deal with chronic bronchitis, to be definite. Assume, further, that the patient is infected by a virus against which there is no antibody in his organism. In this case the cloning process of plasma cells starts, which produces specific antibodies against new virus. During this period of the formation of antibodies the viruses are actively reproducing. As a result, with regard to the organism's immunodeficiencies, the gravity of the viral disease aggravates sharply. The virus becomes the main enemy of the organism, all resources of its immune system are directed to the struggle against this enemy, and the immune surveillance over bacterial chronic locus is weakened temporarily, which leads to gradual exacerbation of the bacterial process against the background of the development of viral infection. As a result, the chronic bacterial process may transfer into the acute form against the background of acute viral disease. This is the way viral-bacterial pneumonia begins.

Here the question arises: which enemy is more dangerous: bacteria or viruses? Viruses prove to be more dangerous because their redoubling time of their concentration in the organism, as was mentioned above, is much shorter than that for bacteria. Therefore the process of the development of viral infection can lead to a lethal outcome, and the viral component of bacterial-viral disease demands principal attention. It is necessary to slow down the progressive development of viral infection, and then to eliminate viruses completely in the organism with the help of antiviral therapy. If the bacterial process develops into a serious form, it is necessary to use antibacterial drugs on the early stage of the disease, but these drugs must be effective with respect to the given bacterial infection and should not depress the immune system, since it exacerbates the viral disease which is serious anyway.

Of course, this approach to the medical treatment of viral-bacterial and bacterial-viral infections is not universal. Sometimes the bacterial component proves to be more dangerous, especially when the destruction of tissue is possible or when a viral infection is not acute being suppressed, even if not completely, by a protective immune means. In these cases we must first of all apply antibacterial therapy with persistent control of viral infection dynamics, which can exacerbate the disease during several hours up to the most serious form with unpredictable outcome.

The process of recovery from viral-bacterial infection is so complex that no simple approach had been found until now. This approach should be developed on the basis of a quantitative estimate of a disease's gravity with the acuteness of inflammatory process and the immunological status of a patient taken into account.

1.8. Stimulation of Immune System. Hypotheses and Problems

1.8.1. Biostimulation of Immune System. As was noticed above, the treatment of chronic diseases requires the normalization of immune processes that are weakened and proceed extremely slowly. Mathematical models enabled us to find a mechanism that helps an organism to escape from the chronic state. It turned out that the exacerbation of disease, i.e., conversion of chronic disease into an acute one may serve as a highly effective method of treatment. This method is well known to clinicians and is sometimes called "treatment by exacerbation". Let us consider the immunological aspect of this approach.[1]

Since chronic states of disease are very stable, the aim of therapy is to substantially increase the antigen concentration in an organism (sometimes by an order of 2–3), which leads to effective stimulation of the immune system as related to a given antigen.

Under normal conditions this situation is observed, for instance, in the case when a viral disease complicates an existing chronic one which provokes the exacerbation of the latter. Complete recovery in this case is not always achieved, since the antibiotic therapy which is usually used causes the immunodepression that limits the possibilities of immune surveillance and permits either the existence of a weakened antigen population of the chronic disease in the organism or the penetration of new infestants of the same specificity into the organism because the immune barrier is low. Thus under natural conditions the possibility of disease relapses is raised, as a rule, after the exacerbation of chronic disease.

For this purpose the polysacharide pyrogen has been used for a long time in clinical practice, which, together with its biostimulating effect, has the capacity of raising the body temperature. The first use of

[1]Besides the defects in the immune system, chronic forms of disease can be characterized by disorders in other protective systems.

this drug goes back to the nineteenth century. At that time it was used effectively for the treatment of syphilis. In recent years the applications of this drug have been growing continuously: from tuberculosis [92] and pneumonia [197] to hepatitis. The first systematic investigations related to the treatment of chronic diseases (mainly pneumonia) with the use of polysacharides as biostimulators had been carried out by Ermol'eva [92]. This treatment can be described by the model of disease based on immunological conceptions as follows.

During the chronic course of the disease there is a stable equilibrium between the processes of proliferation and elimination of antigenes. All immune components in this case are in some stationary state slightly different from a norm. In order to lead an organism out from this state of "equilibrium" it is necessary to obtain the extended reproduction of specific antibodies. However, this level of equilibrium is very stable due to nonlinearity in the immune processes. Therefore the activation of the immune system can be achieved only if the balance is broken. This goal can be reached with the help of "distracting" biostimulation. Assume that an increasing amount of polysacharides or other biostimulants of antigen nature are introduced into the organism. An immune reaction against the new antigen (biostimulator) starts in the organism. Due to the increase in the dose of biostimulators the organism is "distracted" from its former chronic enemy, decreasing the production of antibodies spent before to neutralize it, and concentrates its own forces on the struggle against the new antigen. This occurs because the concentration of VF-complexes of biostimulator in organism becomes much higher than the concentration of analogous complexes for chronic disease antigen, and, therefore, the recognition of the latter becomes difficult. All this create favorable conditions for intensive multiplication of chronic disease antigen, and, therefore, its concentration increases. When biostimulators injections are over, a greater number of VF-complexes of chronic antigen is generated in the organism than before the injections. This causes effective stimulation of the immune system and, as a result, the infestants of chronic disease are quickly eliminated from the organism, and recovery begins (Fig. 15). This is one of approaches to the treatment of chronic diseases on the bases of biostimulating (more precisely, immunostimulating) therapy. Of course, this scheme embraces only general features of the process and works when the immune system is in relatively good condition and the increase in chronic antigen concentration does not lead to serious pathological changes.

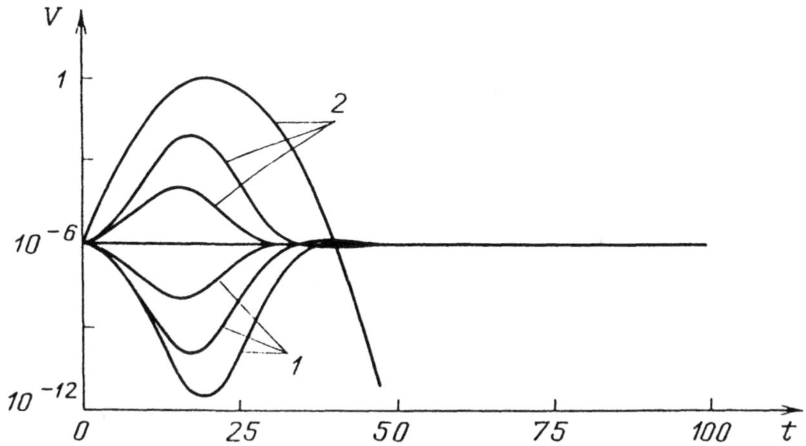

Fig. 15. Treatment of chronic disease. Passive (1) and immunostimulating (2) therapy.

Another variant of treatment of chronic disease, as we believe, is possible. As a result of gradual biostimulation, the levels of lymphocytes, macrophages, immunoglobulins and other vitally important components grow higher in an organism. Some of these resources may be used directly against chronic antigen and its after effects. In this case macrophages may play an especially important role utilizing the after effects of antigen activity ("waste" lymphocytes, antigen–antibody complexes, etc.). This leads to a certain normalization in the functioning of the affected organ, and finally, it increases the effectiveness of the immune system in its struggle against the chronic antigen. This process can, in certain cases, result in complete recovery. If the process of self-healing is combined with antibacterial or antiviral therapy, then the destruction of chronic antigen proves to be more effective under conditions of a high level of immune defence. It should be noted that antibiotics frequently act as immunodepressants together with their bactericidal property, and they can be used for the treatment of chronic diseases together with biostimulation of the immune system. The main way of treating chronic diseases seems to be in the combination of drugs biostimulating the immune system, antiviral drugs, and antibiotics. This approach is being developed now actively for the treatment of patients with acute form of pneumonia, chronic bronchitis, and bronchial asthma [197, 198] on the basis of complex immunostimulating and antibiotic therapy.

The development of methods that enhance the organism's resistance is of ever greater importance for the prevention of chronic diseases. The progress in immunology made it possible to interpret the organism's resistance as the capacity of an infected organism to provide the immune response conditioned by initial immunological status of a patient.

Various biological stimulators (so-called immunocorrectors, immunoprotectors, immunomodulators) are used to stimulate the immune system. The autohemotherapy and its variants, injections of drugs such as aloes, corpus vitreum, FIBS, ginseng, pentoxyl, etc., have been most widely adopted. The enumerated drugs provoke rather short nonspecific immune response and do not stimulate the immune memory; therefore adequate immune and, hence, clinical effects are not always observed after such treatment. Moreover, collateral substances in these drugs lead quite often to local irritation or to general reaction of sensibilization, which limits the repeated injections of the drugs. Most of the enumerated drugs have been used in practice until now, but, unfortunately, with no immunological control.

At present, all economically developed countries immunize their population by live or dead influenza vaccine during the periods of expected epidemic influenza outbreak and for the prophylaxis of an influenza pandemic and prevention of serious hypertoxic forms of influenza [169, 170, 318, 349, 350]. It has been established that the rapidity of immune system adjustment to the production of specific antibodies is the crucial factor responsible for the gravity of a disease. Therefore it is important, that the memory cells specific to potential (possible) antigen should be present in an organism. The immunization creates the lymphocyte populations that serve as the memory cells, which immediately adjusts the immune system for the struggle against an antigen after it has penetrated into an organism. The proliferation (adjustment) of memory B-cells in this case occurs very rapidly and, as a result, the antibodies (specific immunoglobulins) appear which neutralize the antigenes in the stage of subclinical or mild clinical form of disease. If there are no memory cells (immunization has not been done) an organism adjusts the immune system against the unknown antigen, and, if the adjustment of the immune system occurs slowly, then during this period of time the viruses (or bacteria) may create large populations of new generations of viruses and provoke the damage of considerable portion of organs, i.e., a serious form of the disease will develop. The same situation may occur in case of immunodeficiency if it is genetically determined.

The investigations revealed that the immune status of a human being is influenced just partially by immunization, since it provokes the production of antibodies strictly specific to the injected vaccine. The effectiveness of the immunization proves to be higher and it does not lead to after effects if the immune status of population is taken into account. The usage of vaccines permits carrying out both nonspecific and specific stimulation of the immune system.

The most effective immunostimulators known at present are the lipopolysacharide complexes extracted from the prodigiozum, E.coli, S.typhi or P.aeruginosa, pneumococcus or bacterial polysacharides (pyrogen, prodigiozan, salmazan, etc.).

The bacterial polysacharides are believed to stimulate the hormonal activity of the hypophysis-adrenal cortex system and to intensify the antibacterial factor's activity in an organism and the production of intracellular interferon and also to decrease the negative influence of antibiotics on the immune system. The enhanced immune response to bacterial polysacharides injection is due to the stimulation of humoral immunity strengthening the macrophages and neutrophiles phagocytic activity, and to the opsonizing action on the leukocytes as related to bacteria from which the polysacharides have been prepared. The bacterial polysacharides enhance the activity of all T-lymphocyte subpopulations (effectors, suppressors, and helpers) simultaneously with the intensification of macrophage–killer activity if these immunological indices were at low level.

In the case of a normal immune status of a patient (with no deficiency) the application of biostimulating therapy is useless and sometimes leads to the decrease in T-lymphocyte level in blood. It was found that of all the above biological stimulators, pyrogen (prodigiozan to a smaller degree) is the polyclonal stimulator which provokes both specific and nonspecific immune response. At present there is some information about the correcting influence on the immune system of many other medicines besides pyrogen and prodigiozan and about positive clinical experience of their applications in the cases of various diseases which are accompanied by secondary immunodeficiency. In the case of immunodeficiency T-activin, thymalin, thymomodulin, B-activin, bronchovaxom, etc. are used.[2]

1.8.2. Immunotherapy. A more than thirty-year period of wide application of antibiotics and sulphamides has shown that these drugs not always can cope with acuteness of an inflammatory process. There is just one possible reason: immunodeficiency. The process of plasma cell cloning is weak and does not ensure the production of immunoglobulins at the level necessary for complete suppression of bacterial infection. If this situation remains unchanged for a long time the process usually turns into chronic one, and immune status and homeostasis reform to ensure the continuous production of specific antibodies neutralizing new chronic inflammatory process.

There is no question about complete elimination of bacterial antigenes in the organism, the question is about the control of its amount at a certain level in the place of its localization. It appears that energetically it is more advantageous for an organism to maintain a constant level of pathological antigen concentration than to eliminate it completely from the organism since this would require the expansion of antibodies production which can not be provided by the organism.

In this connection, when an acute inflammatory process develops in the organism of an immunodeficient patient and the dynamics is not positive, it is advisable to use the immunotherapy based on the injection of prepared immunoglobulins or donor (passive) antibodies, which may

[2]See Marchuk, Berbentsova [197].

"tip the scale" against the background of intensive antibacterial therapy in favour of convalescence before the process becomes chronic. It is possible that this is the only way at present to eliminate the pathogenic bacteria from the organism and cure a patient.

The immunotherapy is most important in the cases when acuteness of an inflammatory process continues to grow and turns into a serious clinical form with unpredictable outcome in spite of measures taken against the infection. In these cases even though the injected passive antibodies are not sufficient for complete recovery, they can at least transform the acute process of disease into a more mild or chronic form.

There is another situation when the application of immunotherapy is very effective in the treatment of a bacterial process. It happens when chronic infectious disease turns into an acute form and the usual intensive methods do not stabilize a patient's state. Usually this situation occurs when a patient has considerable immune deficiency.

The immunotherapy is particularly important in the case of severe viral or bacterial-viral disease. Rapid growth of the disease gravity in these cases is usually connected with considerable deficiency of immune system, whose protective means are forming slower than the viruses do their damage to the target organ. This process is still more aggravated when at the same time chronic bacterial infection turns into acute form. Recall that in the case of viral damage of the target organ the infected cells are destroyed by the organism's lymphoid or macrophage systems that possess the killer effect. The organism has no alternative (except for cellular interferon) to fight with the viruses penetrating into a cell. Thus mass infection of the cells leads to their mass destruction by the organism's immune system. Naturally, the disease's gravity increases. In many cases, however, viruses, having destroyed a cell, go out into extracellular space and infect new cells. If there is a deficiency of the organism's own antibodies in plasma, the process of damaging new cells will develop threatening with lethal outcome. Such a situation is characteristic of the hypertoxic form of a viral disease. Therefore it is necessary to conduct immunotherapy directed to the suppression of the viruses found in extracellular plasma in the case of growth of the disease's gravity. Depending on the aetiology of viral process, anti-influenza and anti-measles gammaglobulins or normal human immunoglobulins are used for this purpose. Immunotherapy by passive antibodies is also expedient in the cases where secondary (acquired) immunodeficiency is diagnosed or assumed.

1.9. Immunophysiological Reactions of Organism

1.9.1. Temperature reaction of organism. It is generally known that the rise in temperature is a powerful defence reaction of an organism in the struggle with viral or bacterial disease. Almost all bacteria and viruses are subjected to depressing influence of the rise in temperature. The immune process activity increases simultaneously with temperature. There is a generally known fact that confirms this: the case of gastrointestinal poisoning when vigorous temperature reaction develops, although there are no pathogenic bacteria or viruses in an organism. The temperature reaction of the organism is necessary for the intensification of the production of macrophages, albuminous, and other components promoting the most rapid neutralization and removal of the toxic substance from the organism.

It can be said with no overstatement that the positive role of the temperature reaction has been noticed in the life of nearly every human being. This becomes especially pronounced in case of acute inflammatory process of panaritia when, as a result of damage and subsequent infection, fingers and toes abscess. It is long since known that such suppurative inflammation is treated effectively by hot baths. Such a procedure suppresses or entirely destroys the bacteria population that provoked the pathological process, which usually leads to complete recovery. It is not by mere chance that practical experience of people has resulted in the usage of a quite effective remedy against acute respiratory diseases, namely – drinking hot milk with butter and household soda. Such a procedure leads to the local rise in temperature of the larynx and suppresses the viruses or bacteria and sometimes leads to their complete elimination. Many other analogous examples can be cited here.

Of course, medicine has more powerful means now to fight such diseases, but the experience accumulated by humanity allows one to perceive objective laws in this therapy which we would like to explain in immunological terms. Pyrogen, for instance, is a powerful means of raising a temperature, and its application in practice have already been discussed above. The capacity to raise temperature is the most important mechanism of self-defence and self-healing of an organism. Therefore the capacity of an organism to increase the effectiveness of treatment is very important in therapy. Meanwhile, unnecessary use of febrifuges leads to the decrease of an organism's resistance, thus creating more favorable conditions for viruses and bacterial populations to develop.

Unfortunately, the temperature reaction of an organism is generally diminishing with age and, therefore, it is more and more difficult for the organism to cope with an illness. The most pronounced temperature reaction is observed in children of early age who often fall ill because of their still imperfect immune system, and also recover quickly due to exceptionally pronounced temperature reaction. Children run a temperature often but usually are not taken seriously ill.

Here the question arises, of whether it is possible to stimulate the rise in temperature in ' aged patients. The answer to this question is affirmative. Furthermore, these methods have been used in many cases in medical practice, they should only be analyzed in the light of general immunological conceptions and then they should be confirmed by mathematical modelling. However, the basis of health and of rapid recovery after infections is still the normal immune status of a man. The control of this immune status is the main aim of clinical examinations, which becomes generally accepted at present.

We have already noticed the important role of pyrogen therapy which is the antigen stimulation of the entire immune system including the phagocyte one. We would like to consider here the capacity of the pyrogen therapy with the help of biostimulation to raise the temperature of an organism. It should be noted that in this sense pyrogen (an extract of dead P.aeruginosa) has long been used in medicine for treating the lues, whose stimulant, spirochaeta pallida, dies at the temperature of 40°C. The pyrogen therapy of lues consists of intramuscular injections stimulating the rise in temperature up to 41°C.

At present this therapy for treating the lues is not so actual. However, artificial stimulation of the rise in body temperature will evidently find more applications in medical practice not only in cases of chronic diseases but also in cases of acute diseases during the initial period of disease development when the temperature reaction of a patient is weakly expressed. Naturally, the set of physiotherapy means is in practice considerably greater and we will not consider the immunologic significance of each of them. Our task has been to attract attention to their immunological action on an organism, since such an aspect of the analysis of physiotherapy helps to reach deeper understanding of their application in various situations of medical practice.

Young people have a very close correlation between their body temperature and pulse rate. But this correlation is broken with age: the temperature reaction of an organism on the antigen invasion is weakened, but the tachycardia increases. This is, evidently, some compensation of weakened temperature reaction by the increase in the pulse rate. Indeed, if the temperature reaction is not sufficient, the depressing temperature influence on the virus and bacteria population development is not sufficient either. This can be compensated partially by more intensive circulation of blood supplying the organs damaged by the infection with additional albuminous, enzymatic, and immune components. In any case, during the infection the increase of pulse rate is clearly observed, and as a result the organism's defence functions are enhanced and viral and microbial toxins disturbing normal vital activity of an organism are eliminated.

1.9.2. Oedema reaction and arterial tension. While the rise in temperature and increased pulse rate are observed in the course of both viral and bacterial diseases high arterial tension under conditions of infection is believed to be connected with viral infection. The virus, penetrating the cell, provokes the oedema as the reciprocal organism reaction, which, in the first place, localizes the seat of viral damage of tissues preventing rapid infection of the still healthy cells, and, in the second place, the lymphoedema attracts the defence components of immune and macrophagal systems to fight the viruses moving out of a cell into extracellular space.

Therefore the oedema is extremely important in the defence of an organ, preventing vast damage. The role of oedema in the cases of respiratory virus infection have been reported in a series of papers [170, 275, 350, 354].

Indeed, so far as the oedema is formed, normal arterial tension is not sufficient to provide the seat of viral damage, that is blocked up by the oedema, with the oxygen. Just for normal provision with oxygen the arterial tension is raised. Therefore the capacity of an organism "to regulate" arterial tension is the most important component of the defence reaction.

The rise in arterial tension is a defence reaction of an organism connected with forming lymphoedemata of tissues damaged by viruses. If the tension is decreased artificially the blood will not supply the organs blockaded by oedema with oxygen, and as a result the conditions of tissues' oxygen starvation and feasible infarction develop. The infarctions of myocardium, lung, and kidney happen this way. The arterial tension should be decreased only in the cases when the high pressure can have grave consequences for a patient's life (see 3.3).

1.10. Some Problems and Hypotheses of Clinical Immunology

We will consider in this section some important problems connecting immunology and medicine. Some of these problems are well advanced. Others are just being formulated in literature as hypotheses. But both give an impetus to scientific research, though some of them still raise discussions. Perhaps new results in immunology and mathematical simulation can provide medicine with answers.

Of course, recent achievements in immunology will exert and are already exerting direct influence on the methods of medical treatment and on clinical practice in general (see Petrov [273]).

Surgeons were the first, apparently, who encountered problems of noninfectious immunology during both postoperative healing of wounds and the renewal of activity of separate operated organs. Postoperative healing proceeds differently in individuals and depends essentially on the status of their immune system. Therefore this status in the preoperative period together with its subsequent dynamics constitute the most important components of the tactics of surgical operation.

Transplantation operations led to understanding the genetic peculiarities of individual organism and to the problem of its incompatibility with foreign organs and cells. It is well known that the transplantation of tissue or of an organ from a donor to a patient is followed by heavy immune reaction of rejection, where T_E-system of effector lymphocytes plays a key role. The task of an immunologist in this case is to depress the cellular immunity in the preoperation period. Although medicine has been enriched in recent years with first-class elaborations of such operations, nevertheless, the problem of transplantation has not yet been solved completely. It is believed that success in grafting is ensured first of all by appropriate selection of a donor genetically close to a recipient (Petrov [273]). This raises a problem of creating an information database of the population's immunologic status. There can be no doubt that very soon the immunologic status tests of an individual will be just as usual as, for instance, the blood group or Rh-factor tests.

It is known that antibodies IgM, IgG, IgA, IgD, and IgE play an important role in the immune process. Widespread diseases, such as acute respiratory diseases, are associated, to a great extent, with the antibodies IgA which protect, as a rule, the mucous membranes of organs. Therefore if the deficiency of this immunoglobulin occurs, for one or another reason in an organism, the resistance of the organism is diminished and the probability of a disease increases. Such immunodeficiency usually leads to frequent diseases of the nasopharynx, lungs, Eustachian tubes, and so on. Therefore frequent pharyngitis, otitis, and tonsillitis are not so much the result of disease of the throat, nasopharynx, or middle ear, as the consequence of a disease of the immune system. Local treatment of such diseases leads only to a short-term remission. Radical improvement in the state of health can only be achieved in the case of the restoration of normal qualitative and quantitative levels of IgA in an organism. We can assume, therefore, that the main task of the therapy directed to the restoration of IgA level is the liquidation of chronic disease, i.e., complete elimination of the associations of staphylococci, streptococci, and other bacteria. Only under this condition can normal immunological level of antibodies be restored in an organism. However, it is still not clear whether this level can be restored to a norm by itself or whether specific biostimulators are required for this purpose. At present, many immunologists all over the world are looking for biostimulators to raise the level of IgA. (It is believed that Indo-Tibetan "herbal" medicine has some collection of biostimulators of this kind. Obviously, this question requires special investigation.)

Another very important branch of medicine deals with the diseases of the immune system itself. We mean the autoimmune reactions when the system of T-cellular immunity that distinguishes a self cell from a foreign one is in disorder. In this case some organism's cells are identified by the T_E-system as foreign and the organism's immune system starts to fight erroneously against the host. It has been established that rheumatoid diseases of joints, cardiac muscles, and so forth, are examples of a disease of this kind [273, 297]. Apparently, in this case the suppressing component of the immune system responsible for the sensitivity of T_E-system becomes weakened, which activates the T_E-system. It is not a coincidence that clinicians treating an autoimmune disease usually use immunodepressants decreasing the T_E-killer activity. However, the nature of autoimmune process has been investigated insufficiently, and this hypothesis at present is attracting many supporters.

Immunology opens new ways in therapy where optimal strategy of the treatment will be based more and more on the immunological status of a patient. Particular attention will be paid to individuals with congenital or acquired immunodeficiency. They are the individuals who will need special prophylaxis against the diseases which are difficult to treat because of immunodeficiency of one or another component. Special attention will be focused on children during the period of immune system formation, when an organism fighting with the infectious disease undergoes the greatest stress and the probability of immune system disorders is particularly high. This large area of clinical medicine demands exceptional attention.

Gerontology assumes ever greater importance: it will be also based on information on the immune system status that becomes weak at ages from 70 to 80. Taking this into account, it is necessary to change the conventional methodology of treating infections and especially chronic diseases proceeding under condition of immunodeficiency of the system; therefore the treatment of old people must differ essentially from that of the analogous disease in young and middle-aged people who have a relatively good immune system.

And now just a few words about cancer diseases. Immunologists believe that certain types of cancer are caused by congenital or acquired immunodeficiency of an organism. In particular, there is a number of experimental data obtained while treating the children suffering from T-deficiency. These data analyses allow the conclusion that the probability of canceration for a child with congenital T-deficiency is many thousand times higher than for a child with a normal immune system.

Other observations show that immunodepression caused by radiation, chemical and biological drugs leads to the increase in frequency of cancer diseases.

Finally, there exists a maximal age-related tumoral diseases rate. It falls at the end of early childhood when the immune system is only forming and at the age of 60 to 70 when the immune system is weakening as a result of partial or complete thymus atrophy (Petrov [273]).

These and other facts make one conclude that tumors appear in an organism when the immune surveillance over genetically degenerated cells by effector T-lymphocytes is weakened. It means that the immunodepression of a T-system increases the probability of a tumor, since it decreases the effectiveness of T-effectors in the recognition of degenerated cells.

It is believed that in case of a malignant tumor surgical management, radiation therapy, and chemotherapy, applied with regard to the tumor characteristics, restores at a certain stage a balance between the population of cancer cells, T-lymphocytes, and macrophages. In this case, when the tumor is removed, the quantity of T-lymphocytes and macrophages is sufficient to destroy the remainder of the degenerated tissues and to restore the immune cellular surveillance in an organism.

At present, various antibiotics applied in large dosages are the most active immunodepressants. The conception on dependence of allergic, autoimmune, and cancer diseases on the immunologic status suggests that there is a danger in boundless applications of antibiotic therapy, especially for old patients who have weak immune system (Chazov et al. [54]). We should also take into account the possibility of cancera-tion caused by contacts with carcinogenic chemicals, since these substances increase significantly the probability of genetic mutations of cells so that their number may exceed the maximal amount eliminated by populations of T-lymphocyte and macrophages.

It seems possible that in future the immune system of old persons will be prophylacticly stimulated by biologically active substances that can ensure normal functioning of all immune system components for long period of time. Perhaps it will be also an effective prophylaxis against acute infectious diseases.

We do not consider here the aspects of molecular genetics of cancer which are actively investigated at present. The basic hypothesis and research in these directions are aimed at the determination of cancer genome structure. This genome actively "works" as is known, in embryonic development, and then, as the superrapid growth of foetus cell is not necessary any more, it ceases to function until the time when this mechanism is called on again to complete the vital activity of an organism.

CHAPTER 2

Survey of Mathematical Models in Immunology

A large amount of observations on the course of various infectious diseases has been accumulated by immunologists and clinicians up to the present, and fundamental results have been obtained related to the mechanisms of interaction between pathogens and the immune system at different levels of consideration: from the macroscopic to the intracellular genetic. These results make it possible to construct mathematical models of immune processes of corresponding levels of detailing.

The work by Hege and Cole [128] is to be considered as one of the first in the field. These authors constructed an equation describing the change in the number of circulating antibodies depending on the number of plasma cells. Then Jilek [157–161] suggested a series of probabilistic models of interaction of an antigen with an immunocompetent B-cell, and simulated the process of clone formation from one B-cell using the Monte-Carlo method.

The first work of Bell was published in 1970 [18]; in this work, using the basic hypotheses of Burnet's clonal-selection theory, he constructed a mathematical model of the humoral immune reaction to a nonmultiplying monovalent antigen. In his next works [19, 20] Bell modified his model into one taking into account heterogeneity of antibodies with regard to affinity and multivalent antigens. In 1974, he suggested simple model of immune reaction to multiplying antigen [21], where the antibody–antigen interaction was described in terms of a "predator–prey" system. The qualitative properties of this model were investigated by Pimbley [276, 277] and, after introduction of an equation for the B-cells into the model, by Pimbley [278], Hsu and Kasarinoff [143].

Similar approaches were simultaneously developed by Smirnova [317] and Romanovski with coauthors [298].

In 1974 Bruni, Giovenco, Koch, and Strom [46, 47] suggested a model of humoral immune reaction describing the heterogeneity of immunocompetent cell population with continuous functions of two arguments: affinity and time. The main distinctive feature of this model is the description of immune reaction in terms of bilinear systems. This work developed further in two directions. On the one hand, Mohler [225, 228] modified the model to describe more wide range of phenomena (production of various classes of antibodies, cooperation between T- and B-systems of immunity, etc.). On the other hand, there are works that try to solve the identification problem of the original model [224, 226].

In 1975 this author constructed a simple mathematical model of an infectious disease [193] which is a system of nonlinear delay-differential equations. Besides the humoral immune reaction this model describes the influence of damage of a target organ caused by an antigen on the immune system processes. A qualitative analysis of this model has been made by Asachenkov and Belykh [7]. In 1978 the author and Asachenkov suggested a more complete model of infectious disease that describes the cooperation between T- and B-systems of immunity and the production of immunoglobulins of two classes.

In 1975 Richter [295] and Hoffmann [133] suggested original models of immune system based on the network theory of Jerne [156]. In these models the authors pay the main attention to the description of various events occurring in a network.

In 1976 Waltman [341] described a model of immune reaction using the idea of threshold switching of B-lymphocytes from one state to another. The thresholds are described by model equations as delays in time that are the functions of the system's state. Further development of the model was made by Gatica [109, 110].

In 1976 De Lisi published several papers [74–76] on the mechanisms of immune interactions on the surface of a lymphocyte, and suggested a model of tumor growth in an organism [77, 78] resembling in essence Bell's simple model.

Dibrov, Livshits, and Volkenstein [80–82] considered a simple model of humoral response where special attention was paid to the analysis of influence of time lags on the immune process dynamics.

Early works of Perelson [258, 266] should be mentioned, where an immune reaction was considered from the viewpoint of optimal control theory, and the works of Merrill [215, 216], where the author suggests describing an immune reaction in terms of catastrophe theory.

Immunology, as a science, studies an extremely wide spectrum of processes that occur at various levels of organization and control: genetic, cellular, in separate organs or tissues, and in an organism as a whole. The problems under study are remarkable for their diversity: from regularities of the functioning of IR-genes, action of cytokines, to the defence mechanisms and immunopathology in virus infections and autoimmune diseases.

The last decade is characteristic of the widening sphere of application of mathematical methods and models in theoretical and experimental immunology for the investigation of regularities in immunoregulation, recognition and learning in immune system, dynamics of an idiotypic immune network, spatial organization of immune processes, immune mechanisms of infectious diseases. A rather complete contemporary state of investigations in mathematical modeling of immune processes is presented in the reviews made by Lumb [187], Mohler et al. [227], Perelson [259], Perelson and Weisbuch [268], conference proceedings: [53, 137, 208, 229, 293, 330–332], and monographs: Belykh [26], Zuyev [356], Kuznetsov [168], Macken and Perelson [190], Asachenkov et al. [8].

It is difficult to perform a comparative analysis of the main directions in the modeling of immune processes, since each direction was formed in the course of investigations of a certain circle of problems that arise in different disciplines of immunology: cellular immunology and immune control, molecular and structural immunology, clinical immunology, etc. Therefore mathematical models differ by approximation of processes under study, methods of formalization of interaction characteristics of immune system components, temporal scales of analysed processes, methods of fitting a model to experimental data, etc.

From the mathematical standpoint this brings about the models of various dimension and linearity, formulated for various classes of differential equations: ordinary differential equations, differential algebraic equations (Bell [18, 21]), delay differential equations (Behn et al. [17], Dibrov et al. [80–82], Farooqi and Mohler [98], Marchuk et al. [202, 204], Nelson and Perelson [235], Sidorov and Romanyukha [312]), integro-differential equations (Bruni et al. [46], De Boer et al. [71], R.Mohler et al. [227]), partial differential equations (Kuznetsov [174]).

Great interest was attracted recently to discrete methods of modeling in immunology (Agur [3, 4]) on the basis of cellular automata (Brass et al. [42], Celada and Seiden [52], De Boer et al. [72], Ermentrout and Edelstein-Keshet [91], Moiser and Sieburg [231], Pandey [252], Parisi [253], Sieburg et al. [313]), spin glasses (Lefevre and Parisi [179]). The conventional approach to the construction of models is based on the separation and description of the key interactions (McLean and May [213]), which can be supplemented with certain requirements on the structure of phase portrait of a model (Segel and Jaeger [305]).

Two complementary conceptions are used in the analyses of the mechanisms of organization and regulation of immune processes: the clonal theory of Burnet [49] which considers the immune system as a set of large number (about $10^6 - 10^7$) of independent clones of lymphocytes of different specificity activated by antigens selectively, and the theory of idiotypic network suggested by Jerne [156], which represents the immune system as a set of clones interacting between themselves due to the presence of idiotypes on antibodies and immunoglobulin receptors of B-cells.

Respectively, two directions in mathematical modeling in immunology are usually separated: the investigation of the dynamics of immune reactions caused by various antigens on the level of separate clones, and the investigation of properties of immune system as a idiotypic network of interacting clones (De Boer [59], Mohler et al. [227], Perelson [260, 261], Perelson and Weisbuch [268]).

At the same time, currently the conception has been formed on the basis of experimental data, suggesting that both types of the organizations are present independently in the immune system: 80–90% of clones form an autonomous block of lymphocytes reacting to external antigens, and 10–20% of clones take part in idiotypic interactions, which determines the autonomous activity of immune system (De Boer and Perelson [69], Perelson and Weisbuch [268]). This point of view was reflected in recent theoretical studies by the concepts of a "peripheral" and "central" immune systems (Stewart and Varela [322]).

One of the primary functions of the immune system is the defence of the organism against various infections. The study of the mechanisms of viral and bacterial infections is a focus of attention of clinical and experimental investigations in immunology. At the same time there are no doubts that the main problems in understanding the mechanisms of immune system are caused by difficulties of the conceptual analysis of immune system *in vivo* (Doherty [85], Langman and Cohn [178], Paul [255], Zinkernagel et al. [352, 353]). Theoretical analysis of immune reaction mechanisms on the basis of mathematical modeling permits one to formulate an integrative description of organization and kinetic regularities of immune processes *in vivo.*

Mathematical models of immune processes during infectious diseases are the main content of this book. Therefore the review of investigations on mathematical modeling in immunology will be associated first of all with the consideration of models of immune reactions to antigens according to the following features: peculiarities of processes under investigation; detailing of the description of immune processes; methods of mathematical formalization of characteristics of interactions and activation of cells; representation of specificity; parameter analysis of models; interpretation of results of modeling. We will consider just modern models, since there are detailed reviews of various directions in mathematical modeling in immunology (Mohler et al. [227], 1980, Lumb [187], 1987).

Let us introduce a number of notions which will be useful for the discussion of results of qualitative analysis of mathematical models of immune processes. A process is referred to as *deterministic* if its future course and the past are uniquely determined by the current state. A set of all possible states of the process is called a *state space.*

Points of a state space where a vector function of the right-hand sides of model equations is equal to zero, are called equilibrium states (or stationary points, or singular points, or steady states) of a system.

The rearrangement of the qualitative picture of dynamics with the change of a system's parameters are studied by the *bifurcation* theory. Currently the investigation of transition scenarios for dynamical system towards the chaos through a sequence of bifurcations due to changes of some parameter is the actively developed method in the analysis of dynamical systems. These investigations in mathematical immunology are presented in De Boer et al. [63, 64].

2.1. Mathematical Models of Humoral Immune Response

Mathematical models of humoral immune response represent the most numerous class of models in immunology. They are used for the investigation of properties of immunity both at a level of separate clones and in the framework of idiotypic network. These models were formulated initially in the framework of clonal selection theory and were used, first of all, for the analysis of quantitative characteristics of immune reaction to an antigen. These are the models with discrete description of the affinity (Bell [18–20]), the models with continuous description of the affinity (Bertuzzi et al. [30], Bruni et al. [46, 47]), and the models taking into account the existence of antibodies of various classes (De Lisi [75], Mohler et al. [227], Perelson et al. [266]).

Then, in order to investigate the properties of idiotypic immune systems, a set of mathematical models of humoral immune response was formulated and used actively by various groups. These models differed with the level of detailed description of processes but were formulated in the framework of unified approach. Let us consider these models first of all, since they are the mathematical tool which is currently used most actively for the investigation of intimate processes and properties of immune system as a whole (De Boer [71], Perelson and Weisbuch [268], Faro and Velasco [95–97], Stewart et al. [322, 323]).

Since the content of this monograph is related to investigations of immune response to pathogens, we concentrate our attention upon the modeling of clonal dynamics. Parameters of these models that characterize phenomenologically the processes of interaction during the development of immune response can be estimated on the basis of the data of experimental clinical immunology.

2.1.1. B-model (Perelson and Weisbuch [268].). The B-model describes the dynamics of a set of M clones of B-lymphocytes. the corresponding equation for ith clone, $B_i(t)$, $i = 1, \dots, M$ has the form:

$$\frac{dB_i}{dt} = m + B_i(pf(h_i) - d_B), \tag{2.1.1}$$

where m describes the influx of B-cells from the bone marrow; p is the division rate of activated lymphocytes; d characterizes the death rate of cells. The activation function $f(h_i)$ is a key element of the model, determining the intensity of stimulation of the clone's cells as a function of the strength of idiotypic interactions. The latter are characterized by the value h_i called a "field" which is formed by another clone, j. The value of a signal affecting ith clone as a result of idiotypic interactions with other M clones of immune system, whose total number was assumed to be fixed, is chosen as $h_i = \sum_{j=1}^{M} J_{ij} B_j$. Elements of the matrix J characterize the degree of complementarity (for instance, the affinity) of immunoglobulin receptors of the clones B_i and B_j. The choice of the matrix J is related to the topology of the network of interactions. As a rule, the values 0, 1, or 0.1 are used as the elements J_{ij}. The activation function $f(h_i)$ is assigned as follows:

$$f(h_i) = \frac{h_i}{\theta_1 + h_i} \frac{\theta_2}{\theta_2 + h_i},$$

where $\theta_1 \ll \theta_2$. This log bell shaped activation function is the phenomenological description of regularities of the activation of B-cells which takes into account the phenomena of high- and low-zone tolerance, as well as qualitative character of dependence of the degree of cross-linked receptors on the quantity of multivalent antigen.

Here are characteristic estimates for the model's parameters taken from [268]: $p = 1\, d^{-1}$, $d_B = 0.5\, d^{-1}$, $\theta_1 = 10^2$, $\theta_2 = 10^4$.

2.1.2. AB-model. This model is a generalization of B-model, most frequently used in theoretical studies of an immune network [307, 322, 323]. AB-model describes the dynamics of B-lymphocytes ($B_i(t)$), and antibodies ($A_i(t)$) for each i-th clone ($i = 1, \ldots, M$). The corresponding dynamics equations have the form:

$$\begin{aligned} \frac{dB_i}{dt} &= m + B_i(pf(h_i) - d_B), \\ \frac{dA_i}{dt} &= sB_i f(h_i) - d_C h_i A_i - d_A A_i. \end{aligned} \tag{2.1.2}$$

Idiotypic interactions in the framework of this model, unlike the preceding model, are mediated by antibodies, and therefore the model is more realistic. The manifestation of the anti-idiotypic field h_i is determined by concentration and affinities of antibodies A_j capable of reacting with B-cells and antibodies of ith clone:

$$h_i = \sum_{j=1}^{M} J_{ij} A_j.$$

Additional parameters have the following meaning: s is the secretion rate of antibodies; d_A is their decay rate; d_C is the elimination rate of antibody–antibody complexes. These equations were used to study the attracting sets (attractors) of a system of two interacting clones, various dynamical regimes: oscillatory, chaotic, etc. Standard estimates of parameters (in addition to above estimates) are as follows: $d_A = 0.05\, d^{-1}$, $d_C = (0.01 - 0.001)\, d^{-1}$, $s = 1.4 \cdot 10^{-15}$ mole/d.

2.1.3. ABG-model. The ABG-model, which is a further generalization of AB-model, describes for each clone the dynamics of B-lymphocytes ($B_i(t)$), antibodies ($A_i(t)$), and the function $G_i(t)$ characterizing the state of differentiation of activated B-cells and, therefore, their capacity for the production of antibodies (De Boer et al. [66, 67, 69]).

The introduction of this "gearing up" function allows us to take into account phenomenologically the time needed for the differentiation of a B-cell into plasma cells synthesizing antibodies.

The corresponding equations take, therefore, the following form:

$$\frac{dB_i}{dt} = m + B_i(pf(h_i) - d_B),$$

$$\frac{dA_i}{dt} = sB_iG_i - d_c h_i A_i - d_A A_i, \tag{2.1.3}$$

$$\frac{dG_i}{dt} = k[f(h_i) - G_i].$$

The parameter of differentiation rate k is estimated as $k = 0.2\,d^{-1}$.

The B- and AB-models were used to analyse the dynamics of the system of idiotypically interacting clones under the assumption that the total number of clones in a system is fixed. A classification for various equilibrium states was suggested, and characteristics of their stability were investigated.

The ABG-model was used to study the "metadynamics" of an immune network, i.e., of the total number and repertoire of clones with regard to continuous influx of clones, which possibly possess a new structure of variable regions of immunoglobulins, from the bone marrow into secondary lymphoid organs. The investigation of the dynamics of the model network has revealed, in particular, the following properties of self-regulation: the total number of clones in the network tends to some equilibrium value despite continuous influx of new clones; the number of clones in the network is in direct proportion to the intensity influx of clones and is inversely proportional to the probability of the event that random clones are complementary to each other; the number of cells and antibodies of each clone fluctuates in time; a particular repertoire of the network is a consequence of the "selection" promoted by idiotypic interactions with another clones.

It should be noted that the main problem of the verification of the results of modeling the immune networks is the absence of necessary data on the structure and parameters of idiotypic interactions in a real immune system. A very interesting result of modeling the idiotypic networks was the development of the methods of formalized representation of stereochemical structure of the antigen-recognizing part of immunoglobulins and receptors of B-cells (shape) (De Boer et al. [69, 71],

Segel and Perelson [308, 309], Weisbuch [343]). Mathematical modeling of the dynamics of immune idiotypic network represents an intensively developing branch in mathematical immunology.

A simple mathematical model of viral infection is presented in chapter 3 of this monograph. The model describes the humoral reaction to foreign antigen. Equations of this model rewritten in terms of the AB-model have the form:

$$\frac{dB(t)}{dt} = \alpha\xi(h)A(t-\tau)h(t-\tau) + m - d_B B(t),$$

$$\frac{dA(t)}{dt} = sB(t) - d_C h(t)A(t) - d_A A(t),$$

where τ is the delay equal to the time of the differentiation of B-cell into plasma cell and $h(t)$ stands for the amount of antigens. The system of equations of the simple model of humoral immune response has a number of terms common with the AB-model, but describes clonal dynamics for B-cells on the basis of differential equation with a delay. The suppressive effect of high concentrations of antigen (virus) is taken into account phenomenologically by the function $\xi(\cdot)$ decreasing with the growth of antigen (see Chapter 3).

2.2. Mathematical Models of Cellular Immune Response

The above mathematical models of humoral immune response are based on generalized phenomenological schemes of the processes of activation, proliferation, and differentiation of B-cells. Immune reactions to the majority of multiplying antigens (viruses, bacteria) develop as a result of intercellular interactions which may be contact or mediated by secreted cytokines. Antigen presenting cells, T-helper cells, B-lymphocytes, and others participate in these reactions. The antigen-presenting cells, in particular macrophages and dendritic cells, have the function of absorption, processing, and presentation of antigen to T-lymphocytes. Current understanding of mechanisms of development of humoral immune response to T-dependent antigens and cellular immune response allows us to construct mathematical models of cellular immune reaction with different levels of details. The investigations on modeling the regularities of T-cellular reactions were initiated rather recently. Let us discuss typical models of this type.

2.2.1. Model of T-helper cell reaction (McLean [211]). This model describes the dynamics of T-helper cell reaction to multiplying antigen involving the processes of initial activation of T-cells by antigen, subsequent proliferation of activated cells controlled by the interleukin-2, and the generation of memory cells. Suppression mechanisms are incorporated phenomenologically by the restriction of T-helper cell proliferation rate. The process of low-level background activation is also taken into account, based on cross-reactive stimulation or presentation of captured antigen. The model's variables are the quantities of: antigen (A); resting T-helpers (W), activated T-helpers (X), T-helper memory cells (M), and interleukin-2 (I). The model equation have the following form:

$$\frac{dA}{dt} = (r - \gamma X)A,$$

$$\frac{dW}{dt} = \Lambda - \alpha AW - \mu W + \sigma M,$$

$$\frac{dX}{dt} = \alpha AW - \frac{\rho IX}{1 + \xi X} + (\delta\alpha A + C)M - \mu X,$$

$$\frac{dM}{dt} = \frac{2\rho IX}{1 + \xi X} - (\delta\alpha A + C)M - \mu M - \sigma M,$$

$$\frac{dI}{dt} = \phi X - \beta IX - \psi I.$$

Three variants of interactions between multiplying antigen and immune system were separated out as a result of investigations: very slowly multiplying antigens that are injected in low doses may lead to persistent infections; the antigens with low rate of replication are eliminated, generating the immune memory; antigens with high rate of replication lead to the persistent infection characteristic of low level of T-cell population. The immune memory is explained in the framework of the model as a dynamic state that is attained when the antigen stimulation increases the number of activated T-helper cell subpopulations and memory cells up to some threshold level.

The model was used in various modification (McLean et al. [212, 213]) to study possible types of reactions to multiplying antigens, the mechanisms of long-term maintenance of immune memory, and to analyse the data on activation and division of Tcells.

2.2.2. Model of interaction between T-helper and antigen-presenting cells (Fishman, Perelson [100]). It is established that the antigen-presenting cells (APCs) that can be represented by macrophages, dendritic cells, and B-lymphocytes play a key role in initial activation of T-lymphocytes. A mathematical model was suggested describing the interactions between $CD4^+$ T-cells and APCs during immune response to multiplying antigen for the study of induction processes of T-helper reaction. The model has the following variables: Ag, population of foreign microorganisms which is a source of antigens; C_0, population of antigen-presenting cells possessing n independent sites for the presentation of an antigen by T-helpers; T, resting T-helpers; T^*, activated T-helpers. For the sake of simplicity the concentrations of free and bound sites on APC (S_0, S_1) were used rather than quantities of antigen-presenting cells C_j with varying numbers of occupied sites j, $j = 1, 2, \ldots, N$. The systems of model equations has the form:

$$\frac{dAg}{dt} = (q - eT^*)Ag,$$

$$\frac{dS_0}{dt} = nbC_PAg - k_bS_0T + (k_a + k_d)S_1 - d_CS_0,$$

$$\frac{dS_1}{dt} = k_bS_0T - (k_a + k_d)S_1 - d_CS_1,$$

$$\frac{dT}{dt} = a - d_TT - k_bS_0T + k_dS_1 + \mu k_rT^*,$$

$$\frac{dT_1^*}{dt} = k_aS_1 - k_rT^*.$$

It is assumed in the model that the elimination of antigen is proportional to the number of activated T-cells. The model was used for the analysis of delayed-type hypersensitivity reactions, which are the main type of defence mechanism in the infections caused by intracellular bacteria *Listeria monocytogenes, Mycobacterium tuberculosis*, parasites inducing such infections as malaria, helminthosis. The model parameters were estimated on the basis of experimental data.

The comparison was performed of the roles of two types of APCs: macrophages and dendritic cells in the induction of delayed hypersensitivity reactions. It was concluded that dendritic cell have higher antigen-presenting efficiency, which is caused by the differences in duration of the expression of antigens on the surface of cells.

This model and more simple modifications were used for the analysis of stability of equilibrium states and various dynamical regimes. The model permits the existence of two types of stationary states: state of no infection, with no antigen, and the state of coexistence, where an antigen persists against the background of immune reaction. It was found that the state of no infection is asymptotically unstable. In this connection, in order to preserve biological adequacy of the model, the number of antigens was zeroed when it reached a certain threshold level, (effective zero). Situations were found with the model, when the development of cellular reactions may lead not to the elimination of antigen, but to their suppression and low concentration value, which is interpreted as a carrier state. It was shown that low doses of infection by a pathogen may lead to chronic variants of infection, and that high doses may lead to the elimination of infectious agent, which is connected with different level of activation of cellular immune response.

2.2.3. Model of regulation of Th1/Th2 reactions. This model (Fishman and Perelson [101]) is a large-scale detailing of the above T-helper–APC model. This model was constructed to analyse the regularities of the development of humoral and cellular immune reactions on the level of Th1- and Th2-helper cell subpopulations. The model describes the processes of interactions between Th1- and Th2-cell populations mediated by the cytokines IFN-γ and IL-10 they respectively secrete. This interaction is described in the model as cross-suppression, so that IFN-γ suppresses the proliferation of Th2 and IL-10 suppresses the synthesis of IL-2 by Th1-cells and, as a result, the division intensity of Th1.

The dendritic cells perform the functions of antigen-presenting cells; it is assumed that each dendritic cell has n independent sites for the interaction with T-cells. In order to decrease the number of model's variables, the concentrations of free and bound sites are considered rather than APC with various number j of occupied sites.

The model describes the dynamics of 12 variables characterizing the concentrations of: the antigen $Ag(t)$; free sites on antigen-presenting cells $S_0(t)$; the sites on APC occupied by Th1-cells, $S_1(t)$; the sites on APC occupied by Th2-cells, $S_2(t)$; resting Th1-cells, $T_1(t)$; resting Th2-cells, $T_2(t)$; activated Th1-cells secreting IFN-γ, $T_1^*(t)$; activated Th2-cells secreting IL-10, $T2^*(t)$; activated Th1-cells which secrete no IFN-γ, $T_1^{*ns}(t)$; activated Th2-cells which do not proliferate $T_2^{*np}(t)$; the Interferon-γ, $IFN(t)$; the interleukin-10, $IL_{10}(t)$. The model's equations have the following form:

$$\frac{dAg}{dt} = [q - e_1 T_1 - e_2(T_2^* + T_2^{*np})]Ag,$$

$$\frac{dS_0}{dt} = nbC_pAg - k_{b_1}S_0T_1 + (k_{a_1} + k_{d_1})S_1 - k_{b_2}S_0T_2 + (k_{a_2} + k_{d_2})S_2 - d_CS_0,$$

$$\frac{dS_1}{dt} = k_{b_1}S_0T_1 - (k_{a_1} + k_{d_1})S_1 - d_CS_1,$$

$$\frac{dS_2}{dt} = k_{b_2}S_0T_2 - (k_{a_2} + k_{d_2})S_2 - d_CS_2,$$

$$\frac{dT_1}{dt} = a_1 - d_TT_1 - k_{b_1}S_0T_1 + k_{d_1}S_1 + (\mu k_p + k_r)T_1^* + k_{r_{ns}}T_1^{*ns},$$

$$\frac{dT_2}{dt} = a_2 - d_TT_2 - k_{b_2}S_0T_2 + k_{d_2}S_2 + (\mu k_p + k_r)T_2^* + k_{r_{np}}T_2^{*np},$$

$$\frac{dT_1^*}{dt} = k_{a_1}S_1 - (k_p + k_r + k_{ns}\frac{IL_{10}}{K_{IL_{10}} + IL_{10}})T_1^*,$$

$$\frac{dT_2^*}{dt} = k_{a_2}S_2 - (k_p + k_r + k_{np}\frac{IFN}{K_{IFN} + IFN})T_2^*,$$

$$\frac{dT_1^{*ns}}{dt} = k_{ns}\frac{IL_{10}}{K_{IL_{10}} + IL_{10}}T_1^* - k_{r_{ns}}T_1^{*ns},$$
$$\frac{dT_2^{*np}}{dt} = k_{np}\frac{IFN}{K_{IFN} + IFN}T_2^* - k_{r_{np}}T_2^{*np},$$
$$\frac{dIFN}{dt} = p_{IFN}T_1^* - d_{IFN}IFN,$$
$$\frac{dIL_{10}}{dt} = p_{IL_{10}}(T_2^* + T_2^{*np}) - d_{IL_{10}}IL_{10}.$$

While describing the suppressive action of IFN-γ and IL-10, the effect of saturation under high concentrations of these factors is taken into account phenomenologically. The estimations of the model's parameters were constructed using experimental data.

The model has four types of stationary solutions: one corresponds to the state with no infection, and another three correspond to persistent infection characterized by pronounced immune reaction on the level of Th1-cells, or Th2-cells, or Th1- and Th2-cells simultaneously. The state of no infection and the state of antigen persistence along with high levels of Th1- and Th2-cells are unstable in the framework of the model.

It was shown that there exists a dominant type of reaction on the level of separate populations of Th1 and Th2 rather than combined. The model and it's special cases were used to analyse the dynamics of humoral and cellular immune reactions to multiplying antigen with various initial doses of antigen and various relationships between the parameters that determine the activation rate of Th1- and Th2-cells by antigen-presenting cells and the efficiency of the elimination of a pathogen by humoral and cellular components of the reactions.

The parameters that determine if Th1- or Th2-cellular reaction predominate were isolated. The questions of treatment were discussed in the cases of parasite infections and HIV-infection whose pathogenesis was considered to be associated with the inadequacy of the components of dominant type of immune reaction on the level of Th1- or Th2-cells.

2.3. Mathematical Models of Immune Response During Viral Diseases

Let us discuss the basic processes of antiviral immunity (Doherty [84, 85], Picker [274], Zinkernagel [352, 353]). Having got into the organism of a host and infecting the cells of sensitive tissue, viruses begin to proliferate. The damage to cells of the target organ is the basis for the clinical manifestation of infection. Viruses give a signal to the immune system through certain antigenic structures on the surface of infected cells or antigen-presenting cells. Macrophages play a key role in the induction and increasing specific immune reaction. Cellular and humoral immune responses are activated during viral infections. The cellular immunity is represented basically by cytotoxic T-lymphocytes (CTLs), which perform their functions through contact with the cells of the target organ. The humoral response is mediated by B-cells, plasma cells and antibodies. T-helper cells are the most important regulatory elements for the development of cellular and humoral reactions. The activation and triggering the effector's functions of T-lymphocytes are mediated by the antigens presented by the lymph node macrophages or tissue macrophages together with the molecules of major histocompatibility complex of the class I or II. It orders the interactions in lymph tissue which contains the greatest number of macrophages expressing class II MHC. The Interleukin 2 and other factors secreted by T-helpers control locally the proliferation and differentiation of T- and B-lymphocytes. The antigen-specific CTL leave the lymphatic tissue after the proliferation and differentiation and are transported by the lymph and blood flux to a target organ where viruses are multiplying. CTL interrupts the replication process of viruses by delivering a lytic signal to the infected cells. The antibodies neutralize free viruses and enhance the elimination of viral particles from fluids. The defense against the secondary infection by the same virus strain is determined basically by quantity of memory B-cells and by neutralizing antibodies.

This organization of immune processes is common for all acute viral infections over an interval of 10–14 days. It is worth bearing in mind that nonspecific mechanisms of defence such as the interferon system, inflammation, etc., may make an essential contribution to the defence during some viral infections (influenza) over certain intervals of time.

Most of the papers on mathematical modeling of viral diseases in an organism are devoted to the investigations of infection caused by human immunodeficiency virus (HIV) (De Boer and Boerlijist [61], Essunger and Perelson [93], Intrator et al. [152], McLean and Kirkwood [212], McLean and Michie [214], Merill [217], Mosier and Sieburg [231], Nee [233], Nelson and Perelson [235, 236]). Mathematical models are used also in the studies of hepatitis B (Marchuk et al. [202, 203], Payn et al. [256], Romanyukha et al. [299]), influenza (Bocharov and Romanyukha [39]), bacterial infections (Karpov and Romanyukha [163]), mixed infections (Marchuk et al. [206]), parasitic infections (Agur [4], Hellriegel [129], Maizels et al. [192]).

Though any model of immune reaction may be considered as a model of viral infection, if antigen is understood as a virus population, we include into this set of models just the models that reflect it in their names and were used mostly for the analysis of interactions between the viruses and immune system. Let us consider the most frequently used models.

2.3.1. Models of interaction between HIV and immune system. The infection caused by human immunodeficiency virus (HIV), has a number of peculiarities:

– the viruses damage $CD4^+$ T-cells of the immune system;

– there is the phase of initial virus-positive blood (up to 2–4 weeks) which is characterized by the development of immune response and by the appearance of neutralizing antibodies against various HIV antigens, and then this phase goes into a latent period;

– long incubation period of various durations (up to 10 years), when the level of $CD4^+$ T-cells in blood of patients with AIDS decreases slowly but steadily;

– the presence of genetically different strains of HIV both in one patient and in different individuals related to the replication errors;

– HIV strains may manifest essential differences in their biological properties such as the replication rate in $CD4^+$ T-cells, cytopathicity, cellular tropism, antigenic characteristics;

– the diversity of strains increases during the incubation period, though it decreases in the last phases of HIV infection.

It should be noted, as a whole, that the phenomenology of the infection caused by HIV viruses and the mechanisms (molecular, cellular, and systemic) of its development are, no doubt, much more complex and diverse, and are the subject of extended investigations.

It is possible to separate a number of directions in mathematical modeling of the dynamics of HIV infection that study, in particular, the effect of antigen diversity of virus mutants and mechanisms that help some of them escape the immune control at the initial period of infection.

Genetic diversity of HIV is considered in the series of works by Nowak et al. [242–243, 245–248] as the principal mechanism of the development of AIDS. A number of mathematical models of similar structure was suggested that describe the interactions between an ensemble of viral strains ("quasi-species") and the immune system represented by T-helper populations of various specificity. The variables of the basic model are respectively the densities of virus strains, $v_i(t)$ and clones of T-helpers, $x_i(t)$ that are strain-specific or could be cross-reactive too. The structure of model equations reflects the following processes, determining presumably the dynamics of HIV: continuous appearance of new antigenic variants of viruses; viruses of each strain can infect and destroy any $CD4^+$ T-cell, no matter what is their specificity; the immune reaction against HIV includes specific reaction against individual strains and cross-reaction neutralizing all strains. There are several versions of the model that differ by details in the description of T-helper population. Equations of the basic model have the form:

$$\frac{dv_i}{dt} = v_i(r - px_i),$$

$$\frac{dx_i}{dt} = kv_i - uvx_i, \quad i = 1, \ldots, n, \quad v = \sum v_i.$$

The main result of the investigations with this model is the formulation of a threshold condition for the number of different strains ("diversity"). Unlimited growth of the total quantity of viral strains begins when the number of strains exceeds this threshold. This condition takes the following form under the assumption that quasi-species differ just with the antigenicity:

$$n > n_C = \frac{pk}{ru}.$$

In the more realistic case, when the parameters of HIV strains are different, this condition is transformed as follows:

$$\sum_{i=1}^{n} \frac{r_i u_i}{p_i k_i} > 1.$$

The nonlinear character of interactions between the viruses and immune system is emphasized, which stipulates the appearance of the threshold conditions of this kind. In the framework of the model, fast-multiplying viruses with a high level of cytopathicity or strains poorly recognized by the immune system contribute considerably to the threshold condition. It has been shown that the interactions between the strains are of a competitive character when they pass over the threshold.

The central result of the model is a hypothesis that the increase in the antigenic diversity of HIV mutants is rather the cause than a consequence of AIDS, i.e., there is a threshold of diversity up to which the immune system is capable of neutralizing the replication of viruses, but it is not capable of restricting the growth when the threshold is exceeded. As a result, the viral population escapes the immune control, which leads to the decrease in total number of $CD4^+$ T-cells. The model was used for the interpretation of results of long-term observations of infected patients. In particular, a conclusion was made about the higher efficiency of immunotherapy and chemotherapy at the initial periods of infection. The qualitative character of model predictions is emphasized, since the data for precise estimation of model's parameters are absent.

2.3.2. Model of immune reaction to HIV. The threshold condition of diversity contains the parameters associated with cytopathicity of viruses, replication rate, and antigenicity. It is possible, in this sense, to consider the threshold condition as a threshold condition of virulence. A mathematical model (De Boer and Boerlijst [61]) close in structure to the above model was used to determine the threshold conditions linking the diversity and a generalized characteristic of "virulence" that combines infectivity and antigenicity of strains. The variables characterizing the quantity of the ith strain of a virus, $v_i(t)$, and of the clone $CD4^+$ T-cells specific to ith strain, $x_i(t)$, are included in the model. The model's equations have the form:

$$\frac{dv_i}{dt} = v_i(r - cx_i),$$

$$\frac{dx_i}{dt} = \sigma + x_i\left(\frac{pv_i}{k + v_i} - 1 - Iv_i\right), \quad i = 1, \dots, n\ .$$

The existence of a stable steady state was shown, which is interpreted as a persistence regime of a virus, and a domain was separated in the space of model's parameters corresponding to AIDS. A dimensionless parameter (of the virulence) was introduced, determining local and global bifurcations in the model, $\nu = kI$. The threshold condition for the bifurcations is $n\nu = (p^{1/2} - 1)^2$. Therefore the number of quasi-species which is sufficient for HIV to avoid the immune control is $n > (p^{1/2} - 1)^2/\nu$. In the framework of this approach the virulence and diversity are equivalent parameters from the standpoint of the AIDS mechanism.

There exist mechanisms permitting slowly multiplying strains to escape the neutralizing action of immune system at the beginning of HIV infection.

2.3.3. Model of interaction between HIV, macrophages and $CD4^+$ T-cells. Differences in the replication rates permit us to conditionally divide the whole virus population into groups of fast replicating and slowly replicating virus strains, and to investigate their dynamics in organism with the help of models. A model was constructed by Nelson and Perelson in [235] for the study of regularities of the escape of immune control by slowly replicating HIV strains. The following characteristics of infections are described in the model: free viruses, $V(t)$; antigenic material, $A(t)$; macrophages, $M(t)$; infected macrophages,

$M^*(t)$; antigen-presenting macrophages, $\tilde{M}(t)$; nonspecific and specific $CD4^+$ T-cells, $T(t)$ and $\tilde{T}(t)$. The model was formulated as a system of nonlinear delay-differential equations:

$$\frac{dV}{dt} = k_{MV}M^* + NL_T - (\omega_V + k_{VT}(T + \tilde{T}) + (k_{VM} + k_u)M)V,$$

where

$$L_T(t) = k_{VT}V(t-\tau)(T(t-\tau) + \tilde{T}(t-\tau)) + k_s p_p p_t \tilde{T}(t-\tau)M(t-\tau),$$

$$\frac{dA}{dt} = Ag_L NL_T + Ag_V\omega_V V - (k_u + \omega_A)A,$$

$$\frac{dM}{dt} = \sigma_M - (k_{VM} - k_u(V + A) + \omega_M)M + \omega_{\tilde{M}}\tilde{M},$$

$$\frac{dM^*}{dt} = k_{VM}V(t-\tau)M(t-\tau) - \omega_{M^*}M^*,$$

$$\frac{d\tilde{M}}{dt} = k_u(V + A) - \omega_{\tilde{M}}\tilde{M},$$

$$\frac{dT}{dt} = \sigma_T - (k_{VT}V + \omega_T)T,$$

$$\frac{d\tilde{T}}{dt} = \sigma_{\tilde{T}} - (k_s p_p p_t M^* + k_{VT}V + \omega_T)\tilde{T}.$$

The interaction between viruses, macrophages and T-helpers are described in detail, corresponding to modern conceptions. Estimates for all the model's parameters were obtained. The results of modeling allow one to suggest a hypothesis (which can be verified experimentally) about the properties of strains that escape the immune control: tropism to macrophages, weak replication in macrophages, suppression of T-cell activation. A model of similar structure was used in [236] for theoretical study of the methods of therapy against HIV-infection by means of defective interfering viruses (DIV) created by the methods of gene engineering and suppressing or completely blocking the replication of HIV-1 viruses in a jointly infected cell. the quantitative character of dynamical analysis of the interaction between HIV, DIV and $CD4^+T$-cells should be emphasized.

In addition to the above mathematical models of virus infections, a direction in the investigations should be noted connected with the modeling of cellular immune reactions to tumor cells. The structures of these models are very diverse, but have many elements common to the above models; the dimension of the system of equations varies from 2 to 8. Since models of tumor growth are not discussed in this monograph, we just note the works containing the description of immune reactions to tumor antigens (Smirnova [317], De Boer and Hogeweg [65], Kuznetsov et al. [174–176], De Boer and Hogeweg [62]).

We represent below the mathematical models of viral diseases with different degrees of detailing, that are used in this monograph, in order to compare them with the above models. They have in common the description of interactions between the infectious agent, immune system, and target organ. It permits study of both laboratory and clinical characteristics of infectious disease and "to close" the description of processes which is necessary for the application of models in clinical immunology.

2.3.4. Simple model of infectious disease. A simple model of infectious disease (see Chapter 3) is one of well-known mathematical models in this field (Hairer et al. [123], Adomani and Adomani [2], Srinivas and Pattabhi Ramacharyulu [319]). The basic factors of infectious process, i.e., the model's variables, are: the quantities of viruses, $V(t)$; of antibodies, $F(t)$; of plasma cells, $C(t)$; and the mass of affected target organ, $m(t)$ (generalized characteristic of a damage which was inflicted to an organ by viruses). The introduction of the mass of a damaged organ into the system of model's variables allows one to close the description of immune reaction of organism during infectious disease, regarding it as a simple model of infectious disease where the basic processes are taken into account. The simple mathematical model of infectious disease is formulated as a system of four nonlinear delay differential equations subdivided into blocks of equations representing separate processes:

1. Block of the virus dynamics

$$\frac{dV}{dt} = \beta V - \gamma F V.$$

2. Block of the humoral immune response dynamics

$$\frac{dC}{dt} = \alpha\xi(m)F(t-\tau)V(t-\tau) - \mu_C(C - C^*),$$

$$\frac{dF}{dt} = \rho C - \eta\gamma V F - \mu_F F.$$

3. Block of the target organ dynamics

$$\frac{dm}{dt} = \sigma V - \mu_m m.$$

The function $\xi(m)$ decreasing with increase of m in the equation for plasma cells permits us to describe phenomenologically the decrease in productivity of immune system as a function of the degree of damage to target organ. The model was used for the study of *qualitative* regularities in the course of an infectious process and for the interpretation of some results of clinical observations, in particular, on the stimulation of the immunity during chronic infections and for the study of regularities in the development of mixed infections (Belykh [26], Marchuk [194] these questions are discussed in detail in Chapter 4).

The problem of *quantitative* modeling of specific virus and bacterial infections requires the detailed description of antiviral immunity for the analysis of mechanisms of defence and immunopathology, the role of specific and nonspecific mechanisms of defence, the contribution of humoral and cellular components to the struggle against infection, sensitivity of the infection dynamics to variations of parameters of antigen presentation, size of specific clones, replication rate of viruses, virus cytopathicity, etc.

2.3.5. Mathematical model of antiviral immune response. Foreign pathogens (viruses and bacteria) that get into organism initiate a cascade of protective reactions: the activation of the interferon system, the division and differentiation of lymphocytes, the inflammation, etc. The intensity of antiviral and antibacterial reactions is determined by the intercellular interactions in peripheral lymphoid tissues whereas the efficiency of defence is a consequence of a "number game" that characterizes the kinetics of infectious agents and immune processes (Doherty [85], Zinkernagel [352], Zinkernagel et al. [353]). One of approaches to the determination of these "constants" is associated with the study of indices of immune status (Nisevich et al. [240], Marchuk and Berbentsova [197, 198]).

Another approach to the determination of these factors as quantitative parameters of mathematical models is described in this book. For this purpose multiparameter mathematical models of immune response were proposed which are quantitatively correlated with experimental and clinical data and describe in a generalized way typical or normal course of infectious diseases (Marchuk et al. [200–202, 206]). Corresponding estimates of model's parameters characterize the kinetics of evolutionarily fixed defence reactions on a population level and provide for further transit to individual estimation of parameters of immune defence by the study of deviations as related to these characteristic values of parameters.

A mathematical model of antiviral immune response is used to this end (see Chapter 4). The model was formulated in the framework of clonal selection theory and of the principle of MHC-restricted antigen recognition by T-cells; it deals with the three-cellular schemes of the induction of immune response. The variables of this mathematical model of antiviral immune response describe the quantities of the following populations: free viruses, $V_f(t)$; activated (antigen-presenting and interleukin-I producing) macrophages, $M_V(t)$; T-helpers of cellular immunity $Th1$, $H_E(t)$; T-helpers of humoral immunity $Th2$, $H_B(t)$; cytotoxic T-lymphocytes, $E(t)$; B-lymphocytes, $B(t)$; plasma cells, $P(t)$; the antibodies specific to viruses, $F(t)$; infected cells of a target organ, producing viruses, $C_V(t)$; damaged (destroyed) cells of the target organ, $m(t)$. The model is formulated as a system of nonlinear delay differential equations:

1. Block of the virus dynamics

$$\frac{dV_f}{dt} = \nu C_V + nb_{CE}C_V E - \gamma_{VF}V_f F - \gamma_{VM}M^* V_f - \gamma_{VC}V_f(C^* - C_V - m).$$

2. Block of the humoral immune response dynamics

$$\begin{aligned}\frac{dH_B}{dt} &= b_H^B\left[\xi(m)\rho_H^B M_V(t-\tau_H^B)H_E(t-\tau_H^B) - M_V H_B\right]\\ &\quad - b_p^{H_B} M_V H_B B + \alpha_H^B(H_B^* - H_B),\end{aligned}$$

$$\begin{aligned}\frac{dB}{dt} &= b_p^B\left[\xi(m)\rho_B M_V(t-\tau_B)H_B(t-\tau_B)B(t-\tau_B) - M_V H_B B\right]\\ &\quad + \alpha_B(B^* - B),\end{aligned}$$

$$\frac{dP}{dt} = b_p^P \xi(m)\rho_P M_V(t-\tau_P)H_B(t-\tau_P)B(t-\tau_P) + \alpha_P(P^* - P),$$

$$\frac{dF}{dt} = \rho_F P - \gamma_{FV} F V_f - \alpha_F F.$$

3. Block of the target organ dynamics

$$\frac{dC_V}{dt} = \sigma V_f(C^* - C_V - m) - b_{CE}C_V E - b_m C_V,$$

$$\frac{dm}{dt} = b_{CE}C_V E + b_m C_V - \alpha_m m, \quad \xi(m) = 1 - \frac{m}{C^*}.$$

4. Block of the antigen-presenting macrophages dynamics

$$\frac{dM_V}{dt} = \gamma_{MV} M^* V_f - \alpha_M M_V.$$

5. Block of the cytotoxic immune response dynamics

$$\begin{aligned}\frac{dH_E}{dt} &= b_H^E\left[\xi(m)\rho_H^E M_V(t-\tau_H^E)H_E(t-\tau_H^E) - M_V H_E\right]\\ &\quad - b_p^{H_E} M_V H_E E + \alpha_H^E(H_E^* - H_E),\end{aligned}$$

$$\begin{aligned}\frac{dE}{dt} &= b_p^E\left[\xi(m)\rho_E M_V(t-\tau_E)H_E(t-\tau_E)E(t-\tau_E) - M_V H_E E\right]\\ &\quad - b_{EC}C_V E + \alpha_E(E^* - E).\end{aligned}$$

The most important features of the model are: the description of the interactions in a target organ, the development of cellular and humoral immune reactions in terms of variables corresponding to the basic characteristics of immune status, and of the suppressive influence of severe damage to the target organ on the efficiency of immune processes.

This model is a generalization of a simple model; the above blocks correspond to individual equations of the simple model and demonstrate additional processes taken into account by the generalized model.

The model was used for quantitative investigation of the kinetics of the following diseases: infection caused by hepatitis B viruses, and by influenza A viruses (Marchuk et al. [205], Romanyukha and Bocharov [38, 39, 299], Marchuk et al. [202–206], Sidorov and Romanyukha [312]). The choice of infections was motivated by the quest for the study both common quantitative regularities and the peculiarities of the kinetics of immune reactions during diseases caused by two extreme infectious agents. Results of this modeling are discussed in Chapter 7.

2.3.6. Mathematical model of antibacterial immune response. Pathologic potential of bacterial infection is determined by the virulence of bacteria, dose of infection, and state of defence systems of organism. Normally small populations of bacteria are controlled by ordinary mechanisms of antibacterial defence including physical factors (mucociliary apparatus), phagocyting cells (alveolar macrophages, neutrophils in lungs), and humoral factors (complement, sufractants, immunoglobulins). However, in some conditions bacteria overcome these mechanisms. In these cases the development of humoral immune response and nonspecific reaction of inflammation is necessary to clear out the bacteria (Marchuk and Berbentsova [197, 198]).

Mathematical modeling of the dynamics of immune response to bacterial infections is dealt with just in isolated works [163, 206]. A mathematical model of antibacterial immune response usually describes the following characteristics of bacterial infection and immune reaction: the number of bacteria in a target organ, $W(t)$; the number of activated macrophages, $M_W(t)$; the number of T-helpers $Th2$ for B-lymphocytes, $B(t)$; the number of plasma cells, $P(t)$; the number of antibodies specific to bacteria, $F(t)$; damaged part of a target organ, $m(t)$. The system of equations for mathematical model of antibacterial immune response is a combination of the blocks 1, 2, and 4 of the antiviral model and of the 3th block of simple model. The mathematical model of antibacterial immune reaction is used for quantitative description of the dynamics of acute and chronic bacterial infections in lungs (see Chapter 8).

The diversity of mathematical models of immune processes reflects the complexity of the immune systemm, Perelson [268]. It is obvious that the approaches to the analysis of regularities of immune reactions on the basis of mathematical models with low level of details (micro models) and with high level of details (macro models) complement each other. As a rule, models of the first type are used for the analysis of qualitative fundamental laws of immune reactions that are realized on the long time intervals. Models of the second type are constructed as a rule for quantitative organization of conceptions and observational data and for *quantitative* predictions on concrete time interval. Furthermore, in the framework of idiotypic network models and multi-clonal dynamics models the processes are investigated of the same nature associated with variations in active repertoire of immune system, which can be manifested clinically as the formation of secondary immunodeficiencies.

CHAPTER 3

Simple Mathematical Model of Infectious Disease

In this chapter a simple mathematical model will be constructed and investigated. The mathematical model, of course, is extremely approximate and requires further detailed elaboration. However, even in this form it allows one to include in the system various factors essential for infectious disease dynamics, and this model seems to be useful for investigating the general picture of a disease's course and for finding an explanation of some results of observations. It is also possible that some results of the theory will be used for finding effective methods of treatment. Finally, it should be noted that we shall try to construct a model averaged over a large group of patients. We shall not consider the question of description of individual characteristics of disease dynamics.

The simple mathematical model will be constructed on the basis of an equilibrium relation for each component that participate in an immune response. This conception implies that partial characteristics of the immune systems functioning are not essential for the analysis of disease dynamics, but principal features of the course of the organism's immune reaction play a key role. Therefore, while constructing the simplest model we shall not distinguish cellular and humoral components of immunity fighting the antigens that have penetrated into an organism. We shall just assume that an organism does have such defence components. We shall call them *antibodies*, no matter whether we deal with a cellular-lymphoid system or a humoral-immunoglobulin system of immunity. In this model we shall also assume that an organism has sufficient resources of macrophages utilizing the products of immune reaction and also other nonspecific factors necessary for normal functioning of immune system. In this connection we confine ourselves to the consideration of three components: antigen, antibody, and plasma cell that produces antibodies. Antigenes will be represented by either pathogenic bacteria or viruses in our model.

It is worthy of notice that during disease the degree of damage of an organ subjected to antigen attack is of great significance, since it leads finally to lowering of the immune system's activity. This phenomenon, naturally, must be taken into account in mathematical models.

Notice that the simple model in this interpretation permits variations which can help us obtain probable explanations of some important peculiarities in the functioning of the immune system, such as formation of subclinical, acute, and chronic processes of disease, to clear out the role of temperature effects, to investigate the conditions favourable for the development of complications of disease, and also to find the mechanisms of biostimulation, etc. These questions will be considered in this chapter.

3.1. Construction of Simple Model of Disease

We shall assume that the basic acting factors of an infectious disease are as follows:

1) *Concentration of pathogenic multiplying antigens*, $V(t)$.

2) *Concentration of antibodies*, $F(t)$. By *antibodies* we mean substrates of immune system that neutralize antigens (immunoglobulins, cell receptors).

3) *Concentration of plasma cells*, $C(t)$. This is the population of carriers and producents of antibodies (immunocompetent cells and producents of immunoglobulins).

4) *Relative characteristic of affected organ*, $m(t)$.

Let us proceed now with the construction of the model's equations. The **first equation** will describe a change in the number of antigens in an organism:

$$dV = \beta V dt - \gamma F V dt. \tag{3.1.1}$$

The term on the left-hand side of this equation describes the increase of antigens dV during the interval dt. Naturally, this increment is proportional to V and to a value β which we will call the *coefficient of virus multiplication*. The term $\gamma FV dt$ describes the number of antigens neutralized by the antibodies F during the time dt. Indeed, the number of such viruses will be obviously proportional to the number of antibodies in organism as well as to the number of antigens; γ is a coefficient connected with the probability for an antigen to be neutralized after encountering the antibodies. Dividing (3.1.1) by dt we obtain

$$\frac{dV}{dt} = (\beta - \gamma F)V. \tag{3.1.2}$$

Now proceed with deriving the **second equation**, which will describe the growth of plasma cells. To this end we use the simplest hypothesis on the formation of cascade populations of plasma cells. As was shown in Section 1.1, an immunocompetent B-lymphocyte is stimulated by the complex of an antigen with Ig receptor in presence of a signal from specific T-helper activated by antigen-presenting macrophages, and starts the cascade process of formation of the cells which synthesize the antibodies neutralizing the antigens of this kind. Since in our model the antibodies are the substrates capable of binding the antigens, the number of lymphocytes stimulated in this way will be proportional to VF. Thus we obtain the relationship describing the increase of plasma cells over a normal level C^* that is the constant level of plasma cell in a healthy organism:

$$d(C - C^*) = dC = Q(t - \tau)dt, \tag{3.1.3}$$

$$Q(t) = \alpha FV. \tag{3.1.4}$$

A more complete equation has the form:

$$dC = \alpha F(t - \tau)V(t - \tau)dt - \mu_c(C - C^*)dt. \tag{3.1.5}$$

The first term on the right-hand side of (3.1.5) describes the formation of plasma cells; τ is the time of formation of the cascade of plasma cells; α is a coefficient allowing for the probability of antigen–antibody collision, the stimulation of cascade reaction, and the number of newly generated cells. The second term in this formula describes the decrease in the number of plasma cells due to aging; μ_c is a coefficient equal to the inverse value of plasma cell life time. Divide (3.1.5) by dt to obtain

$$\frac{dC}{dt} = \alpha F(t-\tau)V(t-\tau) - \mu_c(C - C^*). \tag{3.1.6}$$

In order to obtain the **third equation** we calculate the balance of the number of antibodies reacting with antigens, starting from the relationship

$$dF = \rho C dt - \eta\gamma F V dt - \mu_f F dt. \tag{3.1.7}$$

The first right-hand term $\rho C dt$ describes the generation of antibodies by plasma cells during the time dt; ρ is the rate of antibodies production by one plasma cell. The second term $\eta\gamma F V dt$ describes the decrease in the number of antibodies during the time dt due to the binding with antigens. As was noted while deriving equation (3.1.2), the number of eliminated antigens during the time dt that were neutralized by antibodies was equal to $\gamma F V dt$. If the neutralization of one antigen requires η antibodies, we arrive at this very term. The third term $\mu_f F dt$ describes the drop in antibodies population due to aging, where μ_f is a coefficient inversely proportional to the decay time of antibodies. Divide (3.1.7) by dt to obtain

$$\frac{dF}{dt} = \rho C - (\mu_f + \eta\gamma V)F. \tag{3.1.8}$$

The equations we have constructed do not account for weakening of an organism during disease due to decrease in the activity of organs providing the delivery of immunologic materials: leukocytes, lymphocytes, antibodies, etc., that are necessary for the struggle with multiplying antigens. Let us adopt as a hypothesis that the productivity of such an organ depends on the degree of damage of the target organ. To this end, introduce an equation for relative characteristic of the target organ damage. Let M be corresponding characteristic of a healthy organ (mass or area), and let M' be corresponding characteristic of healthy part of damaged organ. Let us define m as follows:

$$m = 1 - \frac{M'}{M}, \tag{3.1.9}$$

that is, m is relative characteristic of damage to the target organ. For an intact organ m is equal to zero naturally, and for completely destroyed one m equals 1. We introduce the **fourth equation** governing this value:

$$\frac{dm}{dt} = \sigma V - \mu_m m. \tag{3.1.10}$$

The first term in the right-hand side of (3.1.10) characterizes the degree of damage to an organ. It is assumed that during the time dt the increase in relative value of damaged organ is proportional to the number of antigens described by the term σV, where σ is a constant that is special for each disease.

A decrease in this characteristic is caused by the recovery activity of an organism. This term depends on m with the proportionality coefficient μ_m that is the inverse of the value of the period of recovery of restoration of an organ by e times.

It is absolutely clear that under conditions of serious damage of the vitals the antibody production decreases. This process is fatal for an organism and leads to lethal outcome. In many cases lethal outcome is caused by this very circumstance. It is possible to describe the factor of vitals damage with equation (3.1.6) if we substitute the product $\alpha\xi(m)$ for the coefficient α.

A typical scheme for the function $\xi(m)$ is presented in Fig. 16. In this figure the curve $\xi(m)$ is equal to 1 within the interval $0 \le m \le m^*$. It means that the efficiency of the immune system organs in this interval does not depend on the heaviness of disease. But when $m^* \le m \le 1$, their productivity decreases rapidly, which corresponds to the linear segment of the curve on this interval.

It is natural that in real conditions the $\xi(m)$ diagram may have more complex shape, but qualitatively this curve will always consist of the constant part $\xi = 1$ in the beginning of the argument m changes and decreasing (maybe in a nonlinear way) part when the argument's value increases further. For various diseases the steepness of this part of the curve as well as the value m^* will be different.

Thus we arrive at the following system of nonlinear delay differential equations:

$$
\begin{aligned}
\frac{dV}{dt} &= (\beta - \gamma F)V, \\
\frac{dC}{dt} &= \xi(m)\alpha V(t-\tau)F(t-\tau) - \mu_c(C - C^*), \\
\frac{dF}{dt} &= \rho C - (\mu_f + \eta\gamma V)F, \\
\frac{dm}{dt} &= \sigma V - \mu_m m.
\end{aligned}
\tag{3.1.11}
$$

Specify the initial values for $t = t^0$ to system (3.1.11). Initial conditions for the delay-differential equations that are usually set on the interval $[t^0 - \tau, t^0]$. However, as follows from the biological sense of the process under consideration, there were no viruses in an organism until the moment of infection t^0: $V(t) \equiv 0$ for $t < t^0$, and therefore, the initial values can be assigned in the point t^0. Below, when considering initial values for this type of equations, their assignment in a point t^0 will mean that $V(t) \equiv 0$ for $t < t^0$. We have

$$
\begin{aligned}
V(t^0) = V^0, \;\; C(t^0) = C^0, \\
F(t^0) = F^0, \;\; m(t^0) = m^0.
\end{aligned}
\tag{3.1.12}
$$

Since the model is autonomous we may assign $t^0 = 0$ with no loss of general applicability. Let us call the system of equations (3.1.11) with initial values (3.1.12) as *simple mathematical model of infectious disease.*

One must to keep in mind that our simple model deals with joint population of immunocompetent and antibody-producing cells $C(t)$. When there are no viruses in an organism, $C(t) = C^* > 0$, i.e., C^* is, in fact, the normal level of immunocompetent cells in a healthy organism. In the case when these cells are absent $C^* = 0$ and an organism is tolerant (nonreceptive) to a given antigen. However, it may occur that an organism has no information about a given antigen and, therefore, has no immunocompetent cells against it. In such cases it is possible that the reaction involves immunocompetent cells with similar specific receptors capable of provoking an immune response to this antigen. Let us assume that an organism has a nonzero level of cells C^* with their own receptors F^*, capable of provoking an immune reaction, and let us identify this with above case. A more precise starting mechanism of immune reaction can be traced only with more complex mathematical models. Let us make several remarks.

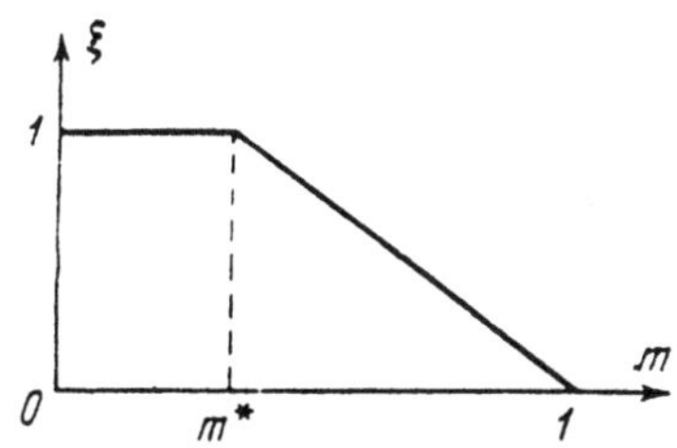

Fig. 16. Diagram of the function $\xi(m)$.

First, $0 \leq m(t) \leq 1$ by definition, but this fact is not expressed explicitly in the model. It is assumed that in the case of complete destruction of an organ, pathogenic antigens have nothing to damage in this organ, i.e., $dm/dt = 0$ if $m(t) = 1$ $(t > t^0)$.

Second, we take $t^0 = 0$ as the initial value in (3.1.11) and assume that the initial values are furthermore nonnegative and that all parameters of the model are positive and constant values.

3.2. Qualitative Analysis of Simple Model of Infectious Disease

3.2.1. General results. The system of equations (3.1.11) with initial conditions (3.1.12) describes the dynamics of pathologic infection development during immune response. We assume that all the constants in the system of equations are nonnegative; it follows directly from their biological meaning. The function $\xi(m)$ is continuous and nonnegative. Then it is easy to show that under nonnegative initial conditions for $t^0 = 0$

$$V^0 \geq 0, \quad C^0 \geq 0, \quad F^0 \geq 0, \quad m^0 \geq 0, \tag{3.2.1}$$

there exists a unique solution to problem (3.1.11), (3.1.12) for $t = t^0 = 0$ for all $t \geq 0$.

As was proven in [26], under given assumption for all $t \geq 0$ the solution of problem (3.1.11), (2.1.12) is continuous and nonnegative:

$$V(t) \geq 0, \quad C(t) \geq 0, \quad F(t) \geq 0, \quad m(t) \geq 0, \tag{3.2.2}$$

The existence and uniqueness theorem for the solution of initial value problem (3.1.11), (3.1.12) allows us to use the simple model of an infectious disease for the interpretation of clinical situations.

3.2.2. Stationary solutions. The system of equations (3.1.11) has stationary solutions. To find them we equate all the derivatives to zero. Thus we obtain

$$\begin{aligned}&(\beta - \gamma F)V = 0,\\ &\xi(m)\alpha V F - \mu_c(C - C^*) = 0,\\ &\rho C - (\mu_f + \eta\gamma F)V = 0,\\ &\sigma V - \mu_m m = 0.\end{aligned} \tag{3.2.3}$$

Since variables V and F do not depend on time $V(t-\tau) = V = \text{const}$, $F(t-\tau) = F = \text{const}$. Notice, that the value C^* is connected with F^* by the relationship

$$C^* = \mu_f F^* / \rho,$$

where C^* and F^* are C and F values for a healthy organism for $V = 0$. It is easy to see that one of the stationary solutions is trivial and describes the healthy state of an organism:

$$V = 0, \quad C = C^*, \quad F = F^* = \rho C^*/\mu_f, \quad m = 0. \tag{3.2.4}$$

Here the concentration of antigen population and damaged mass of an organ are equal to zero, but the quantities of plasma cells C and antibodies F correspond to the values of immunological status of a healthy man.

We shall prove that this state is asymptotically stable for $\beta < \gamma F^*$. Consider small deviations of unknown functions from the equilibrium state (3.2.4) assuming that

$$V = V', \quad C = C^* + C', \quad F = F^* + F', \quad m = m'. \tag{3.2.5}$$

We substitute this expressions into the system of equations (3.1.11) and, assuming that V', C', F', and m' are small, neglect small variables of the second order of smallness to obtain the following system:

$$\begin{aligned}
&\frac{dV'}{dt} - (\beta - \gamma F^*)V' = 0,\\
&\frac{dC'}{dt} + \mu_c C' = \alpha F^* V'(t-\tau),\\
&\frac{dF'}{dt} + \mu_f F' = \rho C' - \eta\gamma F^* V,'\\
&\frac{dm'}{dt} + \mu_m m' = \sigma V'.
\end{aligned} \tag{3.2.6}$$

This system will be solved under the following initial conditions:

$$V' = \varepsilon_1, \quad C' = \varepsilon_2, \quad F' = \varepsilon_3, \quad m' = \varepsilon_4 \quad \text{for} \quad t = 0. \tag{3.2.7}$$

Solving the first equation of (3.2.6) for $V' = \varepsilon_1,\ t = 0$, we obtain the expression for V':

$$V' = \varepsilon_1 e^{(\beta - \gamma F^*)t}. \tag{3.2.8}$$

If the following inequality is satisfied,

$$\beta < \gamma F^*, \tag{3.2.9}$$

then it is necessary and sufficient condition for solution (3.2.8) to tend to zero. Let us rewrite solution (3.2.8) as follows:

$$V' = \varepsilon_1 e^{-(\gamma F^* - \beta)t} = \varepsilon_1 e^{-\beta_1 t}, \tag{3.2.10}$$

where $\beta_1 = \gamma F^* - \beta > 0$ by virtue of (3.2.9).

Since V' differs from zero only for $t \geq 0$, then $V' = 0$ on the interval $[-\tau, 0)$ and, therefore, $V'(t-\tau) = 0$ for all $t < 0$. This means that a solution of the second equation from (3.2.6) for $t < \tau$ will have the form

$$C' = \varepsilon_2 e^{-\mu_c t}. \tag{3.2.11}$$

For $t \geq \tau$, taking into account the initial condition $C'(\tau) = \varepsilon_2 e^{-\mu_c \tau}$ obtained from (3.2.11), and the fact that according to (3.2.10) $V'(t-\tau) = \varepsilon_1 e^{-\beta_1 (t-\tau)}$, the solution for $C'(t)$ has the form:

1) in the case $\beta_1 \neq \mu_c$

$$C'(t) = \varepsilon_2 e^{-\mu_c t} + \frac{\alpha F^*}{\mu_c - \beta_1}\varepsilon_1\left[e^{-\beta_1(t-\tau)} - e^{-\mu_c(t-\tau)}\right]; \qquad (3.2.11a)$$

2) in the case $\beta_1 = \mu_c$

$$C'(t) = \varepsilon_2 e^{-\mu_c t} + \alpha F^* \varepsilon_1 (t-\tau) e^{-\beta_1(t-\tau)}. \qquad (3.2.11b)$$

It is easily seen that in both cases, if the inequality (3.2.9) holds, i.e., when $\beta_1 > 0$, a small deviation from the stationary solution (3.2.5) tends to zero with time. The general solution of the third equation is

$$F'(t) = e^{-\mu_f t}\left[\int \rho C'(s) e^{\mu_f s} ds - \int \eta\gamma F^* V'(s) e^{\mu_f s} ds + A\right], \qquad (3.2.12)$$

where the constant A is determined through corresponding initial conditions. Considering, as has been done for the second equation, all possible relationships between the model's coefficients resulting in various expressions in (3.2.12), one can be assured that $F'(t) \to 0$ for $t \to \infty$, if $\beta_1 > 0$. The same statement is true for a solution of the fourth equation from (3.2.6) which has a structure analogous to (3.2.11a), (3.2.11b):

a) in the case $\beta_1 \neq \mu_m$

$$m'(t) = \varepsilon_4 e^{-\mu_m t} + \frac{\sigma}{\mu_m - \beta_1}\varepsilon_1\left(e^{-\beta_1 t} - e^{-\mu_m t}\right);$$

b) in the case $\beta_1 = \mu_m$

$$m'(t) = \varepsilon_4 e^{-\mu_m t} + \sigma\varepsilon_1 t e^{-\beta_1 t}.$$

Thus all small deviations from stationary solution (3.2.4) for $\beta < \gamma F^*$ tend to zero with time, which means asymptotical stability of this solution.

Later on we shall be concerned with the infection of a healthy organism by a small dose of an antigen $V^0 = \varepsilon_1$, i.e., at the moment $t = 0$ only the variable V will deviate from stationary solution (3.2.4). It appears that in this case one can estimate the "smallness" of the infection dose that does not lead to the loss of stability. Such a dose satisfies the inequality

$$0 < V^0 < \frac{\mu_f(\gamma F^* - \beta)}{\beta\eta\gamma} = V^*. \qquad (3.2.13)$$

Let us call the value V^* an *immunological barrier.* Let us say that the immune barrier is *exceeded* when an infection dose V^0 satisfies the condition $V^0 > V^*$, and is not exceeded otherwise. Thus our investigation may have the following biological interpretation. If in the case of infection of a healthy organism by a small dose of an antigen the immunological barrier cannot be exceeded, then regardless of the infectious dose the disease does not develop, i.e., the number of antigens in the organism decreases with time tending to zero, and the damaged organ is restored. In addition, the elevation of C^*, that is of the level of immunocompetent cells in a healthy organism (for instance, on account of memory cells under the vaccination) increases the immunologic barrier (since $F^* = \rho C^*/\mu_f$) and, therefore, it is an effective method of prophylaxis and, possibly, of disease treatment.

Let us pass now to the study of a stationary solution that simulates the chronic process of disease. This process is described also by the system (3.2.3) for $V > 0$ and $\xi(m) \equiv 1$. The system permits the solution:

$$\begin{aligned}
\overline{V} &= \frac{\mu_c \mu_f(\beta - \gamma F^*)}{\beta(\alpha\rho - \mu_c\eta\gamma)},\\
\overline{C} &= \frac{\alpha\mu_f - \eta\mu_c\gamma^2 C^*}{\gamma(\alpha\rho - \mu_c\eta\gamma)},\\
\overline{F} &= \frac{\beta}{\gamma},\\
\overline{m} &= \frac{\sigma\overline{V}}{\mu_m}.
\end{aligned} \tag{3.2.14}$$

The bar shows that this is a nontrivial stationary solution. Notice that solution (3.2.14) was obtained assuming that $\alpha\rho \neq \mu_c\eta\gamma$. Otherwise, i.e., for $\alpha\rho = \mu_c\eta\gamma$, a nontrivial stationary solution exists only when $\beta = \gamma F^*$; then $\bar{V}$ can take any positive value, $\bar{C} = C^* + \alpha F^* \bar{V}/\mu_c$, $\bar{m} = \sigma\bar{V}/\mu_m$, $\bar{F} = \beta/\gamma = F^*$. Apparently it is hard to interpret this solution from the biological viewpoint, and it is of interest only for mathematicians. Anyway, we will assume hereafter that $\alpha\rho \neq \mu_c\eta\gamma$ and consider the solution (3.2.14).

Let us show that the stationary solution (3.2.14) is stable (Belykh [23]). For this purpose we will seek for a solution of system (3.1.11) in the form

$$V = \bar{V} + V', \quad C = \bar{C} + C', \quad F = \bar{F} + F', \quad m = \bar{m} + m', \quad (3.2.15)$$

where $\bar{V}$, $\bar{C}$, $\bar{F}$, $\bar{m}$, defined by formulas (3.2.14), satisfy the system of equations (3.2.3).

Substituting expressions (3.2.15) into (3.1.11) and taking into account that stationary solution satisfies system (3.2.3), we have for $\xi(m) = 1$, neglecting small quantities of the second order:

$$\begin{gathered} \frac{dV'}{dt} - \beta V' + \gamma \bar{V} F' + \gamma \bar{F} V' = 0, \\ \frac{dC'}{dt} + \mu_c C' - \alpha[\bar{V} F'(t-\tau) + \bar{F} V'(t-\tau)] = 0, \\ \frac{dF'}{dt} + (\mu_f + \eta\gamma\bar{V})F' - \rho C' + \eta\gamma\bar{F}V' = 0, \\ \frac{dm'}{dt} + \mu_m m' - \sigma V' = 0. \end{gathered} \quad (3.2.16)$$

We will investigate this system on the basis of harmonic analysis, assuming that

$$V' = ve^{\lambda t}, \quad C' = ce^{\lambda t}, \quad F' = fe^{\lambda t}, \quad m' = me^{\lambda t}. \quad (3.2.17)$$

Substitute expressions (3.2.17) into (3.2.16) to obtain

$$\begin{gathered} (\lambda - \beta)v + \gamma\bar{V} f + \gamma F v = 0, \\ (\lambda + \mu_c)c - \alpha\bar{V} F' e^{-\lambda\tau} f - \alpha\bar{F} e^{-\lambda\tau} v = 0, \\ (\lambda + \mu_f + \eta\gamma\bar{V})f - \rho c + \eta\gamma\bar{F} v = 0, \\ (\lambda + \mu_m)m - \sigma v = 0. \end{gathered} \quad (3.2.18)$$

The investigation of roots of a characteristic equation of the system (3.2.18) revealed that the real part of the parameter λ was negative, i.e., the system's solution was asymptotically stable for $\alpha \to \infty$ under the following conditions (see [26]):

$$\mu_c \tau \leq 1$$

$$0 < \beta - \gamma F^* < \left(\tau + \frac{1}{\mu_c + \mu_f}\right)^{-1} \approx 0.33 \text{ (day}^{-1}\text{)} \tag{3.2.19}$$

Since the coefficient α in the model is a value characterizing the sensitivity of the immune system, and $\beta - \gamma F^* = d \ln V/dt|_{t=0}$, these conditions may be interpreted biologically in the following way. First, a small nonzero antigens population can exist in the organism, that has arbitrary high sensitivity of immune system. Second, antigens, causing stable chronic forms of disease in the organism that have sufficiently high sensitivity of immune system, have weak dynamics.

3.2.3. Possible forms of disease dynamics and their classification. Analysis of the model of disease (3.1.11) allows the evaluation of the qualitative behavior of the solution $V(t)$, the antigen's concentration with that or another values of coefficients. Before classifying the solutions consider two limit cases which, in essence, are the boundaries of the solution $V(t)$.

Assume that an organism produces no antibodies of given specificity i.e., $F(t) = F^0 = 0$ for all $t \geq 0$ and $\rho \equiv 0$. The equation for $V(t)$ has then the form

$$\frac{dV}{dt} = \beta V.$$

A solution of this equation is given by the formula

$$V(t) = V^0 e^{\beta t},$$

where V^0 is initial antigen concentration (infection dose) at the moment $t = 0$. As for the dynamics of an organ's damage, it is described by the equation

$$\frac{dm}{dt} + \mu_m m = \sigma V^0 e^{\beta t},$$

whose solution, under the condition that $m = 0$ for $t = 0$, is given by

$$m = \frac{\sigma V^0}{\beta + \mu_m}(e^{\beta t} - e^{-\mu_m t}).$$

It is easily seen that in the absence of recovery processes in the affected organ, i.e., when $\mu_m = 0$

$$m = \frac{\sigma V^0}{\beta}(e^{\beta t} - 1),$$

and for all $t \geq 0$

$$V = V^0 e^{\beta t}, \quad F = 0, \quad m = \frac{\sigma V^0}{\beta}(e^{\beta t} - 1). \tag{3.2.20}$$

Apparently such a solution corresponds to a disease with lethal outcome, since there are no factors compensating for the growth of the antigens. This case is a limiting one in a certain sense. Such cases are extremely rare in practice. However, sometimes the immune response to an antigen is so weak that the above ideal case is a good approximation. Such a situation happens, for example, to some old patients whose immune system fails to react to an antigen, or to people with acquired or congenital immune defects.

The second limiting case is realized when there is strong response of the immune system, i.e., when the level of antibodies in an organism specific to a given antigen is sufficient for neutralization of all the antigens penetrating the organism. The antibody-producing mechanism is not switched on in this case and the equation for $V(t)$ has the form

$$\frac{dV}{dt} = (\beta - \gamma F)V,$$

where $\beta \ll \gamma F$. Assuming that the infection dose V^0 is small we can consider F as a constant determined by normal level of antibodies F^*. Then above equation can be rewritten in the form

$$\frac{dV}{dt} = (\beta - \gamma F^*)V,$$

and its solution will have the form:

$$V = V^0 e^{-(\gamma F^* - \beta)t}.$$

It means that the antigen population in an organism will decrease exponentially. In the limiting case of $\beta = 0$ we obtain

$$V = V^0 e^{-\gamma F^* t}.$$

Thus we have found two limiting solutions corresponding to a lethal outcome and a high immunological barrier respectively. For given values of coefficients and initial conditions, obviously, the whole family of various disease dynamics lies in the shaded area in Fig. 17.

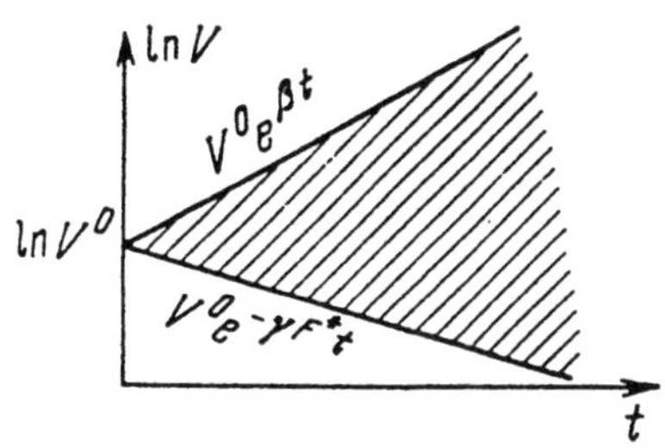

Fig. 17. Area, containing admissible solutions of the model.

Our next aim is to investigate a behavior of other, less trivial disease dynamics. Assume that at the time $t = 0$ the infection of the healthy organism by an initial dose of an antigen V^0 happens and that $\beta > \gamma F^*$ holds. Then initial conditions for the model (3.1.11) have the form

$$V(0) = V^0, \quad C(0) = C^*, \quad F(0) = F^* = F^*, \quad m(0) = 0.$$

The antigen concentration begins to increase for $t > 0$ since the derivative $dV/dt > 0$ in the neighborhood of the point $t = 0$ due to $\beta > \gamma F^*$. At the moment $t = t_1 > 0$ $V(t)$ reaches its maximum, i.e., $V(t_1) = V_{\max}$ and $F(t_1) = \beta/\gamma$. When $t > t_1$, $F(t)$ exceeds the level of β/γ and $V(t)$ decreases while $F(t) > \beta/\gamma$ holds, since $dV/dt < 0$.

A situation is possible when $F(t) > \beta/\gamma$ holds within an interval of time (t_1, t_2) that is large enough and $V(t)$ drops to small values (to zero, practically) on this interval. This case is shown in Fig. 18. Let us call the solution of this type an *acute form of disease.*

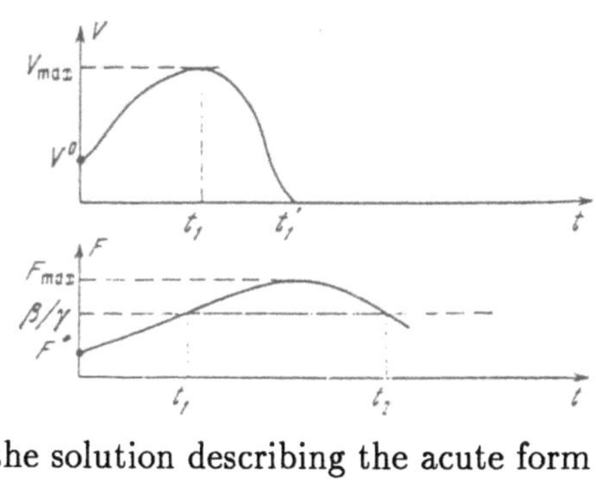

Fig. 18. Behavior of the solution describing the acute form of disease.

If the interval (t_1, t_2) is rather narrow, then $F(t_2) = \beta/\gamma$ in the point $t = t_2$ and $V(t)$ reaches its minimum $V_{\min}$ before it drops to small values, but $V(t)$ begins to increase again with $t > t_2$, since $dV/dt > 0$ for $t = t_2 + \varepsilon$ in view of $F(t) < \beta/\gamma$, where ε is some small positive quantity. The process does not change further qualitatively, and the alternation of local minima and maxima follows, which is shown in Fig. 19. Let us call the solution of this type the *chronic form of disease.*

Thus a ratio between the lengths of intervals $\Delta t = t_1' - t_1$ and $\Delta T = t_2 - t_1$ (see Fig. 18) determines an outcome of disease. In the case $\Delta T > \Delta t$ we deal with the acute form (Fig. 18). If $\Delta T \leq \Delta t$ we deal with the chronic form (Fig. 19). It is clear that the higher is the maximal quantity of the antibodies $F_{\max}$, the larger is $\Delta T = t_2 - t_1$, and, therefore, the smaller is the probability for chronic form to develop.

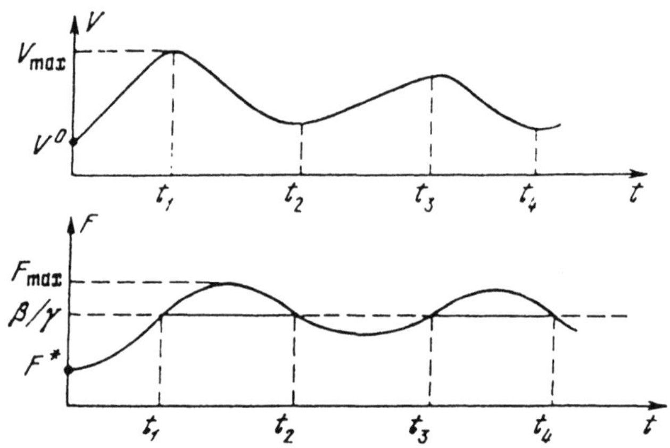

Fig. 19. Behavior of the solution describing chronic form of disease.

Biological interpretation. Chronic forms of disease are caused by an insufficiently effective reaction of immune system. This follows from the fact that when the immune system's reaction is weak (small coefficient α, large delay τ, considerable damage of an organ), on the one hand $V_{\max}$ increases, and on the other hand $F_{\max}$ drops, and as a result the interval $(t_1,\ t_2)$ becomes more narrow.

Thus within the framework of the model (3.1.11) an outcome of disease depends on the derivative of the antigens concentration dV/dt: if it can become negative, and how long it does not change the sign. It is obvious that $dV/dt < 0$ if $V(t) > 0$ and $F(t) > \beta/\gamma$. The value

$V(t) \geq 0$ always in our model; more than that, the value zero is attained just when $V^0 = 0$. Since we assume that the infection of an organism has happened, i.e., $V^0 > 0$, then the validity of the inequality $F(t) > \beta/\gamma$ is the necessary and sufficient condition for dV/dt to be negative. If the immunological barrier is not to be exceeded ($V^0 < V$), then the case 1 of the four following cases is possible.

Case 1. $dV/dt < 0$ in an infinitely large interval of time. We have called a solution of this type the *subclinical form of disease.*

In case of $\beta > \gamma F^*$ $dV/dt > 0$ for t close to zero, and $V(t)$ increases. Assume that $V(t)$ reaches a maximum in the point $t = t_1$ and then decreases. We distinguish the following two cases:

Case 2. $dV/dt < 0$ within sufficiently large interval of time (t_1, t_2). This is the *acute form of disease* (see Fig. 18).

Case 3. $dV/dt < 0$ within sufficiently small interval of time (t_1, t_2). This is the *chronic form of disease* (see Fig. 19).

If the point t_1 does not exist, then the following case takes place:

Case 4. $dV/dt > 0$ within infinitely large interval of time. This case corresponds to *lethal outcome.*

3.3. Results of Modeling

3.3.1. Subclinical form of disease. Fig. 20 shows results of computer simulations with the simple model of disease when $\beta < \gamma F^*$. Two cases are selected: effective or "normal" response ($\alpha\rho > \mu_c\eta\gamma$) and weak response or "immunodeficiency" ($\alpha\rho < \mu_c\eta\gamma$). As we have predicted analytically, when infection doses are small, i.e, lower than the immunological barrier value ($V^0 < V^*$), the removal of the viruses from an organism depends neither on the dose of infection nor on the strength of immune response (curves 1, 2). It means that this removal is provided by the antibody level F^* that is present in an organism constantly; this case corresponds to everyday contacts of an organism with low doses of antigens getting into the organism with breathing and food.

When the dose of infection increases to a considerable extent as compared to the immune barrier value, the strength of immune response begins to play an important role. Effective (normal) immune response can prevent the development of infection (Fig. 20a, curve 3) or suppress it rapidly (Fig. 20a, curve 4). In the latter case the course of the disease resembles the acute form with recovery. Weak immune response, when the level of an antigen exceeds the immunological barrier, finally results in a lethal outcome (Fig. 20b, curve 3)

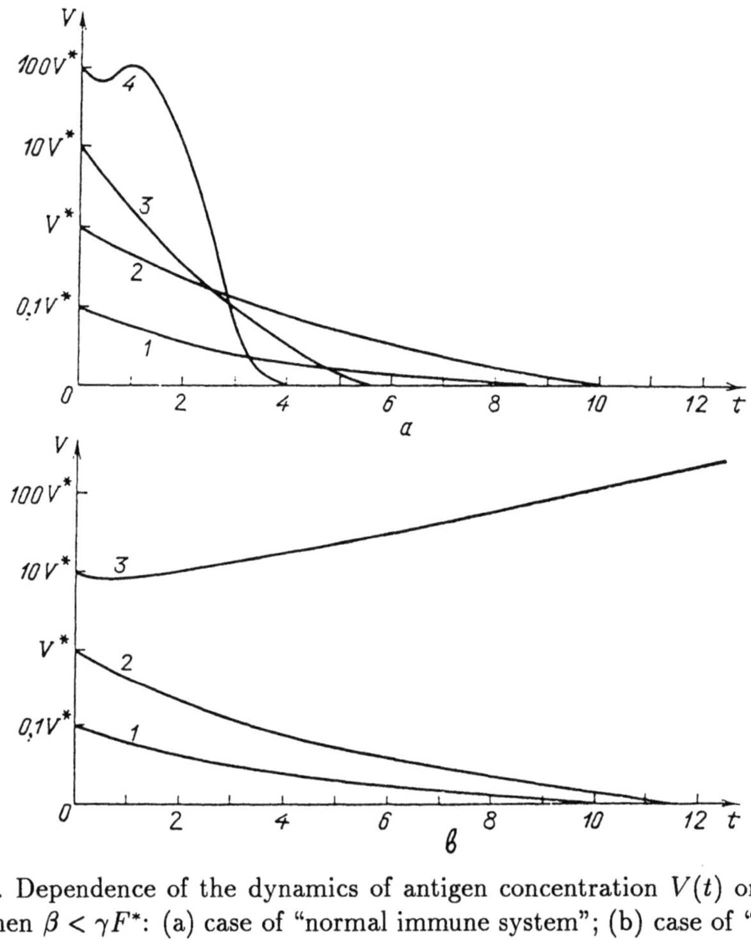

Fig. 20. Dependence of the dynamics of antigen concentration $V(t)$ on infection dose when $\beta < \gamma F^*$: (a) case of "normal immune system"; (b) case of "immunodeficiency". V^* depicts the immunological barrier value.

So, the attraction domain of the state of healthy organism ("recovery zone") with normal immune system ($\alpha\rho > \mu_c \eta\gamma$) is much larger (by several hundred times) than the same domain of an organism with immunodeficiency. Therefore individuals with immunodeficiency are more susceptible to diseases than healthy ones. Medical treatment of these individuals implies diminution of antigen concentration to the immunological barrier value; after that the antigen is removed from organism, irrespective of the immune response strength. It should be noted that in case of normal immune system further increase of infection dose ($V^0 > 100V^*$) leads to lethal outcome.

The case when $\beta < \gamma F^*$ can be interpreted as the vaccination of a healthy organism by weakened antigens. The vaccination is to provoke strong immune response in order to achieve considerable accumulation of memory cells. According to our conception this accumulation is equivalent to the increase in the level C^* of immunocompetent cells that are always present in an organism, and, consequently, it is equivalent to the increase in the immunological barrier. The vaccination effect will be determined by the dose, on the one hand, and by the immune system status on the other hand. It follows from the results of modeling (see Fig. 20), that the injection of small doses (less than the immunological barrier) is ineffective, since in this case antigen can be removed from an organism either with no immune response developed, or with weak immune response. In both cases we will not have the effect of accumulation of memory cells. On the other hand, injection of large doses ($V^0 > V^*$) in individuals with normal immune system ($\alpha\rho > \mu_c\eta\gamma$) results in effective immune response and will lead to lethal outcome, whereas for individuals with immunodeficiency ($\alpha\rho < \mu_c\eta\gamma$) high dose of vaccination may cause serious disease (Fig. 20b, curve 3).

3.3.2. Acute form of disease. Fig. 21 a shows the diagram of the simple model's solutions, which we interpret as the course of the acute form of disease with recovery in the case of a normal immune system. In this case $\beta > \gamma F^*$ and, therefore, the immunological barrier against pathogens does not exist. One can see that the acute form of disease is characterized by a rapid (over several days) increase of antigen quantity in an organism up to the values exceeding the infection dose by several orders, and by rapid elimination of antigens. This character of the disease's course is conditioned by rapid multiplication of antigens which results in rapid accumulation of antigens in an organism, and by strong and effective immune response that leads to production of antibodies

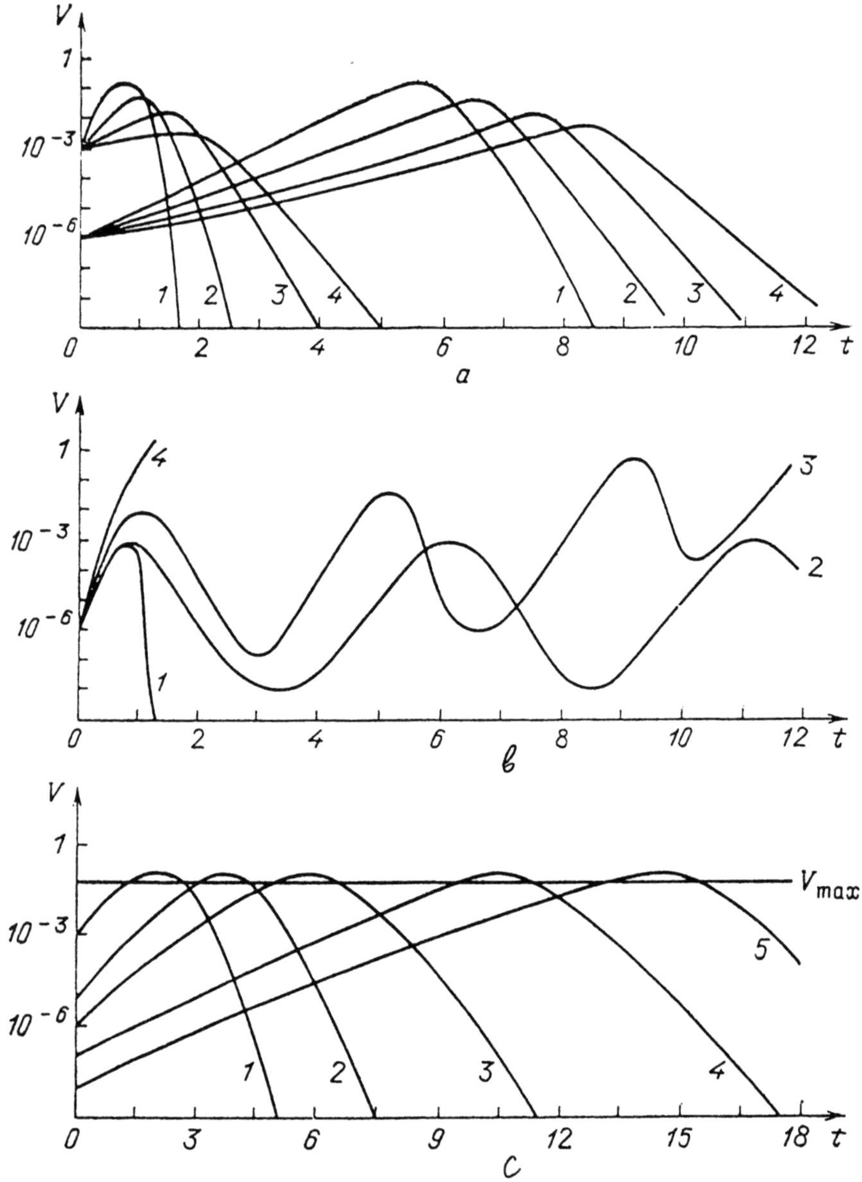

Fig. 21. Dynamics of antigen concentration in the course of theacute form of a disease. (a) Dependence on the rate of multiplication of antigens β ($\beta_1 > \beta_2 > \beta_3 > \beta_4$) for two different values of the infection dose V^0. (b) Transition of acute form of disease (1) into chronic form (2, 3) and lethal form (4) for changing value of organ damage coefficient σ ($\sigma_1 < \sigma_2 < \sigma_3 < \sigma_4$). (c) Dependence on the dose of infection V^0 ($V_1^0 > V_2^0 > V_3^0 > V_4^0 > V_5^0$).

in quantities sufficient for the elimination of antigens. The second is the consequence of the first in the case when an affected organ has a weak influence on the immune system's reactivity, i.e., in the case of weak pathogenic antigen.

Fig. 21a illustrates the course of the acute form of disease depending on the rate of virus multiplication β and the dose of infection V^0. One can see that for a certain dose of infection the higher is the rate of virus multiplication, the higher is the maximum quantity of antigens, and the faster the maximum value of antigen quantity is attained, and the faster this process stops (see the sets of curves 1, 2, 3, 4 for $V^0 = 10^{-6}$ and for $V^0 = 10^{-3}$). On the other hand, for a given rate of virus multiplication, the higher is the dose of infection, the faster the maximum value of antigen quantity is achieved, and the faster this process stops (compare the curves 1 for $V^0 = 10^{-3}$ and $V^0 = 10^{-6}$). This can be partly explained by the fact that a high rate of virus multiplication or a high dose of infection, with other conditions being equal, permits us to reach more quickly the antigen concentrations that stimulate immune system effectively.

One can see in Fig. 21a, that the maximum of antigen quantity, with other conditions being equal, does not change (or, more precisely, hardly changes) with the change of infection dose (cf., for example, curve 1 for $V^0 = 10^{-3}$ and for $V^0 = 10^{-6}$). Fig. 21b illustrates it in more obvious way: $V_{\max}$, an estimate of maximum value of $V(t)$, does not depend on V^0. It was obtained from equations of the simple model with some simplifying assumption when $\beta > \gamma F^*$ and values of σ are small. Thus the value of a "peak of disease" does not depend on the dose of infection in the acute form of disease, but is determined by the immune status of organism as related to given antigen (i.e., by the set of model parameters). The dose of infection affects the time when the peak is reached: the lower is V^0, the later the peak is reached.

In fact, everything we have said above is applicable to weak pathogenic antigens. How would the course of the acute form of disease change in the case of highly pathogenic antigen? The answer is given in Fig. 21b. The acute form of disease (curve 1) turns into the one (curve 2) with the increase in organ damage coefficient σ. It is stipulated by the fact that extensive damage to an organ worsens the general state of organism and, therefore, the immune system's reactivity decreases. Production of antibodies falls, their number is insufficient for complete antigen elimination from the organism. The remaining part of antigens begins to multiply again and the process repeats. Thus the chronic form of disease caused by serious damage of an organ develops. Numerical experiments have revealed that this process is stable and quasi-periodic. Further increase in the value of the damage coefficient σ violates the stability of this process, and a heavy form of disease develops with complications, increasing and alternating with a period of about 40–50 days (this disease ends with lethal outcome if the necessary treatment is not applied (see curve 3). In this case the organ is still more damaged and the production of antibodies is lower than in previous case, accordingly. These antibodies are sufficient only for temporary suppression of infection; the part of the antigen that has not been eliminated becomes considerable and reaches the maximum value rapidly leaving no time for the recovery of organ's viability.

As a result, new outbreak of infection affects the organ anew, which weakens the response of immune system. Each of these outbreaks affects the organ even harder, and the production of antibodies drops even lower. Finally, it causes complete damage of the organ and lethal outcome. In the case where the values of coefficients are still higher, the organ is so strongly affected that the immune system loses its ability to resist the infection which results in fast lethal outcome (curve 4).

The above results enable us to make a conclusion: in order to prevent the transition of the acute form into the heavier form, one should try to lower the antigen pathogenicity.

3.3.3. Hypertoxic form of disease. It is known that in the case of immunodeficiency, when there is no normal development of clone formation of effector lymphocytes and plasma cells, the viral lesion of a target organ can be so extensive that hypertoxic form of disease with unpredictable outcome develops. This happens when the clone formation of plasma cells C is delayed and, therefore, there is a delay in the production of antibodies specific to given antigens.

The case of delay in immune response caused by immunodeficiency is realized in the simple model of disease through large values of delays $\tau_d > \tau$, where τ is normal duration of the clone formation and τ_d is prolonged duration. So we have a mathematical model with a set of parameters, characterized by the value of τ_d larger than τ^*. This model describes the hypertoxic form of disease:

$$\frac{dV}{dt} = (\beta - \gamma F)V,$$

$$\frac{dC}{dt} = \xi(m)\alpha V(t - \tau_d)F(t - \tau_d) - \mu_c(C - C^*),$$

$$\frac{dF}{dt} = \rho C - (\mu_f + \eta\gamma V)F,$$

$$\frac{dm}{dt} = \sigma V - \mu_m m.$$

In order to prevent the development of the hypertoxic form of disease it is necessary to delay the replication rate of viruses in an organism and provide the time needed for plasma cell clones to be formed, that produce protective antibodies. In practice, it can be done with the help of hydrocortisone or other drugs that suppress the viral activity. The injection of such a drug must abolish the virus replication mechanism partly or completely.

This mechanism is realized in the simple model by the decrease in the value of the coefficient β that determines the rate of virus replication. If one manages to diminish β essentially for two or three days, then a sufficient quantity of immunoglobulins will be produced and the hypertoxic form of disease can develop into ordinary acute process.

Results of modelling the treatment of hypertoxic form are presented in Fig. 22. It is seen that suppression of the virus replication through β in the interval $t_1 \leq t \leq t_2$ leads to the formation of sufficient clone of plasma cells and to the production of antibodies, thus preventing the transition of the acute form of viral disease into the hypertoxic one.

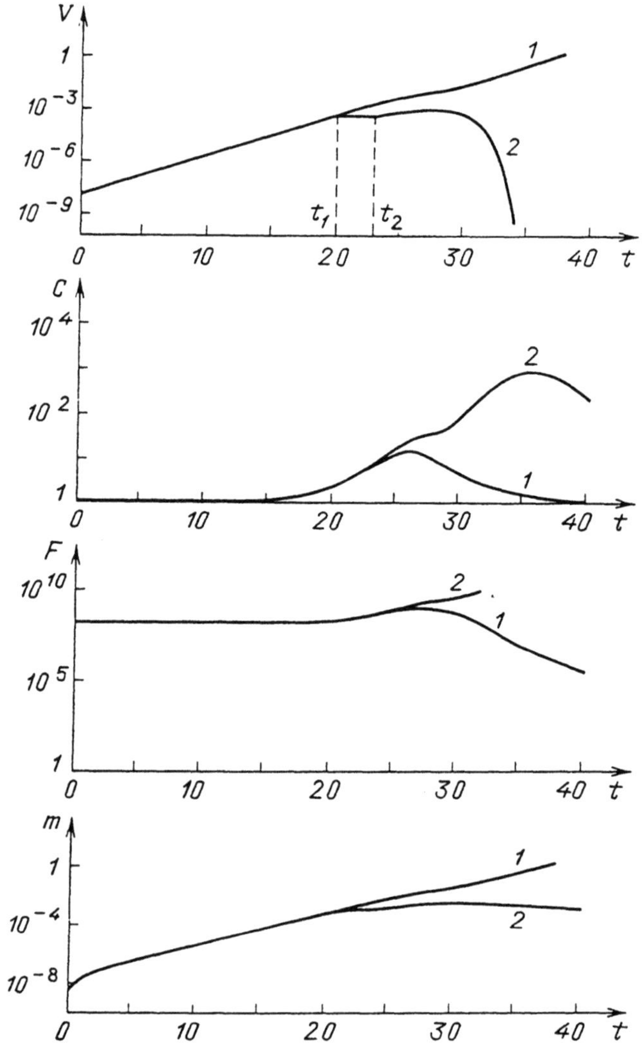

Fig. 22. (a) Simulation of the treatment of hypertoxic form: 1 – disease with no treatment (lethal outcome); 2 – disease with treatment by suppressing the virus replication in the interval $t_1 \leq t \leq t_2$.

3.3.4. Chronic form of disease. We have demonstrated the possibility of transformation of the acute form of disease into the chronic one when an organ is damaged severely (Fig. 21b, curve 2). Here we shall consider another types of stable chronic form, in particular those satisfying the stability conditions (2.2.19). Such typical case is depicted in Fig. 23a (curve 1). One can see that this form is characterized by weak dynamics of stimulants. Their number increases slowly after infection, reaches a maximum, and then decreases slowly down to a minimum. The process repeats until the stationary level of antigens V in an organism is fixed (straight line 4). This results in an equilibrium between replicating stimulants and those eliminated by the immune system. Due to weak dynamics of antigen and its small quantity the immune system is stimulated rather weakly and we cannot observe any appreciable immune response. Let us recall that we consider the case when conditions (2.2.19) hold, i.e., when α, the coefficient of simulation of the immune system, has a large value which guarantees a small value of $\overline{V}$. It was assumed also that the stimulants are weakly pathogenic. The curve 1 can thus also be interpreted as the constant presence of a small quantity of weakly pathogenic microbes in an organism with a normal immune system.

When the infection dose is several times higher than stationary level $\overline{V}$, the antigen dynamics slightly activates by virtue of more active stimulation of the immune system (Fig. 23a, curve 2). Nevertheless, this stimulus usually is not as effective for the immune system activation as to eliminate the antigen from an organism. It only causes considerable decrease in antigen quantity. When the antigen population has reached its minimum, the process becomes analogous to the dynamics depicted by the curve 1 and results in establishing a stationary level V.

With even higher doses of infection ($\log V^0 > \log \bar{V} + p$, where $p \geq 2$), the chronic form turns into the acute form with recovery (curve 3). This is stipulated by the fact that such a dose of infection proves to be effective for the stimulation of the immune system which, in response to infection, produces antibodies in quantities sufficient for the elimination of antigens from an organism. Though, one can always select so high a dose of infection that even weak pathogenic antigens in the case of rapid growth may lead to a lethal outcome (Fig. 23a, curve 5). This means that the chronic form of disease depends on the dose of infection and can develop into the acute form with recovery or lead to a lethal outcome.

The results presented in Fig. 23a can be a basis for the following conclusions. First, the treatment of chronic forms is possible through exacerbation of disease (considerable increase in the antigen quantity as compared to chronic form). Second, it is inexpedient for a normal immune system to react to low doses of antigens with weak dynamics, since it can lead to the chronic form of disease. These conclusions follow from the comparison of curves 1 and 3 in Fig. 23a. We emphasize that they are correct provided that the immune system has high sensibility for the antigen (large value of stimulation coefficient α) and when the antigen has weak pathogenicity (small value of the coefficient of damage σ).

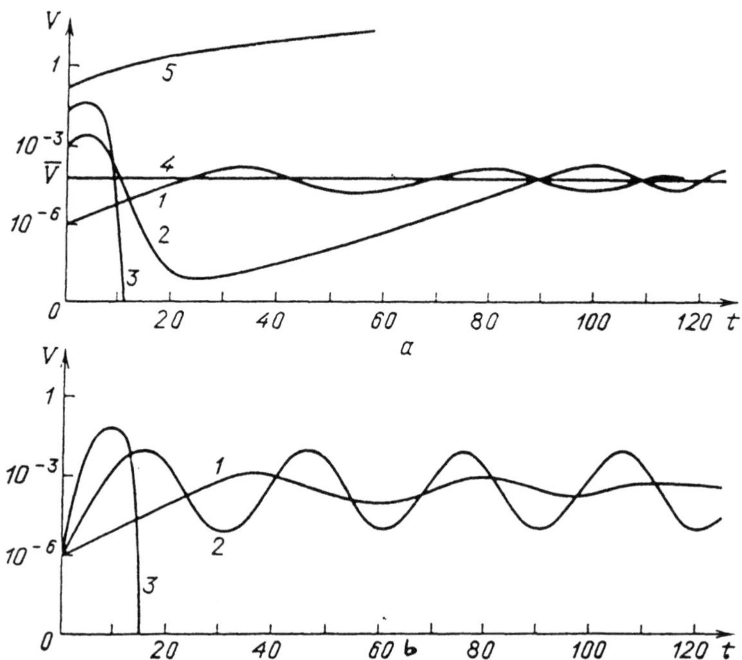

Fig. 23. Dynamics of antigen concentration for chronic form of disease as dependent on: (a) infection dose V^0 ($V_1^0 < V_2^0 < V_3^0 < V_4^0 = \bar{V} < V_5^0$); (b) rate of antigen replication β ($\beta_1 < \beta_2 < \beta_3$).

Fig. 23b depicts the dependence of the course of chronic form of disease on the rate of antigen replication β. When β satisfies the condition (2.2.19), typical chronic form of disease develops (curve 1). If the value of β turns the right-hand side of condition (2.2.19) into equality, stationary periodic solution appears, as numerical experiments show; we interpret it as a limit case of typical chronic form of disease (curve 2). Existence of this type of solution is easily proved. Such a form of disease has more intensive dynamics than any chronic form and can easily develop into the acute form through the exacerbation; the maximum and minimum can then differ by a number of orders, and the period between two neighboring peaks can be as long as a few weeks. Further increase in the rate of antigen multiplication transforms chronic process into acute one with recovery (curve 3). Comparison of curves 1, 2, and 3 shows that the treatment of acute forms of disease by drugs that decrease the rate of virus multiplication promotes the disease becoming chronic.

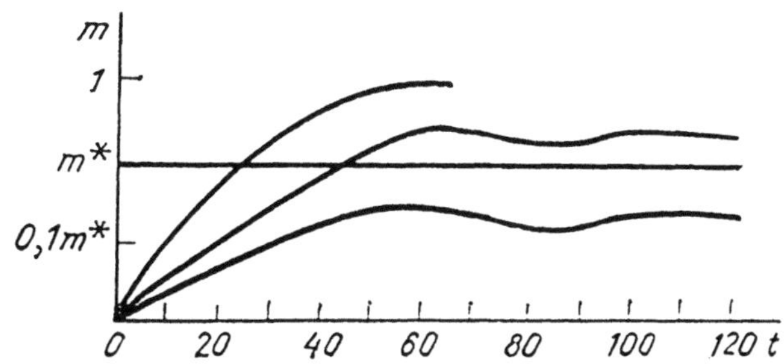

Fig. 24. Dependence of the course of chronic form of disease on damage coefficient.

Fig. 24 demonstrates a dependence of the course of the chronic form of disease on the values of damage coefficient σ, which is illustrated by the dynamics of parameter $m(t)$ characterizing the damage to an organ. Recall that in our model the lesion of an organ can weaken the immune system only when this lesion exceeds certain threshold level m^*, i.e., if $m > m^*$. When antigens are weakly pathogenic (value of σ is small) the typical chronic form of disease with stationary level of damage $\bar{m} < m^*$ appears (curve 1), which does not affect the immune system's response. With the increase of σ the process can move into a new stable stationary state $\bar{m} > m^*$, which we interpret as a serious chronic form of disease (curve 2). The existence and stability of this solution were proved in Belykh, Kalyaev [27]. The lesion of an organ binds partially the immune system's response, and as a consequence, corresponding stationary level of antigen proves to be higher than for typical process. The exacerbation treatment is not advisable here, since the organ has been considerably damaged already. Further increase of σ can lead to a lethal outcome (curve 3).

3.4. Influence of Organism's Temperature Reaction on the Course of Disease

Live culture and vaccine of various types of pathogenic and nonpathogenic bacteria, foreign proteins, products of damage and decay of cells and tissues affect the heat exchange mechanism of an organism, raising its temperature. One connects the pyrogenicity (i.e., the property to raise temperature) of the majority of bacteria types with a high molecular lipopolysaccharide complex that is released with the disintegration of bacteria (endotoxin). The exotoxins and the excretory products secreted by bacteria into the environment also possess a certain pyrogenic effect.

Table 1

Antibody (Ab) dilution in sensibilized rabbits with fever and in control group (by the serum dilutions, N is the number of serum*).

Day of blood sampling	Rabbits with fever					Control group				
	N	With Ab titer				N	With Ab titer			
		1/320 1/160	1/80	1/40	1/20		1/320 1/160	1/80	1/40	1/20
3th	4	1	2	1	–	6	–	3	–	3
7th	4	3	1	–	–	5	–	3	3	–
10th	4	2	1	1	–	5	–	2	3	–
14th	4	–	2	1	1	6	–	–	2	4

* Vesyolkin P.N., *The Fever*, Moscow, Medgiz, 1963 (in Russian)

The influence of the rise of temperature on the dynamics of disease has a dual character. On the one hand, high temperature suppresses most viruses and bacteria; for example, rise of body temperature affects the multiplication and invasion of influenza viruses into the cells of an organism. On the other hand, the rise of temperature within certain limits increases in direct proportion the rate of elementary chemical reactions (*the Arrhenius–Van-Hoff rule*), it raises the activity of enzymes, stimulates immune reactivity and accelerates the rate of physical and chemical reactions of the immunity. Table 1 contains experimental data that directly point out stimulation influence of the rise of body temperature caused by pyrogen on the rate of antibodies production and secretion after the injection of the horse serum to rabbits.

Now, consider a mathematical model of temperature reaction influence on the immune response. Let an infectious disease be described by the simple mathematical model:

$$\begin{aligned}
\frac{dV}{dt} &= (\beta - \gamma F)V, \\
\frac{dC}{dt} &= \xi(m)\alpha_0 V(t-\tau)F(t-\tau) - \mu_c(C - C^*), \\
\frac{dF}{dt} &= \rho C - \eta\gamma FV - \mu_f F, \\
\frac{dm}{dt} &= \sigma V - \mu_m m,
\end{aligned} \tag{3.4.1}$$

with the initial conditions for $t = t_0$:

$$\begin{aligned}
&V = V^0, \quad C = C^0, \quad F = F^0, \quad m = 0, \\
&\varphi(t) = V(t)F(t) = 0, \quad t^0 - \tau \le t < t^0.
\end{aligned} \tag{3.4.2}$$

Here $V(t)$ is the quantity of antigen (viruses, bacteria, etc.) capable of reproduction and affecting a target organ; $F(t)$ is the quantity of specific immune factors (antibodies, cell receptors, etc.) capable of binding with an antigen and neutralizing it; $C(t)$ are immunocompetent cells capable of reception of antigen stimulus and producing specific immune factors in response to this stimulus; $m(t)$ is the relative characteristic of target organ damage ($0 \le m \le 1$); $\xi(m)$ is a piecewise linear function which takes into account the influence of target organ damage on lymphatic organs productivity ($0 \le \xi(m) \le 1$, $\xi(0) = 1$, $\xi(1) = 0$).

Let us assume that the rate of antigen multiplication decreases with the increase of body temperature. In the simplest case this dependence can be described by the formula:

$$\beta = \frac{\beta_0}{1 + \beta_1(\theta - \theta^*)}, \tag{3.4.3}$$

where β_0 is the coefficient of antigen multiplication when the temperature of a body is normal; $\theta = \theta^* = 36.6°\mathrm{C}$; $\beta_1 = \mathrm{const} > 0$. Let us assume further, that the rise of temperature stimulates the synthesis of all proteins including immunoglobulins. Suppose that the coefficient of stimulation α depends on temperature linearly:

$$\alpha(\theta) = \alpha_0(1 + \alpha_1(\theta - \theta^*)), \tag{3.4.4}$$

where α_0 is the coefficient of stimulation for $\theta = \theta^*$, and $\alpha_1 = \mathrm{const} > 0$.

Let us suggest the following hypothesis about the mechanism of temperature changes in the course of an infection disease. The body temperature depends on the concentration of FV-complexes. If their concentration is lower than some threshold value $(FV)^*$, the temperature is not increased. If the concentration of these complexes exceeds $(FV)^*$ the rise of temperature will be proportional to FV. An equation describing changes in temperature during the course of disease can be written as:

$$\begin{gathered} \frac{d\theta}{dt} = \alpha_T VF - \mu_T(\theta - \theta^*), \\ \theta(t^0) = \theta^*, \quad \alpha_T = \begin{cases} 0, & \text{if } \; VF < (VF)^*, \\ \mathrm{const} > 0, & \text{if } \; VF \geq (VF)^*, \end{cases} \end{gathered} \tag{3.4.5}$$

where $(VF)^*$ is the maximum available concentration of FV-complexes that does not yet stimulate the rise of temperature.

Clinical observational data show that the rise of temperature during infection diseases in man does not exceed 40–40.5°C as a rule. Consider in this connection, together with equation (3.4.5), the more complex equation:

$$\begin{gathered} \frac{d\theta}{dt} = \alpha_T VF(1 - \alpha_2 VF) - \mu_T(\theta - \theta^*), \\ \theta(t^0) = \theta^*, \quad \alpha_T = \begin{cases} 0, & \text{if } \; VF < (VF)^*, \\ \mathrm{const} > 0, & \text{if } \; VF \geq (VF)^*. \end{cases} \end{gathered} \tag{3.4.6}$$

Fig. 25 depicts the temperature dependence on the concentration of VF-complexes as follows from (3.4.5), (3.4.6). The following data related to the value of μ_T are presented in [153]. A sudden rise of body temperature by 1°C in a patient weighing 72 kg provokes the reaction that will reestablish an equilibrium with an initial rate of 0.16°C per second. This very value was used in numerical experiments.

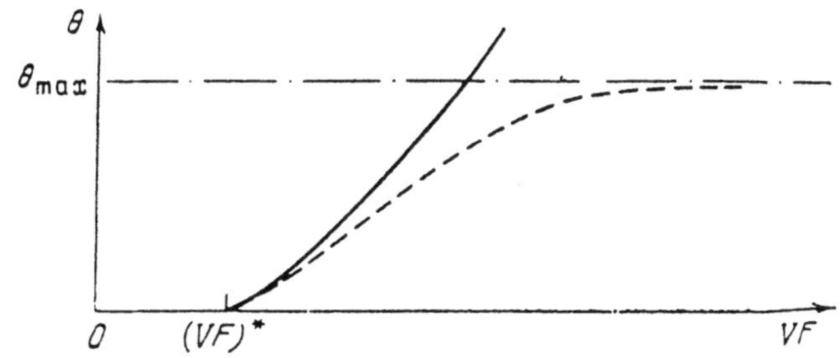

Fig. 25. Dependence of temperature on concentration of VF-complexes as follows from equations (3.4.5) and (3.4.6) (solid and dashed curves respectively).

Qualitative analysis of the model and numerical experiments have revealed (see [7]) that the rise of body temperature decreases the maximum of antigen concentration which decreases the value of the characteristic of the target organ's damage. Such influence can be decisive in the cases of acute forms with unpredictable outcome. This is schematically shown in Fig. 26: if there is no temperature reaction, antigen concentration reaches its maximum at the point t_1 when $F = F^* = \beta_0/\gamma$. The inequality $\beta(\theta)/\gamma < \beta_0/\gamma$ holds as the temperature grows higher, and the rate of antibodies production is higher than that with normal temperature. Therefore the antibodies concentration sooner reaches the level $F^* = \beta(\theta)/\gamma$ above which $dV/dt < 0$.

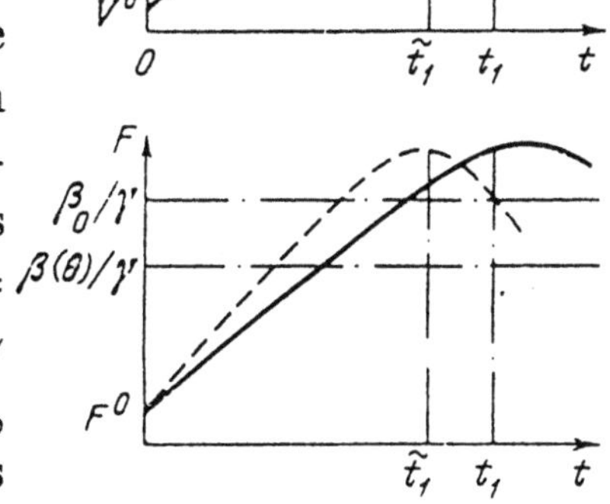

Fig. 26. The lowering of antigen concentration maximum in the case of increase in body temperature.

Results of numerical experiments when the temperature reaction may be of special importance are presented in Fig. 27. The acute form of disease with unpredictable outcome is considered, when the vast damage of a target organ can have an important bearing on the process of disease.

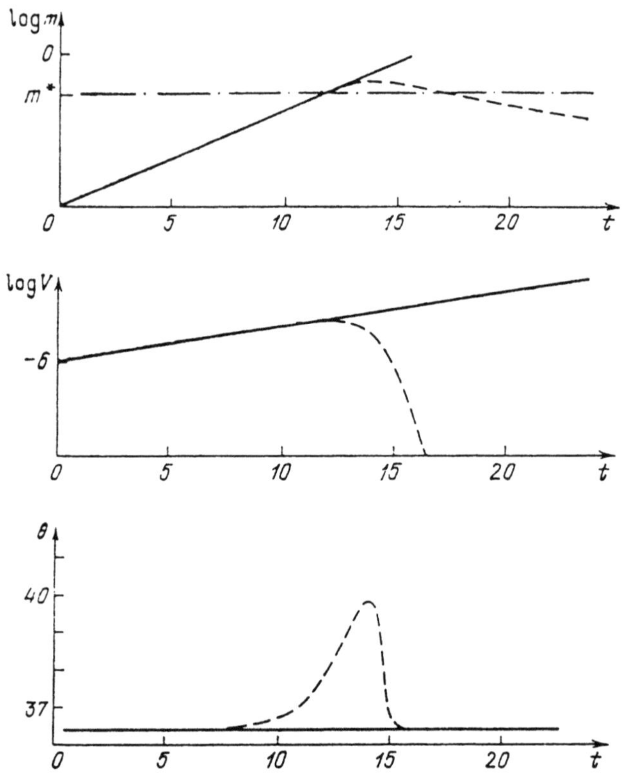

Fig. 27. Influence of temperature on the disease dynamics (solid curve – rise of temperature is not accounted for; dashed line – expressed temperature reaction is taken into account).

As we have noticed in Section 2.3, chronic forms of disease are caused by insufficiently effective reaction of the immune system which leads to the narrowing of the interval (t_1, t_2) where a derivative of $V(t)$ is negative: $dV/dt < 0$.

Pronounced temperature reaction, lowering the maximum of antigen concentration and stimulating antibodies production, enhances the extension of the interval $(\tilde{t}_1,\ \tilde{t}_2)$ which can lead to the transformation of chronic form of disease into acute one with recovery. It is shown schematically in Fig. 28, and Fig. 29 shows the results of corresponding numerical experiments.

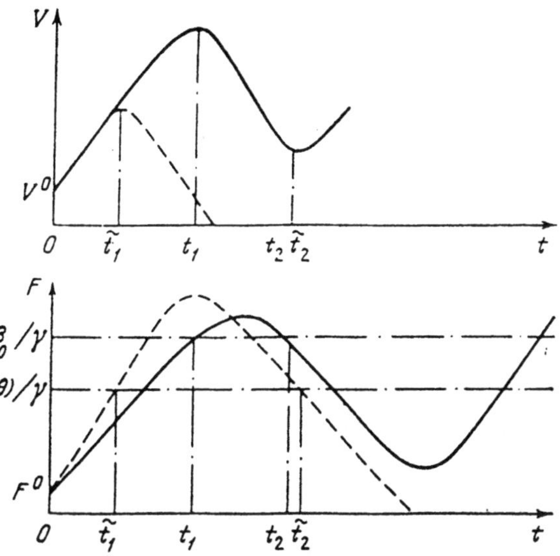

Fig. 28. Transformation of acute form of disease into a chronic one under conditions of depressed temperature reaction.

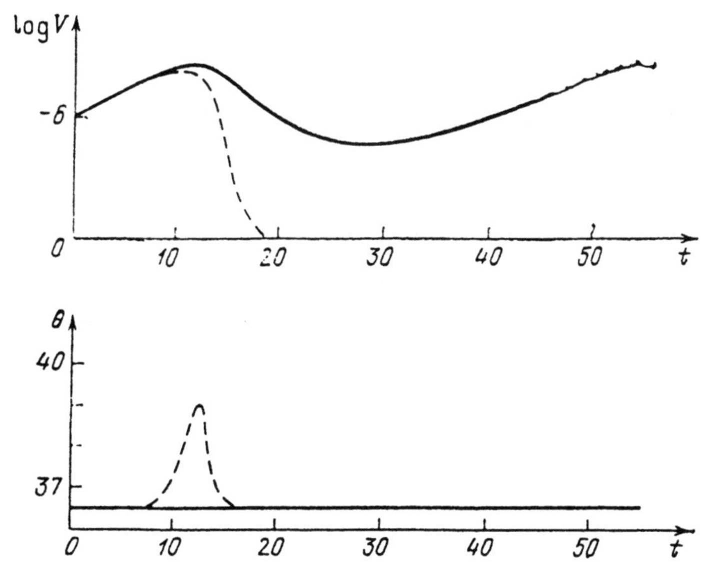

Fig. 29. Transformation of chronic form of disease into an acute one with recovery due to temperature reaction.

Results of numerical simulations emphasize the importance of temperature reaction with respect to chronicisation of the disease. Unnecessary depression of temperature reaction can lead to the transformation of the acute form of disease into a lingering or chronic one. More than that, many febrifuges exhibit the analgetic effectm, thus reducing the expressiveness of the clinical symptoms which may cause underestimation of the clinical gravity of the disease and erroneous treatment [197, 198].

So, the temperature reaction is an important mechanism of self-defence and self-healing of an organism. On the one hand, the rise of body temperature decreases the maximum concentration of an antigen during the course of the disease and, consequently, decreases the degree of damage to the target organ.

On the other hand, the rise of temperature stimulates the production of protein and other components of the immune system being contributory factors for more rapid neutralization and elimination of an antigen from an organism. Therefore, unnecessary usage of febrifuges lowers the resistance of an organism, creating favourable conditions for the development of viruses or bacteria populations. The disease can transform into chronic form as a result.

3.5. On the Question of Antibodies' Level for Chronic Disease

The connection between the level of antibodies and the heaviness of pathological process in the course of a chronic disease manifests itself while analyzing the immunoglobulins. Since the levels of IgM and IgG usually stay normal during stationary chronic disease, then in many cases IgA proves to be the main indicator of a chronic process; its level grows lower with the aggravation of chronic process.

This well established fact brings about the thought that there exists some regularity which can be clarified while analysing the balance equation for the immunoglobulin IgA, whose concentration will be denoted by F. Let us write

$$\frac{dF}{dt} = \rho C - \eta\gamma FV - \mu_f F. \tag{3.5.1}$$

Notice that in the absence of antigens ($V = 0$) we arrive at the stationary balance equation

$$0 = \rho C^* - \mu_f F^*, \tag{3.5.2}$$

where C^* and F^* are respectively the levels of plasma cells and antibodies IgA determined by the homeostasis. It follows from (3.5.2) that F^* and C^* are linked:

$$F^* = \frac{\rho C^*}{\mu_f}. \tag{3.5.3}$$

Assume now that $V \neq 0$ but that the process is stationary, i.e., we deal with the case of chronic disease. Then $dF/dt = 0$ and we arrive at the relationship

$$\rho\bar{C} - \eta\gamma\overline{FV} - \mu_f\bar{F} = 0. \tag{3.5.4}$$

Here the bar marks the elements corresponding to the chronic state.

Introduce a hypothesis: the level of plasma cells is determined in the chronic state by the homeostasis, i.e.,

$$\bar{C} = C^*,$$

which means that the immune system is not stimulated with FV-complexes any more, and the organism stops to react on V-antigens (if their level stays the same for a long time). Then the relationship (3.5.4) turns into

$$\rho C^* - \eta\gamma\overline{FV} - \mu_f\bar{F} = 0,$$

whence it follows that

$$\bar{F} = \frac{\rho}{\mu_f + \eta\gamma\bar{V}} C^* = \frac{F^*}{1 + \frac{\eta\gamma\bar{V}}{\mu_f}}. \tag{3.5.5}$$

Since $\eta\gamma\bar{V}/\mu_f > 0$ it then follows from (3.5.5) that $\overline{F} < F^*$.

Biological interpretation. During chronic disease the level of "active" antibodies (in this case IgA) is to be lower than a norm. The larger is the deficiency of antibodies, the heavier is the form of the chronic process. It may be explained qualitatively as follows. Suppose that the chronicisation of the process led to the development of a tolerance with respect to given antibodies (bacteria or viruses). The organism turned thus to the normal production of plasma cells, corresponding to the level of homeostasis C^*. The plasma cells produce a certain amount of antibodies (see (3.5.3.)). But some part of these antibodies is spent on the blocking of the agents of chronic infection. As a result, the level of antibodies $\bar{F}$ in the body falls as compared with a norm.

We have discussed the immunoglobulins here, but the same seems to be applicable as related to the level of T-lymphocytes effectors. Though, this conclusion must be analysed thoroughly, and verified on large clinical material.

Of course, the antibody deficit condition for chronic diseases seems to be typical, but the antibody deficit can be innate or acquired. Therefore, the immune status of a patient before the disease is of primary importance here. It should be noticed, though, that the deficit of antibodies IgA points out the receptivity to infectious diseases connected with mucosal epithelia of human organs. It is a sign that the patient requires special observation in the process of disease, unless the chronicisation develops.

Though, another simple case is possible, when while the chronicisation the level of antibodies stays normal. Two explanations seem to be possible here: Either the role of IgA is played by T-lymphocytes effectors, or the rise in IgA has occurred on account of some accompanying disease. In the latter case the deficit of IgA will be evident in 2–3 weeks.

3.6. On the Chronicisation of Disease Process

The disease chronicisation process is one of the most urgent problems in contemporary medicine. It is very important for a physician-clinician to know the probable forecast of a disease's outcome at the time when it is possible to change the course of treatment and to avoid a chronic regime. All laboratory clinical information and data on the dynamics of the pathological process must be used to this end. But that is not all. The most informative indices are needed that reflect the state of an organism's defence functions and are the first to react to the changes in the disease's course towards chronicisation.

Actually, in the case of acute forms of infectious diseases with an inflammation process, the level of leukocytes and lymphocytes grows higher, ESR reaction activates, C-reactive protein becomes positive, and so forth. All these characteristics are usually normalized during the treatment, and if the general clinical state of a patient is satisfactory the conclusion is made that the treatment is over with recovery.

But in some cases such a conclusion proves to be premature, since the disease transforms into chronic form, with no manifestations in laboratory or clinical indices. Usually such chronicisation is a consequence of low activity of the organism's immune system because of medicinal immunodepressive preparations (antibiotics, sulfanilamides, and others). The question arises: can one be ever sure of a complete recovery? The question is, of course, multi-plan; but it is possible, for some diseases, to make the conclusion of a possible recovery of a patient with a certain confidence, using the most informative immunological, laboratory, and clinical indices.

So, in the case of infectious diseases of upper respiratory tracts and lungs, the most informative indices are immunoglobulins (IgM, IgG, IgA) and components of the "blood formula" (fibrinogen, leukocytes, ESR, C-reactive protein, lymphocytes, eosinophiles, etc.). If we deal with hepatitis, then, together with immune indices, the analysis is required of indices for free and bound bilirubine, beta-lipopropetids, and ferments.

In the case of other infectious diseases, for example, those of the stomach, intestines, kidneys, and so forth, it is necessary to have, together with immune indices, the data on indices specific and characteristic of these diseases. Complete normalization is necessary in all these cases of immune status and other indicators characterizing the pathological state of an organism.

Let us exemplify this with infectious diseases of the upper respiratory tracts and lungs, whose chronicisation is especially high nowadays. In this case the infection damage of an organism is related to the damage to mucosal epithelia and breathing ways. It is known that the immune defence of mucosal organs is provided finally by the antibodies IgA. If an organism has no inherited or acquired deficit for a given type of immunoglobulins, then, as was shown in Section 3.6, the normalization of IgA is a characteristic test of recovery.

Consider the blood formula. A physician often pays attention to heightened laboratory blood indices that usually characterize a degree of inflammatory process. In the case of acute infectious diseases the indicators that are lower than a norm are not always noticed. But some of these very indices contain valuable information on the chronicisation of a process. A level of leukocytes in blood is one of the most important such indices. In order to justify this conclusion, we will construct a mathematical model of the macrophages (leukocytes) dynamics and discuss the results of numerical simulation.

Let φ be the number of macrophages in an organism at the moment of time t. If φ^* is the number of macrophages in a healthy organism, then their dynamics in the case of pathology can be described with the following equation:

$$\frac{d\varphi}{dt} = q - p - \alpha(\varphi - \varphi^*). \tag{3.6.1}$$

Here q describes the generation of macrophages, whose number in a unit of time is assumed to be proportional to concentration of VF-complexes stimulating the immune system, i.e.,

$$q = aVF, \tag{3.6.2}$$

where V is concentration of antigens, F is concentration of antibodies, and $a > 0$ is a constant.

The second term in the right-hand side of (3.6.1) describes the number of macrophages leaving in a unit of time due to the extraction of products of immune reaction (VF-complexes) from an organism. It is also proportional to VF, i.e.,

$$p = bVF, \tag{3.6.3}$$

where $b > 0$ is a constant.

The third term in the right-hand side of (3.6.1) is responsible for the homeostasis of macrophages and their natural decrease due to aging.

Notice that in the absence of immune reaction ($VF = 0$) it follows from (3.6.1) that $\varphi = \varphi^*$.

Rewrite (3.6.1) with regard to (3.6.2) and (3.6.3) in the form

$$\frac{d\varphi}{dt} = (a-b)VF - \alpha(\varphi - \varphi^*). \tag{3.6.4}$$

Add the initial value

$$\varphi(t) = -\varphi^* \quad \text{for} \quad t = 0. \tag{3.6.5}$$

Introduce the function $\Phi = \varphi - \varphi^*$. Then problem (3.6.4), (3.6.5) is written as

$$\frac{d\Phi}{dt} = (a-b)VF - \alpha\Phi. \tag{3.6.6}$$

Add the initial value

$$\Phi = 0 \quad \text{for} \quad t = 0 \tag{3.6.7}$$

We assume that the quantity VF is known from corresponding tests.

Consider now chronic disease. In this case, due to weakened immune response in the course of prolonged process the generation of macrophages q appears to be much smaller than p, since p is related with the decrease of macrophages on account of the extraction of products of pathological process from an organism. This means that in this case $a < b$, and we have

$$\Phi = -(b-a)\int_0^t VFe^{-\alpha(t-t')}dt' < 0. \tag{3.6.8}$$

Hence it follows that $\varphi < \varphi^*$.

It is important to notice that a chronic process is the heavier, the smaller is a value of Φ, and, consequently, the leukopenia is more pronounced.

It may happen that $a = b$ during a chronic form of disease. Then

$$\varphi = \varphi^*. \tag{3.6.9}$$

It seems that this case is realized in the remission of a chronic process.

As for acute forms of disease, they are characteristic of expressed immune reaction, when $q > p$, i.e., $a > b$. Integrate equation (3.6.6) with condition (3.6.7) to obtain

$$\Phi = (a-b)\int_0^t VFe^{-\alpha(t-t')}dt' > 0, \tag{3.6.10}$$

i.e., $\varphi > \varphi^*$.

More than that, the heavier is the process, the larger is a compared to b, and, consequently, the higher is the level of leukocytes as compared to a norm.

Biological interpretation. The decrease in a level of macrophages is a natural indicator of the chronicisation of a process. And the more expressed is the form of chronic process, the more expressed is the leukopenia.

In case of acute infectious diseases a level of macrophages rises sharply compared to a norm, and the higher, the heavier is the pathological process.

It follows hence, in particular, that the slow decrease in a level of macrophages in an organism during the transition into leukopenia is probably related with the start of the chronicisation of the process (unless this decrease is provoked artificially by depressing the influence of medicinal preparates).

An important conclusion suggests itself, in our opinion, that the immune status and blood formula of a patient must be monitored until all corresponding indices are normalized. If in the case of remission the deficits in the above indices are preserved, this requires more detailed clinical inspection, for example, employing bronchoscopy or other methods of contact diagnosis. A definite conclusion on complete recovery is possible just after such inspection.

One should keep in mind that the simple model of infectious disease, considered in this chapter is a mathematical abstraction. We understand an infectious disease as a reflection of relations established between two members of biocenosis, one of which (the antigen) is capable of existing in another due to pathogenic mechanisms, and this other (organism) is capable of resisting the pathogenic action with the help of the immunity system.

The model does not describe a certain concrete disease induced by a certain concrete antigen. The main task we set for the modeling is the description and search for the most general regularities common to all infectious diseases.

The construction and analysis of this model made it possible to systematise and explain from immunological position the various well known facts concerning the mechanisms of infectious diseases accumulated by immunologists and clinicians. For example, the investigation of the stability of stationary solutions of the model resulted in obtaining the conditions of appearance of various forms of disease and their classification. The investigation of the dependence of model solutions on initial data brought about a mathematically justified method of treatment for the chronic form of a disease that we have called the *exacerbation method.* It was established in the framework of the model in which the appearance and the course of a disease do not depend on the initial dose of infection, and are determined by the immune status of an organism. Besides, various modifications of the model present the possibility of estimating the influence of the temperature reaction of an organism on the dynamics of disease and to explain a possible mechanism of the course of mixed infections.

It is clear that the construction of a model of concrete disease will require, on the one hand, higher detailing of the process, and, on the other hand, wide employment of clinical and laboratory data for the identification of the model's parameters. We believe that regularities obtained in the framework of the simple model will be preserved.

CHAPTER 4

Mathematical Modeling of Antiviral and Antibacterial Immune Responses

We shall construct in this chapter the model of viral disease developed by Marchuk and Petrov [200] on the basis of recent advances in immunology. The cytotoxic action of T-lymphocytes against infected cells has been used as the main point of the model that describes pathological changes in an organism. The destruction of an organism's cells is the essential mechanism of recovery from infection. As for a viral population which gets into lymph and blood plasma from infected cells, it seems to be largely neutralized by the immunoglobulins with further elimination of viruses from organism. Severity of the course of disease, in the context of this model, depends on the degree of damage to target organ caused by viruses, and on the effectiveness of immune response. A modification of the mathematical model of antiviral response will also be considered. This model takes into account local immunophysiological mechanisms of action of cytotoxic T-cells and antibodies in a target organ affected by a virus, which are connected with the oedema development and changes in the circulation of blood in the organ. This modification of the model has been developed by Marchuk in collaboration with Petrov [201].

Then a model of antibacterial immune response will be constructed on the basis of the mathematical model of antiviral immune response. These models as a whole provide a basis for modeling mixed infections.

4.1. Immunological Description of an Antiviral Immune Response Model

Immune response to viral pathogens penetrating into an organism (influenza, measles, poliomyelitis, viral hepatitis, etc.) involves two types: humoral response, when the B-system of lymphocytes produces antibodies, and cellular response, when cytotoxic effector T-lymphocytes accumulate in the organism. The cellular type of immune response providing the defence of an organism is the main one. Antibodies neutralize viral particles circulating in blood, but they cannot clear an organism from pathogens, since virions multiply in the cells of tissues sensitive to the offending virus. As for antibodies, they cannot penetrate into these cells. Cytotoxic T-lymphocytes (CTLs) accumulated during the immune response recognize the cells infected by virus using membrane receptors and destroy them, taking the role of killers of their own body's cells. Thus antiviral immune response seems to be autoimmune in character. However, this is not a cell-mediated true autoimmune reaction that involves pathologic reactions of the immune system against normal (unchanged) cells or normal cellular antigenic substances. In the case of antiviral immunity the CTLs destroy the *virally infected cells of the body.* This, apparently, is the only way to clear an organism from viruses if intracellular defence mechanisms (interferon, enzymes controlling the replication of nucleic acids, etc.) of the cells infected by viruses have not managed to cope with their multiplication by themselves.

The key role of T-cell mediated immune response in antiviral immunity has been demonstrated during previous decade or so. Accumulated effector T-cells that are equipped with receptor apparatus against viral antigens locate the virally infected cells, and, *acting against viral antigens* expressed on the cell membrane, kill infected cells.

In 1974–1975, (Zinkernagel [351]), and later other investigators, had demonstrated that the processes of recognition, initiation of effector T-lymphocyte proliferation, and cytotoxic action of accumulated T-killers are triggered by the virus–transplantation antigen complex rather than by the viral antigen. So called double (MHC-restricted) recognition was discovered. Molecules of transplantation antigens of the major hystocompatibility complex (MHC) (*H*–2K and *H*–2*D* in mice, and *HLA*–*A*, *HLA*–*B* and *HLA*–*C* in human being) are the primary molecules that form complexes with viruses. The term "MHC-restricted recognition" emphasizes the fact that recognizing receptor structures of *T*-lymphocytes are complementary to physically associated virus–transplantation antigen complex rather than to viral particle.

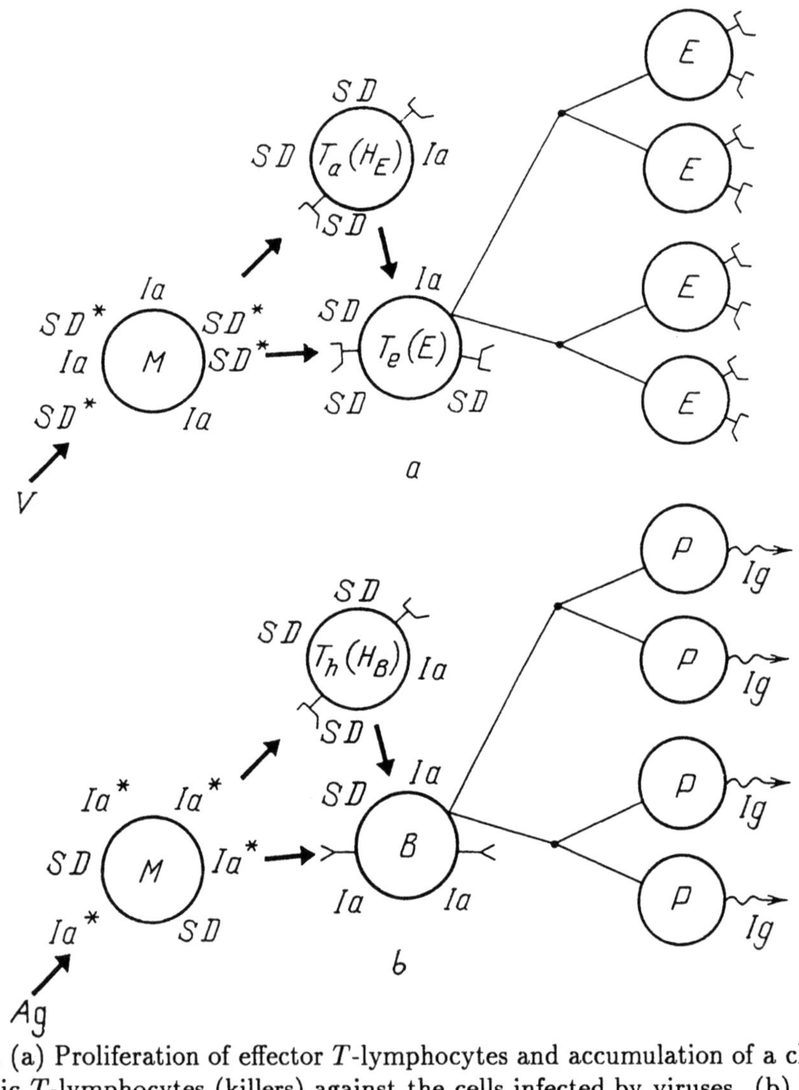

Fig. 30. (a) Proliferation of effector *T*-lymphocytes and accumulation of a clone of cytotoxic *T*-lymphocytes (killers) against the cells infected by viruses. (b) Proliferation of *B*-lymphocytes and accumulation of a clone of plasma cells synthesizing antibodies against viral antigens.

The MHC-restricted recognition principle is true not only for viral antigens, but also for any other antigens, including those stimulating the humoral immune response (Benacerraf, Unanue [28]). Just in these cases,other antigens of the major hystocompatibility complex (Ia in mice and HLA–D in human being) play the role of primary molecules that complexes with antigens. Macrophages are the main cells presenting antigens to the helper T-lymphocytes for further recognition, i. e., they are antigen-presenting cells.

Thus T-lymphocytes recognize antigens only being in a complex with macrophagal antigens of the major hystocompatibility complex; a macrophage is the first cell interacting with foreign antigens, and MHC antigens play the outstanding role in the cell interactions, since they are used by macrophages and lymphocytes in order to recognize each other. If these antigens are not identical, say, they have genetically determined distinctions, then there is no cooperation macrophage–T-lymphocyte, macrophage–B-lymphocyte, T–B, and T–T. Helper T-lymphocytes participating in the antibody response are labelled $Th2$, and $Th1$ in the cellular response. Lymphocytes effectors are denoted as T_e. The antigen–MHC antigen complex of macrophage, which is presented to $Th1$ or $Th2$ is regarded as a switching-on signal. The antibodies against viruses and cytotoxic T-cells recognizing virally infected cells of their own organism are the main factors in clearing the organism from viruses.

Fig. 30 shows the scheme of interaction between the cells in the process of stimulation of cellular immune response to viruses (a) and humoral one (b) to any antigens including viral antigens. Macrophages are denoted by M; the clone of cytotoxic T-cells E and clone of plasma cells P are generated from stimulated $CD8^+$ T-cells and B cells as a result of division and differentiation. Antigens of the major hystocompatibility complex of interacting cells are denoted by Ia and SD; complexes Ia–viral antigen and SD-virus are denoted by Ia^* and SD^*.

Receptors of the double recognition are labelled by the symbol 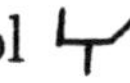, receptors of simple recognition of single antigenic determinant or free antigen are labelled by  . The arrows mark the specific recognition and the direction of the switching on signal.

4.2. Construction of Mathematical Model of Antiviral Immune Response

While deriving the equations simulating the development of antiviral immune response of cellular type, we made the following assumptions:

1. Macrophages M, presenting the antigens of virus V in complex with the SD-molecules denoted by M_V, stimulate helper $Th1$-lymphocytes denoted by H_E and cytotoxic T-lymphocytes denoted by E; accumulated cytotoxic effectors (E) kill the virus-infected cells C_V of a tissue sensitive to given virus (C).

2. There is a sufficient amount of macrophages M in an organism for the appearance of M_V in the number, proportional to the number of viruses.

3. Stimulation of T-helpers (H_E) depends on size of the preceding clone of the same specificity, i.e., bearing the receptors for the complex of Ia with given virus, and on the amount of M_V.

4. H_E stimulated by M_V divide, i.e., their amount increases. If the number of M_V is sufficient, the cells H_E, produced during division, are stimulated again. A helper cell H_E dies after the encounter with E-cells. In other words, the hypothesis of monogamy is suggested for the helper-lymphocytes.

5. The cell E in response to two signals, from M_V and H_E, starts the formation of a clone of effector cells (E) that possess the cytotoxic (killer) activity against C_V. As a result, the number of newly generated cells depends on the preceding clone of E cells of given specificity (i.e., bearing receptors for SD^*), and on the quantity of stimulated H_E.

6. Cells of tissue sensitive to the virus (like all nuclear cells of an organism) bear identical SD-complexes, which become SD^* in the cells infected by virus. Virally infected cells perform their normal functions for a certain time. They die as a result of either the development of irreversible viral damage or the elimination of the cells C_V by cytotoxic T-cells (E). Therefore the damaged mass of a tissue sensitive to virus is the sum of the number of cells killed by the virus and of the number of cells killed by effector-lymphocytes.

While deriving the equations simulating the development of antiviral immune response of humoral type (Fig. 30b), we made the following assumptions.

7. Macrophages, which present viral antigens in association with Ia molecules, stimulate helper $Th2$-lymphocytes (H_B) and B-cells.

8. The same macrophages which express viral particles with SD-structures present complexes of viral antigen with Ia-structures.

9. Stimulation of helper lymphocytes H_B depends on the size of preceding clone of the same specificity (i.e., bearing receptors for the complex Ia with given antigen) and on the amount of M_V.

10. H_B cells are switched on in the same way as H_E (see item 4) and are characteristic of monogamy as related to B-cells.

11. B-cell being affected by two signals: from M_V and from H_B, starts the clone of antibody producing cells (plasma cells P), as E-cell does (see item 5).

12. Generated antibodies eliminate only extracellular viral particles, released form virus-infected cells after their destruction. The number of these viruses is proportional to the number of infected cells killed by viruses and to that of infected cells killed by effector-lymphocytes.

13. All viruses are regarded as "free" in the model, i.e., as circulating freely outside the cells in lymph and blood plasma, assuming that all intracellular processes are described phenomenologically by the coefficient of multiplication of viruses, that get into plasma.

In agreement with the above facts and conceptions on the dynamics of antiviral immune response, we introduce the following model variables.

$V_f(t)$, quantity of viruses circulating freely in an organism;
$M_V(t)$, quantity of stimulated (antigen-presenting) macrophages;
$H_E(t)$, quantity of helper $Th1$-lymphocytes of cellular immunity;
$H_B(t)$, quantity of helper $Th2$-lymphocytes of humoral immunity;
$E(t)$, quantity of cytotoxic T-cells;
$B(t)$, quantity of B-lymphocytes;
$P(t)$, quantity of plasma cells;
$F(t)$, quantity of antibodies;

$C_V(t)$, quantity of cells of a target organ infected by viruses;
$m(t)$, malfunctioning part of a target organ affected by viruses.

Let us now describe the system of equations modelling the immune response to viral invasion. The corresponding equations of the system must, naturally, be a consequence of balance relations between the main components participating in the immune response. The balance relations are written for an instant of time t of an interval Δt, which is so small that the various processes of interaction between the viruses and immune system can be considered as additive. Notice that all quantitative characteristics of different components in the final balance relationships, which with $\Delta t \to 0$ take the form of differential equations, are normalized to a unit of time; this fact will be not mentioned, as a rule, in what follows.

Let us write the balance equation for the number of free viruses:

$$\begin{aligned} \frac{dV_f}{dt} &= \nu C_V + nb_{CE}C_V E - \gamma_{VM} M V_f - \gamma_{VF} F V_f \\ &\quad - \gamma_{VC}(C^* - C_V - m)V_f. \end{aligned} \tag{4.2.1}$$

The term dV_f/dt in the left-hand side of equation (4.2.1) characterizes the rate (velocity) of change of virus population in an organism; the first term in the right-hand side accounts for the increase of viruses population in a unit of time when they multiply in infected cells C_V. The coefficient ν depends on the rate of reproduction of viruses with regard to statistically averaged intracellular factors of defence. It is suggested in this model that new viruses that appear in infected cells get into plasma and become "free". These free viruses stimulate the cells M and transform them into M_V and then they are destroyed by antibodies.

The second term describes viruses which leave infected cells C_V when they are destructed by effector lymphocytes E.

The third and the fourth terms in (4.2.1) describe the decrease in the number of free viruses caused by the interaction with macrophages M and antibodies F. The last term in equation (4.2.1) describes the decrease in population due to the invasion of free viruses into healthy cells which became infected. It is taken into account that quantity of target organ cells accessible to infection is finite. The values of

γ_{VM}, γ_{VF}, γ_{VC}, can be taken constant. They characterize inverse times necessary for the interaction between viruses V_f and macrophages, antibodies, and healthy cells respectively.

Consider the balance equation for the number of macrophages bound with viruses (stimulated macrophages) M_V:

$$\frac{dM_V}{dt} = \gamma_{VM} M V_f - \alpha_M M_V. \tag{4.2.2}$$

The first term in the right-hand side of (4.2.2) describes the increase in the number of macrophages expressing the virus antigens owing to the complexes SD and Ia in a unit of time. Here, as in equation (4.2.1), M is the number of all macrophages in an organism which is meant to be known and determined by the homeostasis. The coefficient γ_{MV} depends on formation time of SD^*- and Ia^*-complexes. The second term in (4.2.2) takes into account the decrease in population of M_V cells due to natural processing or aging. The coefficient α_M is equal to the inverse value of average life time of M_V-cells in an organism.

Consider now the balance equation for the number of helper $Th1$-lymphocytes, that is, H_E-cells which provide for proliferation of E-cells:

$$\begin{aligned} \frac{dH_E}{dt} &= b_H^{(E)} P_H^{(E)}(t - \tau_H^{(E)}) - b_H^{(E)} M_V H_E \\ &\quad - b_p^{(H_E)} M_V H_E E + \alpha_H^{(E)}(H_E^* - H_E), \end{aligned} \tag{4.2.3}$$

where

$$P_H^{(E)}(t) = \rho_H^{(E)} M_V(t) H_E(t).$$

The first term in the right-hand side of (4.2.3) describes the increase in the number of helper T-lymphocytes due to their division in response to the contact (interaction) with stimulated macrophage M_V. Naturally, delay in the process of new H_E-cells generation after the interaction with stimulated macrophage M should be taken into account in this term. Let this delay be equal to $\tau_H^{(E)}$. The coefficient $b_H^{(E)}$ is the value inverse to average time of interaction between H_E-cell and M_V. The second term describes the decrease in the number of activated H_E-cells due to the start of the division cycle after the interaction with macrophages stimulated by viruses. The third term in (4.2.3) describes the decrease of H_E-cells (stimulated by the macrophage M_V) during their interaction with effector E-lymphocytes resulting in the proliferation of effectors. The coefficient $b_p^{(H_E)}$ accounts for the double interaction time, and $\rho_H^{(E)}$ describes the number of generated cells. The last term in (4.2.3) accounts for the maintenance of H_E-cells homeostasis, which is equal to H_E^* in the absence of the immune reaction of organism, and the life-span of cells. The coefficient $\alpha_H^{(E)}$ is equal to a value inverse to average life time for H_E-cells.

It should be noted that terms $b_H^{(E)} M_V H_E$ and $b_p^{(H_E)} M_V H_E E$ in equation (4.2.2) are absent, since we have assumed that after interaction of M_V with H_E and E stimulated macrophage does not die but continues to perform its functions.

Let us write the equation for helper T-lymphocytes providing for the proliferation of B-cells H_B:

$$\begin{aligned}\frac{dH_B}{dt} &= b_H^{(B)} \rho_H^{(B)} M_V(t - \tau_H^{(B)}) H_B(t - \tau_H^{(B)}) - b_H^{(B)} M_V H_B \\ &\quad - b_p^{(H_B)} M_V H_B B + \alpha_H^{(B)} (H_B^* - H_B). \qquad (4.2.4)\end{aligned}$$

Here the first term in the right-hand side is connected with the proliferation of H_B-cells and accounts for the effect of the delayed appearance of new cells after primary contact of H_B-cell with a macrophage stimulated by virus. The second term describes the decrease of H_B-cells due to the start of division cycle. The third term in the right-hand side describes the decrease of H_B-cells as a result of interaction with M_V and B-cells. The last term in the equation describes the homeostasis. The constant $\rho_H^{(B)}$ is equal to the number of cells that appear as a result of division.

Balance equation for the effectors E has the form

$$\frac{dE}{dt} = b_p^{(E)} P_E(t - \tau_E) - b_p^{(E)} M_V H_E E - b_{EC} C_V E + \alpha_E (E^* - E). \quad (4.2.5)$$

The first term in the right-hand side of this equation takes into account the increase in the number of new effectors, which appear per unit of time in the interval of time τ_E due to the division of stimulated effectors, where $b_p^{(E)}$ is a coefficient which accounts for the rate of stimulation of E-cells. The second term describes the decrease of E-cells due to the start of the division process. The third term in (4.2.5) describes the diminution of population of the effector lymphocytes E through the destruction of cells infected by virus. Here b_{EC} is a coefficient which is equal to the inverse value of the average time of interaction between effectors and C_V-cells. The last term describes the homeostasis of E-cells.

Write down a term for effector lymphocytes E stimulated by the double interaction: with stimulated macrophage M_V on the one hand, and with helper T-lymphocyte H_E on the other hand:

$$P_E(t) = \rho_E M_V(t) H_E(t) E(t).$$

This term reflects the fact that the number of stimulated effectors is proportional to the product of probabilities of components of the double interaction of E with M_V and H_E respectively. Here ρ_E is a coefficient which accounts for the number of E-cells that appear during clonal expansion of E-cells.

Balance equation for B-cells has the form

$$\begin{aligned} \frac{dB}{dt} &= b_p^{(B)} \rho_B M_V(t-\tau_B) H_B(t-\tau_B) B(t-\tau_B) \\ &\quad - b_p^{(B)} M_V H_B B + \alpha_B (B^* - B). \end{aligned} \tag{4.2.6}$$

Two first terms in the right-hand side of the equation describe the proliferation of B-cells and their progression through the cell cycle where ρ_B is the number of B-cells generated by one or several divisions. The last term of (4.2.6) describes the homeostasis.

Now write down the balance equation for plasma cells. Let P be a concentration of plasma cells. Their dynamics will then be described by the following equation:

$$\frac{dP}{dt} = b_p^{(P)} \rho_P M_V(t-\tau_P) H_B(t-\tau_P) B(t-\tau_P) + \alpha_P (P^* - P). \tag{4.2.7}$$

The first term in the right-hand side describes the rate of generation of plasma cells from stimulated B-cells with regard to delay of differentiation. The last term in (4.2.7) accounts for the maintenance of homeostasis of plasma cells in the absence of antigenic stimulation.

Next, we write down an equation for antibodies F,

$$\frac{dF}{dt} = \rho_F P - \gamma_{FV} F V_f - \alpha_F F. \tag{4.2.8}$$

Here the first term in the right-hand side describes the production of antibodies by plasma cells P, and ρ_F is a rate of the production of antibodies. The second term describes the expenditure of antibodies spent for the neutralization of viruses, and the last one describes the decrease of antibodies due to natural decay.

Next consider a balance equation for the cells C_V infected by virus:

$$\frac{dC_V}{dt} = \sigma(C^* - C_V - m)V_f - b_{CE} C_V E - b_m C_V. \tag{4.2.9}$$

This equation describes both the infection of healthy cells C by "free" virus, which transforms them into C_V, and the decrease in the number of infected cells which are destroyed by viruses and eliminated by effectors.

Finally, write down the equation for malfunctioning part of an organ affected by viruses:

$$\frac{dm}{dt} = b_{CE} C_V E + b_m C_V - \alpha_m m. \tag{4.2.10}$$

Here the first term in the right-hand side accounts for the effect of elimination of infected cells by effectors, and the second term takes into account irreversible viral damage to cells C_V which cease to function. The last term describes the restoration of damaged cells due to regeneration of tissue; b_{CE}, b_m, α_m are corresponding coefficients.

Few words are to be said about the feed-back relations with the variable m. As was assumed in the case of the simple system of immune response (see Section 3.2.1), when an organ is seriously damaged, the deficiency in the development of immune response occurs. It means that the values of $\rho_H^{(E)}$, $\rho_H^{(B)}$, ρ_E, ρ_B, ρ_P will decrease depending on the increase in m. In order to take into account this factor, it is necessary instead of the values $\rho_H^{(E)}$, $\rho_H^{(B)}$, ρ_E, ρ_B, ρ_P to use in the system of the equations respectively the following values:

$$\rho_H^{(E)}\xi(m),\ \rho_H^{(B)}\xi(m),\ \rho_E\xi(m),\ \rho_B\xi(m),\ \rho_P\xi(m),$$

where the function $\xi(m)$, $0 \leq \xi(m) \leq 1$ is connected with the reduction in the immune system's effectiveness when an organ is damaged by viruses.

Combine the equations we have constructed into the system, which we will call the mathematical model of antiviral immune response:

$$\begin{aligned}\frac{dV_f}{dt} &= \nu C_V + nb_{CE}C_V E - \gamma_{VF}FV_f - \gamma_{VM}MV_f \\ &\quad - \gamma_{VC}(C^* - C_V - m)V_f, \\ \frac{dM_V}{dt} &= \gamma_{VM}MV_f - \alpha_M M_V,\end{aligned} \qquad (4.2.11)$$

$$\begin{aligned}\frac{dH_E}{dt} &= b_H^{(E)}\left[\xi(m)\rho_H^{(E)}M_V(t-\tau_H^{(E)})H_E(t-\tau_H^{(E)}) - M_V H_E\right] \\ &\quad - b_p^{(H_E)}M_V H_E E + \alpha_H^{(E)}(H_E^* - H_E),\end{aligned}$$

$$\begin{aligned}\frac{dH_B}{dt} &= b_H^{(B)}\left[\xi(m)\rho_H^{(B)}M_V(t-\tau_H^{(B)})H_B(t-\tau_H^{(B)}) - M_V H_B\right] \\ &\quad - b_p^{(H_B)}M_V H_B B + \alpha_H^{(B)}(H_B^* - H_B),\end{aligned}$$

$$\begin{aligned}\frac{dE}{dt} &= b_p^{(E)}\left[\xi(m)\rho_E M_V(t-\tau_E)H_E(t-\tau_E)E(t-\tau_E) - M_V H_E E\right] \\ &\quad - b_{EC}C_V E + \alpha_E(E^* - E),\end{aligned}$$

$$\begin{aligned}\frac{dB}{dt} &= b_p^{(B)}\left[\xi(m)\rho_B M_V(t-\tau_B)H_B(t-\tau_B)B(t-\tau_B) - M_V H_B B\right] \\ &\quad + \alpha_B(B^* - B),\end{aligned}$$

$$\frac{dP}{dt} = b_p^{(P)}\xi(m)\rho_P M_V(t-\tau_P)H_B(t-\tau_P)B(t-\tau_P) + \alpha_P(P^* - P),$$

$$\frac{dF}{dt} = \rho_F P - \gamma_{FV}V_f F - \alpha_F F,$$

$$\frac{dC_V}{dt} = \sigma V_f(C^* - C_V - m) - b_{CE}C_V E - b_m C_V,$$

$$\frac{dm}{dt} = b_{CE}C_V E + b_m C_V - \alpha_m m.$$

Initial values must be specified for this system of equations. If an organism is not infected by a virus, then the system permits the following steady state solution:

$$\begin{gathered} V_f = 0, \quad M_V = 0, \quad H_E = H_E^*, \quad H_B = H_B^*, \\ E = E^*, \quad B = B^*, \quad P = P^*, \\ F = \frac{\rho_F P^*}{\alpha_F}, \quad C_V = 0, \quad m = 0. \end{gathered} \tag{4.2.12}$$

As before (see Chapter 3) we are interested in a quite natural situation: the infection of a healthy organism by small dose of viruses V_f^0. We will assume henceforward that before the moment of infection t^0, i.e., for $t < t_o$, a system is in the stationary state (4.2.12), and at the moment $t = t^0$ the infection by the small dose $V_f(t^0) = V_f^0$ occurs. All other components at the moment $t = t^0$ maintain their stationary values. Since the model is autonomous we may assign $t^0 = 0$, not losing the general applicability.

Common properties of the system of model's equations. Assume that all the coefficients of model system of equations (4.2.11) are nonnegative; this assumption is consistent with their biological meaning. Consider the initial conditions for system (4.2.11):

$$\begin{gathered} V_f(0) = V_f^0, \quad M_V(0) = M_V^0, \quad H_E(0) = H_E^0, \quad H_B(0) = H_B^0, \\ E(0) = E^0, \quad B(0) = B^0, \quad P(0) = P^0, \quad F(0) = F^0, \\ C_V(0) = C_V^0, \quad m(0) = m_0, \\ M_V(t)H_E(t) = \varphi_1(t), \quad -\tau_H^{(E)} \le t < 0, \\ M_V(t)H_B(t) = \varphi_2(t), \quad -\tau_H^{(B)} \le t < 0, \\ M_V(t)H_E(t)E(t) = \varphi_3(t), \quad -\tau_E \le t < 0, \\ M_V(t)H_B(t)B(t) = \varphi_4(t), \quad -\tau \le t < 0, \quad \tau = \max\{\tau_B, \tau_P\}, \end{gathered} \tag{4.2.13}$$

where $\varphi_i(t)$, $i = \overline{1,4}$ are continuous functions.

The following statements can be proven (Belykh [26]).

1. There exists a solution of the initial value problem (4.2.11), (4.2.13) for nonnegative values of coefficients in (4.2.11) and for nonnegative initial conditions (4.2.13), which is unique and nonnegative for all $t \ge 0$.

2. The components $V_f(t)$, $M_V(t)$, $C_V(t)$, and $m(t)$ of the vector-function of a solution to the initial value problem (4.2.11), (4.2.13), are for $t > 0$ bounded functions while $0 \leq C_V(t) + m(t) \leq 1$ for all $t > 0$, provided that the values of coefficients and initial conditions are nonnegative and the inequalities $\gamma_{VM}M > 0$ and $0 \leq C_V^0 + m^0 \leq 1$ are satisfied.

3. Sufficient condition for asymptotic stability of stationary solution (4.2.12) is the validity of inequalities

$$(\gamma_{VM}M + \gamma_{VF}F^* + \gamma_{VC}C^*)(b_{CE}E^* + b_m) > \sigma C^*(\nu + nb_{CE}E^*) \quad (4.2.14)$$

and $\alpha_M > 0$, $\alpha_H^{(E)} > 0$, $\alpha_H^{(B)} > 0$, $\alpha_E > 0$, $\alpha_B > 0$, $\alpha_P > 0$, $\alpha_F > 0, \alpha_m > 0$.

Inequality (4.2.14) guarantees the existence of an immunological barrier V^* such that in the case of infection of a healthy organism (4.2.12) by small a dose of viruses V_f^0, $V_f^0 < V^*$ the development of disease does not occur, i.e., $V(t)$ tends to zero with $t \to +\infty$. It is necessary for the disease to develop either to exceed the immunological barrier ($V_f^0 > V^*$) or to break the stability condition (4.2.14).

The system of model equations can have nonnegative stationary solutions $\bar{x} \triangleq [\bar{V}_f, \bar{M}_V, \bar{H}_E, \bar{H}_B, \bar{E}, \bar{B}, \bar{P}, \bar{F}, \bar{C}_V, \bar{m}]^T$, different from (4.2.12) and such that $\bar{V}_f > 0$, $\bar{C}_V > 0$, $\bar{m} > 0$. Solutions of this type describe the chronic form of disease; their existence was shown numerically while simulating chronic forms of viral hepatitis B.

Sufficient conditions for asymptotic stability of stationary solutions $\bar{x}$ of system (4.2.11) have been obtained by Djory and Pertsev [83]. Consider linearized system of equations for system (4.2.11) in the neighborhood of $\bar{x}$:

$$\begin{aligned}\frac{dy(t)}{dt} &= Ay(t) + A_1 y(t - \tau_H^{(E)}) + A_2 y(t - \tau_H^{(B)}) \\ &\quad + A_3 y(t - \tau_E) + A_4 y(t - \tau_B) + A_5 y(t - \tau_P), \quad (4.2.15)\end{aligned}$$

where $y(t)$ is a vector-function with the values in R^{10}; A and $\{A_i\}_{i=1}^5$ are 10×10 matrices.

Let us represent the matrix A as $A = A_0 + B$, where B is a diagonal matrix consisting of diagonal elements of A. Construct the matrix

$G = [g_{ij}]_{i,j=\overline{1,10}}$ following the rule: $G \triangleq B + \sum_{k=0}^{5} A_k^+$, where elements of the matrix A_k^+ are absolute values of corresponding elements of A_k $(k = \overline{0,5})$. As was shown in [83], a sufficient condition for asymptotic stability of stationary solution of system (4.2.15) is the stability of the matrix G, whose stability conditions are determined by Sevastjanov-Kotelyanski criterion (Hantmacher [127]).

We state, omitting intermediate calculations, that the stationary solution of system (4.2.11) is asymptotically stable provided that the following inequalities are satisfied:

$$\frac{(\gamma_{VM}M + \gamma_{VF}\bar{F})\bar{m} + 2\gamma_{VC}\bar{m}(C^* - \bar{m})}{\gamma_{VM}M + \gamma_{VF}\bar{F} + 2\gamma_{VC}\bar{m}} < \bar{C}_V,$$

$$\begin{gathered}(\rho_E\xi(\bar{m}) + 1)(b_P^{(E)}\bar{M}_V\bar{H}_E\bar{E})^2 < \alpha_H^{(E)}\alpha_E H_E^* E^*,\\ \det G_6 > 0,\end{gathered} \tag{4.2.16}$$

$$\begin{gathered}(\rho_B\xi(\bar{m}) + 1)(b_P^{(B)}\bar{M}_V\bar{H}_B B)^2 < \alpha_H^{(B)}\alpha_B H_B^* B^*,\\ \det G > 0,\end{gathered} \tag{4.2.17}$$

where G_6 is 6×6 matrix with the elements:

$$g_{11} = -\gamma_{VM}M - \gamma_{VF}\bar{F} - \gamma_{VC}(C^* - \bar{C}_V - \bar{m}),$$

$$g_{12} = \nu + nb_{CE}\bar{E} + \gamma_{VC}\bar{V}_f, \quad g_{13} = \gamma_{VC}\bar{V}_f, \quad g_{16} = nb_{CE}\bar{C}_V,$$

$$g_{21} = \sigma(C^* - \bar{C}_V - \bar{m}),\ g_{22} = -\sigma\bar{V}_f - b_{CE}\bar{E} - b_m,\ g_{23} = \sigma\bar{V}_f,\ g_{26} = b_{CE}\bar{C}_V,$$

$$g_{32} = b_{CE}\bar{E} + b_m, \quad g_{33} = -\alpha_m, \quad g_{36} = b_{CE}\bar{C}_V,$$

$$g_{41} = \gamma_{MV}, \quad g_{44} = -\alpha_M,$$

$$g_{53} = b_H^{(E)}\rho_H^{(E)}\bar{M}_V\bar{H}_E \left|\frac{d\xi(m)}{dm}\right|_{m=\bar{m}}, \quad g_{55} = -\frac{\alpha_H^{(E)}H_E^*}{\bar{H}_E},$$

$$g_{54} = b_P^{(H_E)}\bar{H}_E\bar{E} + b_H^{(E)}\bar{H}_{(E)}(\rho_H^{(E)}\xi(\bar{m}) + 1), \quad g_{56} = b_P^{(H_E)}\bar{M}_V\bar{H}_E,$$

$$g_{62} = b_{EC}\bar{E}, \quad g_{63} = b_P^{(E)}\rho_E\bar{M}_V\bar{H}_E\bar{E}\left|\frac{d\xi(m)}{dm}\right|_{m=\bar{m}},$$

$$g_{64} = b_p^{(E)}\bar{H}_E\bar{E}(\rho_E\xi(\bar{m}) + 1), \quad g_{65} = b_p^{(E)}\bar{M}_V\bar{E}(\rho_E\xi(\bar{m}) + 1),$$

$$g_{66} = -\frac{\alpha_E E^*}{\bar{E}}.$$

All the remaining g_{ij} are equal to zero. Notice that inequalities (4.2.16) actually describe sufficient conditions for asymptotic stability of the stationary solution for a special case of the model: antiviral immune response of cellular type, whereas inequalities (4.2.17) take into account the components of antiviral immune response of humoral type.

In particular, sufficient conditions for asymptotic stability (4.2.14) can be derived from inequalities (4.2.16) and (4.2.17).

Numerical experiments show that there exist stationary solutions $\bar{x}$ different from (4.2.12), for whom conditions (4.2.16) hold.

The mathematical model of antiviral immune response we have formulated will be used later for simulation experiments.

Notice that the mathematical modelling requires the correct choice of the values of differential equations coefficients which phenomenologically account for various characteristics of the immune response.

4.3. Modeling the Immune Defence Reactions of Organism

The mathematical model of antiviral immune response we have constructed does not account for local mechanisms of the realization of killer action of effector T-lymphocytes and of antibodies in an organ affected by virus. At the same time, the formation of CTLs and antibodies in lymphoid tissues and their circulation in blood cannot yet realize their action by themselves. The transfer of immune defence factors is needed from blood circulation into intercellular tissue glottises through capillary walls. Otherwise killer lymphocytes cannot reach a target cell affected by a virus, and antibodies cannot prevent the infection of healthy cells in the neighborhood of an infected one. Humoral and cellular components of immunity are the constituents of the process of lymph generation and must always retrace the path of lymph:

blood ⇒ intercellular tissue glottises ⇒

⇒ small lymphatic vessels ⇒
⇒ regional lymph nodes ⇒
⇒ large lymph collectors ⇒
⇒ thoracic duct ⇒ blood

Actually, the process of generation of intercellular liquid and its outflow from tissues of any organ (lymph current) is the necessary condition for the realization of antiviral immune defence (Marchuk, Petrov [201]).

Imagine the principal scheme of any organ. Let us call it an abstract organ. It consists of functionally specific cells of given organ (cells of parenchyma), cellular and noncellular structures of connective tissue stroma with macrophages, of organized lymphoid congestions, and of a system of liquid flow. This system is represented by arteries, capillaries, veins, intercellular tissue glottises, lymphoducts. Fig. 31 shows a simplified scheme of an abstract organ with regard to the flow of intratissue liquid:

blood ⇒ intercellular tissue glottices ⇒ lymphoducts.

In normal conditions the amount of generated lymph, its volume in intercellular glottices, and the rate of circulation are optimal for the functions of nutrition, oxygen supply, and immune surveillance. This normal quantity of lymph in tissues can be denoted by Λ^* and defined as "normal oedema" in terms of its physiological meaning. An increased quantity Λ of extracellular liquid in an organ is responsible for one or another degree of oedema. This relative value exceeding the normal one is expressed by $l = (\Lambda - \Lambda^*)/\Lambda^*$. Since the system of lymphoducts has no special mechanism regulating the gaps, the amount of intratissue lymph and the rate of its flow are determined by intensity of the delivery of lymph components (plasma and lymphocytes) from blood. This requires the increase of intracapillary blood pressure and increased penetration of lymphocytes through vascular walls. The rise of intracapillary pressure can be provided by the

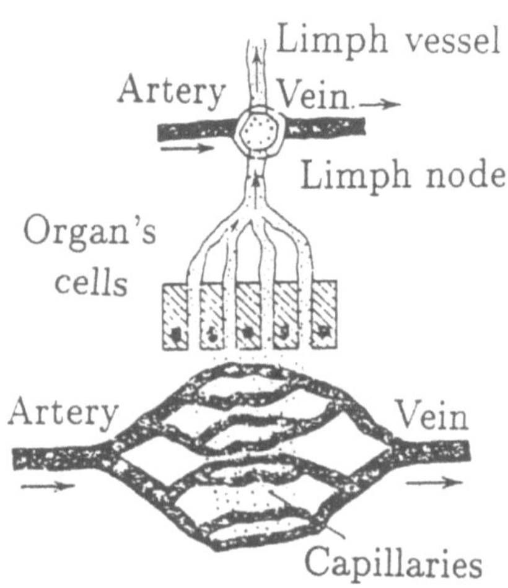

Fig. 31. Simplified scheme of intratissue liquid flow in an abstract organ.

increase of arterial pressure and by slowing down the venous outflow. With the buildup of oedema the increase of arterial pressure is necessary to provide a proper oxygen supply level for the cells which is hampered by increased volume of oedema liquid. Additional outflow of cells from blood ducts is induced by a series of mediators pushed out from damaged or interacting cells of the tissues of given organ.

We consider a certain degree of oedema as an obligatory and necessary component of the defence reaction, which develops in tissues of any organ in the case of viral infection of cells. This component provides increased influx of antibodies and lymphocytes in the locus of damage and requires for its positive realization an increase in arterial tension and some other physiological mechanisms and the outflow of active mediators from affected or interacting with antibodies and lymphocytes cells which provide for increased influx of lymphocytes into the locus of oedema. As we pointed out above, the effector mechanisms of immune defence (antibodies and cytotoxic T-lymphocytes) will not operate unless the process called oedema is realized, since they will not be delivered to target cells affected by virus. The more intensive is the lymph flux and the larger is the quantity of lymphocytes passing through the locus of damage, the higher is the probability of accumulation of sensibilized CTLs of a given clone in this locus, i.e., the clone bearing specific receptors to given targets. More than that, the more intensive is lymphocyte circulation in an organism, the faster is the selection and accumulation in the site of damage of effective quantity of cytotoxic T-lymphocytes of given specificity.

If we have recognized the necessity of oedema, which is accompanied by a rise in arterial tension, the question arises on quantitative and temporal boundaries of their expediency, and on the possible therapeutic effectiveness of their correction (stimulation–depression), and on the search for methods of stimulation of lymphocytes circulation with the intensification of lymph flux (Fig. 32). The problem of "limit of compensation" arises too, i.e., the problem of a level of oedema or arterial tension which, being exceeded, threatens with pathological disorders and dangerous malfunctions of one or another organ. The malfunctions can be manifested by disorders in liver detoxification function, in kidney filtration and concentration function, it may be the development of ischemia of a certain part of the heart or even infarction of the heart, because the provision of myocardium with oxygen is of prime importance. In such a situation the therapy must be directed to lowering the lymph flux and arterial tension.

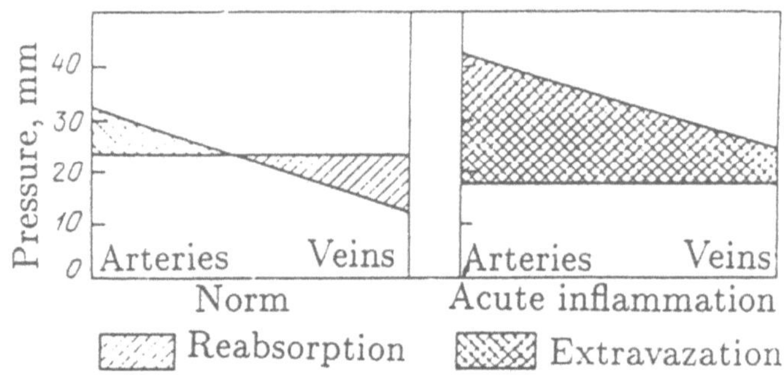

Fig. 32. Blood pressure and colloidal-osmotic plasma pressure (horizontal line) in normal and inflammated tissues.

The notion of the "limit of compensation" must be introduced into the reaction of destructive action of killer lymphocyte against target cells affected by virus. One may conceive two extreme cases. The first one is the case when a virus attack has damaged a small portion of an organ's cells. The failure of this small amount of cells is compensated by the functioning of normal cells. The resulting oedema provides the delivery of a sufficient quantity of antibodies and killer lymphocytes to the organ. Antibodies neutralize the extracellular population of viruses. Cytotoxic T-cells destroy the infected cells. The development of pathology is terminated, the processes of regeneration reconstruct or replace the lost cells. The second case is characterized by uneliminated primary damage because of the delay of immune response. The viruses have propagated and damaged the majority of an organ's cells. Oedema increases, the killer lymphocytes, which have developed with delay, destroy this majority of target cells. Actually, "autoimmune" destruction of the cells of a given type happens. The organ cannot work any more. In other words, when a certain level of damage to cells by the virus (the limit of compensation) is exceeded, it is necessary rather to inhibit than to stimulate the intensity of cellular immune reaction by pharmacological treatment in order to prevent such rate of destruction of virus infected cells, which cannot be compensated by the existing rate of regeneration.

These interrelated processes of the development of damage to cells by viruses, formation of oedema, rise of arterial tension, neutralization of extracellular viruses by antibodies, and destruction of target cells infected by viruses by killer lymphocytes can be modeled mathematically. Mathematical modelling makes it possible to find the limits of compensation for each of the above processes, which can be used to decide whether stimulating or suppressing therapy for given process is required.

The model of viral disease (4.2.11) is based on the assumption that primary viral damage of certain portion of cells of a given organ triggers the immune physiological reaction of defence. The intensity of reactions is determined by the value of this primary damage, i.e., by the number of infected cells (C_V). As was pointed out above, an oedema is characterized by the value of l, a dimensionless parameter which is the deviation of lymph volume in a site of damage (Λ) from the norm (Λ^*), i.e.,

$$l = \frac{\Lambda - \Lambda^*}{\Lambda^*}. \tag{4.3.1}$$

Assume that the value l depends on the quantity of infected cells C_V and on possible physiological or pharmacological influence on the lymph flux denoted by δ. Then

$$l = kC_V + \delta, \tag{4.3.2}$$

where k is a constant.

Suppose, next, that the immune reaction in the site of damage is possible only in presence of oedema ($l \neq 0$) and is proportional to the intensity of the latter, i.e., to the value of l. Therefore we must substitute in our basic model corresponding coefficients of interaction of antibodies F with viruses and of cytotoxic T-lymphocytes E with infected cells with the products of these coefficients by a certain function depending on l:

$$\begin{aligned} b_{EC} &\to b_{EC}f(l), \quad b_{CE} \to b_{CE}f(l), \\ \gamma_{VF} &\to \gamma_{VF}f(l), \quad \gamma_{FV} \to \gamma_{FV}f(l). \end{aligned} \tag{4.3.3}$$

The function $f(l)$ can be considered as linear one, to begin with, i.e.,

$$f(l) = 1 + \mu l, \quad \mu = \text{const.} \tag{4.3.4}$$

Assume further that the function $\xi(m)$ depends not only on m, but on l too, i.e.,

$$\xi(m) \rightarrow \xi(m, l). \tag{4.3.5}$$

Finally, we assume that arterial pressure increases proportionally to the size and degree of oedema, since the arterial tension must provide the intensification of blood circulation which is necessary for the oxygen supply with oedema being increased. If g is blood pressure then

$$g = g^*(1 + \alpha l), \tag{4.3.6}$$

where g^* is a norm and α is a coefficient.

The main system of equations of antiviral immune response with regard to above assumptions will have the form

$$\begin{aligned} \frac{dV_f}{dt} &= \nu C_V + n b_{CE} f(l) C_V E - \gamma_{VF} f(l) F V_f \\ &\quad - \gamma_{VM} M V_f - \gamma_{VC}(C^* - C_V - m) V_f, \end{aligned}$$

$$\frac{dM_V}{dt} = \gamma_{VM} M V_f - \alpha_M M_V, \tag{4.3.7}$$

$$\begin{aligned} \frac{dH_E}{dt} &= b_H^{(E)} \left[\xi(m,l) \rho_H^{(E)} M_V(t - \tau_H^{(E)}) H_E(t - \tau_H^{(E)}) - M_V H_E\right] \\ &\quad - b_p^{(H_E)} M_V H_E E + \alpha_H^{(E)} (H_E^* - H_E), \end{aligned}$$

$$\begin{aligned} \frac{dH_B}{dt} &= b_H^{(B)} \left[\xi(m,l) \rho_H^{(B)} M_V(t - \tau_H^{(B)}) H_B(\tau - \tau_H^{(B)}) - M_V H_B\right] \\ &\quad - b_p^{(H_B)} M_V H_B B + \alpha_H^{(B)} (H_B^* - H_B), \end{aligned}$$

$$\begin{aligned} \frac{dE}{dt} &= b_p^{(E)} [\xi(m,l) \rho_E M_V(t - \tau_E) H_E(t - \tau_E) E(t - \tau_E) - M_V H_E E] \\ &\quad - b_{EC} f(l) C_V E + \alpha_E(E^* - E), \end{aligned}$$

$$\begin{aligned}\frac{dB}{dt} &= b_p^{(B)}\left[\xi(m,l)\rho_B M_V(t-\tau_B)H_B(t-\tau_B)B(t-\tau_B) - M_V H_B B\right]\\ &\quad + \alpha_B(B^* - B),\end{aligned}$$

$$\begin{aligned}\frac{dP}{dt} &= b_p^{(P)}\xi(m,l)\rho_P M_V(t-\tau_P)H_B(t-\tau_P)B(t-\tau_P)\\ &\quad + \alpha_P(P^* - P),\end{aligned}$$

$$\frac{dF}{dt} = \rho_F P - \gamma_{FV} f(l) V_f F - \alpha_F F, \tag{4.3.7}$$

$$\frac{dC_V}{dt} = \sigma V_f(C^* - C_V - m) - b_{CEf}(l) C_V E - b_m C_V,$$

$$\frac{dm}{dt} = b_{CE} f(l) C_V E + b_m C_V - \alpha_m m,$$

with the initial conditions for $t \le t^0$ corresponding to the infection of a healthy organism:

$$\begin{gathered} V_f(t) = V_f^0\theta(t-t^0), \quad M_V(t) = C_V(t) = m(t) = 0,\\ H_E(t) = H_E^*, \quad H_B(t) = H_B^*, \quad E(t) = E^*, \\ B(t) = B^*, \quad P(t) = P^*, \quad F(t) = F^* = \rho_f P^*/\alpha_f \end{gathered} \tag{4.3.8}$$

where $\theta(t) = \begin{cases} 1, & t \ge 0, \\ 0, & t < 0. \end{cases}$

We find it necessary to formulate, in conclusion, a series of general, as well as special and concrete, questions and problems, which are to be experimentally, clinically, and mathematically investigated, modeled, and implemented.

1. Do the lymph vessels actually possess no apparatus regulating their crossection, and do the rate of the lymph flux is determined by the volume of oedema?

2. What are pharmacological and other ways of intensification or inhibition of the lymph flux? Is it possible to shunt a peripheral lymph node or to mount the lymphovenous anastomoses?

3. Is it necessary to introduce into clinical practice the known and to find out new methods of intensification and inhibition of circulation and migration of lymphocytes with elaboration of means to intensify

the discharge of lymphocytes and immunoglobulins reserved in lymphoid tissues into blood? Is it profitable to accumulate and freeze the organism's own lymphocytes isolated from blood in period of remission, with following injection of them into blood in periods of exacerbation?

4. What is an effective regime of increasing the arterial tension pharmacologically of physiotherapeutically at the beginning of disease? When is it profitable to intensify the oedema? Is it possible to increase the tissue's lymph flux by the acceleration or inhibition of venous outflow?

5. It is necessary to elaborate the criteria for determination of limits of the compensation of an oedema degree, of the increase of arterial tension, and of the intensity of immune destruction of target cells, infected by viruses, by killer lymphocytes for specified organ: heart, liver, kidney, etc.

6. It is necessary to elaborate the techniques and regimes for the immunosuppression, which are the most effective as related to killer lymphocytes under conditions when the limit of compensation of the interaction killer–target cell is attained.

7. Detailed immunomorphological investigations are needed of viral diseases based on the above model of immunophysiological reactions to virus attack on an organ.

8. Methods of quantitative estimation of the degree of viral damage of organs are to be developed with determination of cell tropism for the most widespread viruses to given organs or cellular systems. Such an investigation should first of all be concerned with the adeno-, rhino- and enteroviral infections.

9. Extensive virological investigations of concrete organs must be carried out in order to study their susceptibility to DNA-viruses and RNA-viruses and to characterize the viruses of highest tropism for each organ.

10. According to the model, the most early application of immunoglobulins has to provide pronounced clinical effect, since viral-specific antibodies neutralize extracellular viral population and prevent the propagation of viral damage to normal cells. It is known from works since 1940 that the usage of specific antibodies in combination with an

immunizing microorganism (sensibilized vaccine) lowers the immunization effect. New and more precise studies are needed on the influence of joint action of antigen and antibodies, as well as of antibodies that are in immunoglobulin preparations, on the immune system.

11. Most vital problems arise in cardiology in connection with the possible application of the model we suggest (regulation of arterial tension and lymph flux) to the treatment of viral complications in the patients with hypertension, ischemic heart disease, myocardial infarction, and in the states when the arterial tension is low. New limits of compensation for the components of the model must be determined for these patients; the methods of therapeutic administration may prove to be unusual from the stand point of conventional approaches.

12. There are strong grounds to believe that many physiotherapeutical and balneological treatments (massage, certain complexes of therapeutical physical training, mud cure, baths, UHF (ultrasound of high frequency), etc., provide healing effect with amplification of lymph flux and intensification of lymphocytes circulation in an organism. However, this problem has not yet been formulated from this viewpoint, and there is no systematic investigations in this area. It would be efficient to organize these researches in order to establish the criteria for the prescription and evaluation of above procedures.

4.4. Model of Antibacterial Immune Response

The model of antiviral immune response formulated in Section 4.2 describes joint functioning of T- and B-systems of immunity.

Unlike the antiviral immune response, the immune response that provides the defence of an organism against bacterial infection (for instance, the infection of the lung respiratory tract) is basically the humoral response. Specific antibodies (IgM, IgG), produced during the reaction of B-system of lymphocytes to bacterial aggression, bind themselves to the surface antigens of bacteria (opsonization process). Thereafter bacteria are destroyed by various systems of organism's defence: complements system, phagocytosis system, etc.

As was noted in Section 4.1, the principle of MHC-restricted recognition in the cooperation of cells during the development of humoral immune response holds for antigens of any nature, including bacterial ones. Equations of the model of antibacterial immune response represent the part of the model of antiviral immune response that describes the reaction of the B-system of lymphocytes.

While constructing the system of equations which describe the development of the immune response of humoral type in the case of bacterial infection, we have made the following assumptions:

1. Macrophages, which present bacterial antigens with Ia-molecules (class II MHC), stimulate the helper $Th2$-lymphocytes H_B and B-cells.

2. Stimulation of helper $Th2$-lymphocytes H_B depends on the size of the preceding clone of given specificity and on quantity of M_V.

3. A B-cell affected by two signals from M_V and H_B initiates the development of clones of plasma cells producing antibodies.

4. Produced antibodies bind themselves to bacterial particles, destroyed subsequently by the neutrophils, which are believed to be in sufficient quantity, by the complement, and other defence mechanisms.

Here are the variables of the model of antibacterial immune response:

$K(t)$, quantity of pathogenic bacteria in a target organ;

$M_K(t)$, quantity of stimulated macrophages in lymphoid tissue of a target organ;

$H_B(t)$, quantity of helper $Th2$-cells of given specificity;

$B(t)$, quantity of B-lymphocytes of given specificity;

$P(t)$, quantity of plasma cells producing antibodies of given specificity;

$F(t)$, quantity of specific antibodies;

$m(t)$, part of affected cells of a target organ.

Let us construct equations of the model. The balance equation for the number of bacteria has the form:

$$\frac{dK}{dt} = \beta K - \gamma_{KM} MK - \gamma_{KF} FK. \tag{4.4.1}$$

Here the first term in the right-hand side describes phenomenologically the rate of bacteria multiplication in a target organ, the second and the third terms describe the decrease of bacteria population due to the absorption by macrophages M and opsonization by antibodies F with subsequent destruction of bacteria by complement and by phagocytes.

Write the equation for macrophages stimulated by bacteria as follows:

$$\frac{dM_K}{dt} = \gamma_{MK} K M - \alpha_M M_K. \tag{4.4.2}$$

The first term in the right-hand side of (4.4.2) describes the rate of increase in the number of macrophages stimulated by given bacteria, and the second term describes their natural loss due to exhaustion and aging. M is the number of all macrophages in an organism determined by homeostasis.

The equation for helper T-lymphocytes takes the form:

$$\begin{aligned}\frac{dH_B}{dt} &= b_H^{(B)} \left[\rho_H^{(B)} M_K(t-\tau_H^{(B)}) H_B(\tau-\tau_H^{(B)}) - M_K H_B\right] \\ &\quad - b_p^{(H_B)} M_K H_B B + \alpha_H^{(B)} (H_B^* - H_B). \end{aligned} \tag{4.4.3}$$

Here the first term in square brackets represents the appearance of new cells due to the division of stimulated H_B-cells with regard to the effect of delay in the accumulation of cells after primary contact of H_B-cell with the macrophages stimulated by the bacteria. The second term in the brackets describes the entry of stimulated H_B-cells into the proliferation phase of the cell cycle. The third term in the right-hand side in (4.4.3) describes the decrease of H_B-cells as a result of interaction with M_K and B-cells. The last term in the equation describes the homeostasis. The constant $\rho_H^{(B)}$ describes the number of cells that appear in the process of division.

Here is the balance equation for B-cells:

$$\begin{aligned}\frac{dB}{dt} &= b_p^{(B)} \left[\rho_B M_K(t-\tau_B) H_B(t-\tau_B) B(t-\tau_B) - M_K H_B B\right] \\ &\quad + \alpha_B (B^* - B). \end{aligned} \tag{4.4.4}$$

Both terms in the first brackets, as in equation (4.4.3), describe respectively the proliferation of B-cells and their expenditure for the following proliferation, where ρ_B is the number of B-cells generated in the process of division. The last term in (4.4.4) describes the homeostasis.

Write down now the balance equation for plasma cells P. The dynamics of their generation will be described by the following equation:

$$\frac{dP}{dt} = b_p^{(P)} \rho_P M_K(t-\tau_P) H_B(t-\tau_P) B(t-\tau_P). + \alpha_P(P^* - P) \quad (4.4.5)$$

The first term in the right-hand side of (4.4.5) describes the rate of generation and maturation of plasma cells from stimulated B-cells with regard to delay. The last term in (4.4.5) accounts for the maintenance of the homeostasis in absence of antigenic stimulation, and the life-span of P-cells.

Next, write down the equation for the antibodies:

$$\frac{dF}{dt} = \rho_F P - \eta_F \gamma_{FK} K F - \alpha_F F. \quad (4.4.6)$$

The first term in the right-hand side describes here the generation of antibodies by plasma cells P; ρ_F is the number of antibodies that appear per unit of time. The second term describes the expenditure of antibodies for the destruction of bacteria. The last term describes the decrease in the number of antibodies due to natural decay.

Finally, write down the equation for the malfunctioning part of an organ, which is affected by the toxins secreted by bacteria:

$$\frac{dm}{dt} = \sigma K - \alpha_m m, \quad (4.4.7)$$

where the first term in the right-hand side describes the damage to an organ, which is proportional to the bacteria population in an organism; the second term describes the decrease in tissue damage due to regeneration.

If we now introduce the function $\xi(m)$ characterizing the decrease in productivity of immune components which depends on the degree of damage to an organ, we obtain finally the system of equations describing the antibacterial immune response on a basis of the given model:

$$\frac{dK}{dt} = \beta K - \gamma_{KM} M K - \gamma_{KF} F K,$$

$$\frac{dM_K}{dt} = \gamma_{MK} K M - \alpha_M M_K, \quad (4.4.8)$$

$$\frac{dH_B}{dt} = b_H^{(B)}\left[\xi(m)\rho_H^{(B)} M_K(t-\tau_H^{(B)})H_B(\tau-\tau_H^{(B)}) - M_K H_B\right] - b_p^{(H_B)} M_K H_B B + \alpha_H^{(B)}(H_B^* - H_B),$$

$$\frac{dB}{dt} = b_p^{(B)}\left[\xi(m)\rho_B M_K(t-\tau_B)H_B(t-\tau_B)B(t-\tau_B) - M_K H_B B\right] + \alpha_B(B^* - B),$$

$$\frac{dP}{dt} = b_p^{(P)}\xi(m)\rho_P M_K(t-\tau_P)H_B(t-\tau_P)B(t-\tau_P) + \alpha_P(P^* - P),$$

$$\frac{dF}{dt} = \rho_F P - \eta_F\gamma_{FK}(l)KF - \alpha_F F, \tag{4.4.8}$$

$$\frac{dm}{dt} = \sigma K - \alpha_m m.$$

Let us add the initial conditions, corresponding to the infection of a healthy organism for $t^0 = 0$:

$$K(0) = K_0, \quad M_K(0) = 0, \quad H_B(0) = H_B^*, \quad B(0) = B^*,$$

$$P(0) = P^*, \quad F(0) = F^* = \rho_F P^*/\alpha_F, \quad m(0) = 0, \tag{4.4.9}$$

$$M_K(t)H_B(t) = 0, \quad -\tau_H^B \le t < 0,$$

$$M_K(t)H_B(t)B(t) = 0, \quad -\tau \le t < 0, \quad \tau = \max[\tau_B, \tau_P].$$

It is easy to verify that in absence of bacteria ($K = 0$) we will have the steady-state solution:

$$M_K = 0, \quad H_B = H_B^*, \quad B = B^*, \quad P = P^*, \quad F = \rho_F P^*/\alpha_F, \quad m = 0.$$

4.5. Mathematical Model of Immune Response in the Case of Viral-Bacterial Infection

Let us consider a typical case of the development of acute viral and bacterial infection against the background of a chronic bacterial process. A permanent and long lasting weak bacterial process forms in an organism a new level of antibodies specific to the pathogenic bacteria. This state is extremely stable. At the same time, such patients are always threatened by possible exacerbation of the chronic bacterial process due to temporal weakening of the defence mechanisms of their organisms. As we had noticed before, viral infections usually serve as "trigger mechanisms" for the exacerbation of chronic bacterial process. Assume that a patient is infected by viruses against which there are no protecting antibodies. The experimental and clinical data show than under these conditions the virus is characterized by a high rate of multiplication and within two or three days the virus concentration in a target organ increases by $10^6 - 10^7$ times. It leads to the development of acute viral infection and to powerful antigenic stimulation. Since the processes of immunogenesis connected with acute viral and chronic bacterial infections occur in the same lymphoid tissue, say, in lungs, the development of the antiviral immune response reduces the intensity of the immune processes which constrain the chronic bacterial infection.

The mathematical model of viral-bacterial infectious disease combines the above models of antiviral immune response and antibacterial immune response. It is assumed that the interaction between viral and bacterial infection processes is mediated both by the development of damage to sensitive organs (viral and bacterial) resulting in lower immune response effectiveness, and by competition in the process of clone formation between specific lymphocytes realizing antiviral and antibacterial immune responses. We believe that the intensity of general immune reaction of an organism has a limit; if we introduce some characteristic Q which is total productivity of the immunity system in the course of disease, then this characteristic is bounded above ($Q(t) \leq Q_{\max}$) by potentials of the organism. Therefore affected organism cannot produce antibodies and effector T-lymphocytes over certain maximum value. There are reasons to assume that this limit depends on general state of an organism, so that the development of viral or bacterial damage of organs and tissues decreases $Q_{\max}$.

Concrete mechanisms realizing this limitation on the immune system productivity may be different. For example, the processes of immune response associated with acute viral and chronic bacterial infections which take place in the same lymphoid tissue can be characterized by the competition between specific lymphocytes, realizing antiviral and antibacterial immune responses, for a limited number of antigen-presenting macrophages. More than that, the appearance of a large number of viral antigens at the peak of viral disease reduces for some time the presentation level and, therefore, the intensity of reaction to bacterial antigens.

Another mechanism realizing the limitation on the production of immune components may be connected with physical limitations on the size of a lymph node, on its blood provision, on the cells' concentration and, therefore, on the number of proliferating clones of lymphocytes.

Let us formulate principal assumptions that are necessary to construct the mathematical model of viral-bacterial infection.

1. Let m_V be the characteristic of the malfunctioning part of a target organ under virus attack, and let m_K be the characteristic of bacterial damage of corresponding organ. Introduce the value of generalized damage of an organism in case of viral-bacterial infection

$$m = a_V m_V + a_K m_K. \tag{4.5.1}$$

The constants a_V and a_K characterize a drop in the reproduction of immune defence components in the cases when the organs are damaged by viruses and bacteria respectively.

2. Viral-bacterial attack weakens the processes of immunogenesis in an organism. We will describe this influence with the function $\xi(m)$ that was introduced and described in the Sections 3.1, 4.2. It is assumed that the same function $\xi(m)$ of the generalized damage affects in the same way as the development of antiviral and antibacterial immune responses.

3. The total productivity Q of the immune system while reacting to various antigens is limited: $Q(t) \leq Q_{\max}$, and this limit depends on

general damage of an organism, i.e., $Q_{\max} = Q_{\max}(m)$ so that the more serious is the damage, the lower is the limit.

4. It is assumed that the generation of plasma cells which synthesize antibodies and cytotoxic T-cells in an organism is such that there is a certain limit of joint intensity of generation of these lymphocytes during the course of immune reaction to various viral and bacterial antigens.

Introduce the notations $Q_{E_V}(t)$, $Q_{P_V}(t)$, $Q_{P_K}(t)$ for the rates of clone formation of T-effectors and plasma cells specific to the viral and bacterial antigens.

$$Q_{E_V}(t) = \rho_{E_V} b_p^{(E_V)} M_V(t - \tau_{E_V}) H_{E_V}(t - \tau_{E_V}) E_V(t - \tau_{E_V}),$$

$$Q_{P_V}(t) = \rho_{P_V} b_p^{(P_V)} M_V(t - \tau_{P_V}) H_{B_V}(t - \tau_{P_V}) B_V(t - \tau_{P_V}), \qquad (4.5.2)$$

$$Q_{P_K}(t) = \rho_{P_K} b_p^{(P_K)} M_K(t - \tau_{P_K}) H_{B_K}(t - \tau_{P_K}) B_K(t - \tau_{P_K}),$$

where T- and B-lymphocytes specific to viral (V) and bacterial (K) antigens are denoted by H_{E_V}, H_{B_V}, E_V, B_V, H_{B_K}, B_K. By virtue of assumption 3 the total value of the rate $Q(t)$ of the production of plasma cells producing antibodies against viruses and bacteria, and effector T-cells specific for viral antigens is limited:

$$Q_{E_V}(t) + Q_{P_V}(t) + Q_{P_K} = Q(t) \leq Q_{\max}. \qquad (4.5.3)$$

Notice that the choice of restriction as it stands is not, generally speaking, unique and can be modified for a concrete statement of a problem. If the condition (4.5.3) holds then there is no redistribution of resources. Otherwise the rate of generation of clones of plasma cells and cytotoxic T-cells against corresponding antigens is assumed to be changed according to the following formulas:

$$q_{E_V}(t) = \frac{Q_{E_V}(t)}{Q(t)} Q_{\max}, \quad q_{P_V}(t) = \frac{Q_{P_V}(t)}{Q(t)} Q_{\max},$$
$$q_{P_K}(t) = \frac{Q_{P_K}(t)}{Q(t)} Q_{\max}, \qquad (4.5.4)$$

or, in general form:

$$q_{E_V} = Q_{E_V}, \quad q_{P_V} = Q_{P_V}, \quad q_{P_K} = Q_{P_K}, \quad \text{if} \quad Q \leq Q_{\max},$$

$$q_{E_V} = \frac{Q_{E_V}}{Q} Q_{\max}, \; q_{P_V} = \frac{Q_{P_V}}{Q} Q_{\max}, \; q_{P_K} = \frac{Q_{P_K}}{Q} Q_{\max}, \quad \text{if} \quad Q > Q_{\max}. \tag{4.5.5}$$

Now we can close the description of antiviral and antibacterial immune responses combining systems of equations (4.2.11) and (4.4.8) into single mathematical model of viral-bacterial infection. The system of the model's equations takes the following form:

$$\frac{dV_f}{dt} = \nu C_V + nb_{CE}C_V E_V - \gamma_{VM} M V_f - \gamma_{VF} F_V V_f - \gamma_{VC}(C^* - C_V - m)V_f,$$

$$\frac{dM_V}{dt} = \gamma_{VM} M V_f - \alpha_M M_V,$$

$$\frac{dH_{E_V}}{dt} = b_H^{(E_V)} \left[\xi(m)\rho_H^{(E_V)} M_V(t - \tau_H^{(E_V)}) H_{E_V}(t - \tau_H^{(E_V)}) - M_V H_{E_V}\right] - b_p^{(H_{E_V})} M_V H_{E_V} E_V + \alpha_H^{(E)} \left(H_{E_V}^* - H_{E_V}\right),$$

$$\frac{dH_{B_V}}{dt} = b_H^{(B_V)} \left[\xi(m)\rho_H^{(B_V)} M_V(t - \tau_H^{(B_V)}) H_{B_V}(\tau - \tau_H^{(B_V)}) - M_V H_{B_V}\right] - b_p^{(H_{B_V})} M_V H_{B_V} B + \alpha_H^{(B)} \left(H_{B_V}^* - H_{B_V}\right),$$

$$\frac{dE_V}{dt} = \xi(m) q_{E_V}(t) - b_p^{(E_V)} M_V H_{E_V} E_V - b_{EC} C_V E_V + \alpha_E(E_V^* - E_V),$$

$$\frac{dB_V}{dt} = b_p^{(B_V)} [\xi(m)\rho_{B_V} M_V(t - \tau_{B_V}) H_{B_V}(t - \tau_{B_V}) B_V(t - \tau_{B_V}) - M_V H_{B_V} B_V] + \alpha_B(B_V^* - B_V),$$

$$\frac{dP_V}{dt} = \xi(m) q_{P_V}(t) + \alpha_P(P_V^* - P_V), \tag{4.5.6}$$

$$\frac{dF_V}{dt} = \rho_F P_V - \gamma_{FV} V_f F_V - \alpha_F F_V,$$

$$\frac{dC_V}{dt} = \sigma V_f (C^* - C_V - m_V) - b_{CE} C_V E_V - b_m C_V,$$

$$\frac{dm_V}{dt} = b_{CE} C_V E_V + b_m C_V - \alpha_{m_V} m_V,$$

$$\frac{dK}{dt} = \beta K - \gamma_{KM} M K - \gamma_{KF} F_K K,$$

$$\frac{dM_K}{dt} = \gamma_{MK} K M - \alpha_M M_K, \tag{4.5.6}$$

$$\begin{aligned}\frac{dH_{B_K}}{dt} &= b_H^{(B_K)} \left[\xi(m) \rho_H^{(B_K)} M_K(t - \tau_H^{(B_K)}) H_{B_K}(\tau - \tau_H^{(B_K)}) - M_K H_{B_K}\right] \\ &\quad - b_p^{(H_{B_K})} M_K H_{B_K} B_K + \alpha_H \left(H_{B_K}^* - H_{B_K}\right),\end{aligned}$$

$$\begin{aligned}\frac{dB_K}{dt} &= b_p^{(B_K)} [\xi(m) \rho_{B_K} M_K(t - \tau_{B_K}) H_{B_K}(t - \tau_{B_K}) B_K(t - \tau_{B_K}) \\ &\quad - M_K H_{B_K} B_K] + \alpha_B (B_K^* - B_K),\end{aligned}$$

$$\frac{dP_K}{dt} = \xi(m) q_{P_K}(t) + \alpha_P (P_K^* - P_K),$$

$$\frac{dF_K}{dt} = \rho_F P_K - \eta_F \gamma_{FK} K F_K - \alpha_F F_K,$$

$$\frac{dm_K}{dt} = \sigma_K K - \alpha_{m_K} m_K.$$

Initial conditions corresponding to viral infection of an organism subjected to a chronic bacterial process take the form:

$$V_f(0) = V_f^0, \quad M_V(0) = 0, \quad H_{E_V}(0) = H_{E_V}^*, \quad H_{B_V}(0) = H_{B_V}^*,$$

$$E_V(0) = E_V^*, \quad B_V(0) = B_V^*, \quad P_V(0) = P_V^*,$$

$$F_V(0) = F_V^* = \rho_F P_V^* / \alpha_F, \quad C_V(0) = m_V(0) = 0,$$

$$K(0) = K_0, \quad M_K(0) = M_K^0, \quad H_{B_K}(0) = H_{B_K}^0, \quad B_K(0) = B_K^0,$$

$$P_K(0) = P_K^0, \quad F_K(0) = F_K^0, \quad m_k(0) = m_K^0, \tag{4.5.7}$$

$$M_V(t) H_{E_V}(t) = 0, \quad -\tau_H^{(E_V)} \le t < 0,$$

$$M_V(t)H_{B_V}(t) = 0, \quad -\tau_H^{(B_V)} \le t < 0,$$
$$M_V(t)H_{E_V}(t)E_V(t) = 0, \quad -\tau_{E_V} \le t < 0,$$
$$M_V(t)H_{E_V}(t)E_V(t) = 0, \quad -\tau_{E_V} \le t < 0, \quad \tau = \max[\tau_{B_V}, \tau_{P_V}],$$
$$M_K(t)H_{B_K}(t) = \varphi_1(t), \quad -\tau_H^{(B_K)} \le t < 0,$$
$$M_K(t)H_{B_K}(t)B_K(t) = \varphi_2(t), \quad -\tau \le t < 0, \quad \tau = \max[\tau_{B_K}, \tau_{P_K}].$$

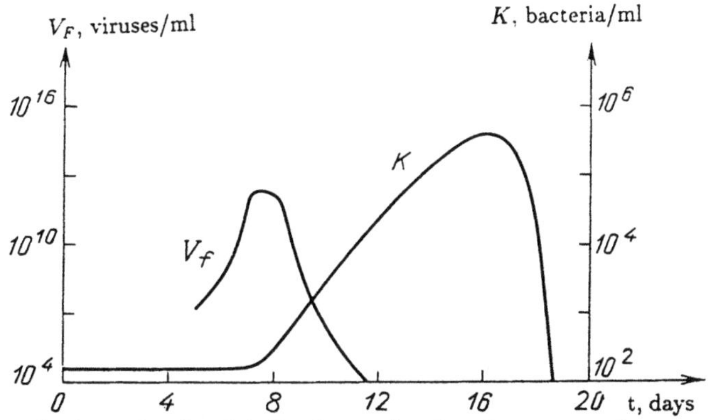

Fig. 33. Simulation of mixed infection: the development of acute viral infection against the background of chronic bacterial infection.

One of the solutions of the mathematical model of viral-bacterial infection which displays qualitatively the dynamics of viral-bacterial process is shown in Fig. 33. In the case of correct estimation of values of coefficient values for viral-bacterial infection model, this model can be used for the simulation of various treatments of mixed infections.

CHAPTER 5

Identification of Parameters of Models

Mathematical modeling in immunology requires above all an extremely precise statement of the original problem. The very statement provides the basis for the most complex processes that are modeled with a computer. All mathematical problems, as a rule, reflect so-called balance relationships written either in algebraic form or in the form of differential, integral, or integro-differential equations. These equations represent an adequate mapping of the processes that we investigate.

It is common for all models we have to deal with that the input data and coefficients are not known. The task for a mathematical researcher is to find those classes of problems and those conditions when the problem is correct.

If the researcher wishes to establish features characteristic of one or another organism's reaction to the invasion of viruses, rather than common statistical laws, he must identify the model's coefficients in order to describe a real process. As a result of statement and solution of such problems it is possible to pass from the mathematical simulation model to real control of a disease process.

the identification problem for the coefficients of mathematical models of immunological processes is dealt with in [30, 263, 265]. Let us consider in brief their main results.

Bertuzzi and coauthors identify in [30] the coefficients of the model of humoral immune response in case of a nonreplicating antigen. In order to obtain a homogeneous set of data an experimental program was planned and carried out. Statistical analysis of measurement errors of quantities observed in experiments have made it possible to justify experimentally the correctness of application of the maximum likelihood method to solve the parameter identification problem. Maximum likelihood estimates for six identified parameters (α) were obtained by the minimization of the corresponding functional $\Phi(\alpha)$ by the method of direct search (the Hook–Jeaves method). Both direct and indirect data reported in literature were used to get an initial guess for the values of parameters. The values of coefficients obtained as a result of minimization of the functional differed from the initial values in the range of 100-500%, and the value of the functional was decreased by 2 orders of magnitude. In order to solve this minimization problem the functional $\Phi(\alpha)$ was calculated about 700 times. The authors noticed that the values found for unknown parameters of the model did not correspond to the point of minimum of $\Phi(\alpha)$; further minimization of $\Phi(\alpha)$ led to negligible decrease in its value while the parameters α took biologically incorrect values. Satisfactory quality of the description of data by the model was attained by graphical correlation of experimental data and solutions of the model for found values of parameters. Later the same authors used the maximum likelihood method to estimate the affinity distribution parameters of antibodies on experimental data.

Perelson and coauthors in [263, 265] solved the parameter identification problem for stochastic models of the kinetics of destruction of target cells by cytotoxic lymphocytes. The least squares method with weighed measurements was used to fit quantitatively the experimental data and model predictions. The values of two unknown model parameters were determined by the minimization of a sum of squares of deviations of measurements, results from the model values of corresponding variables with the Marquard algorithm (a modification of the Newton method). It was noticed that the convergence in the point of minimum was attained by 10–20 iterations. The estimates of standard deviations for each identified parameter were obtained. The values of minimized functional in the points of minimum were used as a basis for the analysis of quality of approximation and, therefore, for the comparison of various models.

Consider the connections between the methods of parameter identification and the character of available information on the modeled system. The system identification and parameter estimations problem are discussed in rather voluminous literature, including [15, 281, 304, 334]. Nevertheless, nowadays there is no universal classification of statements of identification problems and of the methods of their solution [334].

The solution of an identification problem consists of minimization or maximization of certain quantitative criterion of the deviation of model prediction from observed data; we shall call this criterion the residual functional. Available information affects essentially the aim and the method of identification and the choice of the residual functional.

It is known that any observational data are not exact, i.e., contain an uncertainty [15, 162]. This uncertainty may be related to the measurement errors, random effects, nonlinear effects, contribution of unknown processes, etc. If observational data are the results of measurements that display statistical regularity, then standard statistical methods for optimal estimation can be applied to determine the model parameters. In this case optimal estimates of parameters maximize or minimize the chosen criteria: maximum likelihood, Bayesian, the least absolute deviation, etc. [15, 334]. As a rule, the statistical methods of parameter identification allow one to obtain the interval estimates of the model's parameters possessing asymptotically the properties of non-biasness, consistency, and efficiency.

Correct application of these approaches implies a sufficient amount of observation data on the system and the interpretation of their uncertainty as a consequence of fixed probabilistic mechanism, i.e., the possibility of application of standard statistical hypothesis [162]. The parameter identification problem for nonlinear dynamic system is posed as a generalized problem of filtration. The assumption of Gaussian distribution of random variables is essential for the realization of this method. The calculations are rather complex, since it is necessary along with the main system of model equations to solve the Riccatti matrix equation that determines an error of estimation [304].

For a number of modelling problems the observational data either do not permit the use of the standard statistical hypothesis, or represent indirect estimates. In these cases the sum of squares of deviations of observational data from the model variables with corresponding weights over all moments of observations is used as a quantitative criterion for the adjustment of the model to the data [15, 281]. The identification problem is treated in this case as a problem of searching for a minimum of the functional, determined by a simple differential subproblem, for the system of model equations [118].

The experience in solving the parameter identification problems of chemical kinetics can be used for the parameter identification of mathematical models of immune response. Above all, these are the following questions: correlation between the exactness of the solution of direct problem and the algorithm chosen for the minimization of the objective function; the choice of identified parameters on the basis of the sensitivity analysis of the model; analysis of interdependence of the model parameters as related to the identification problem; analysis of various degeneracy situations, instability, that are characteristic of dynamic systems in chemical kinetics [50, 89, 244, 321].

This chapter deals with two approaches to the solution of parameter identification problem for mathematical models of immune response. Each of them takes into account: the goals of concrete problem of modeling; specific features of modeled processes and observational data; properties of mathematical models (dimensionality, complexity, nonlinearity); numerical realization of an identification method.

Section 5.1 presents the approach to identification problems proposed by Romanyukha and Bocharov [40], and Section 5.2 describes the methods of identification developed by Zuev [355].

5.1. The Identification of Model Parameters by Sequential Local Minimization of Deviations

5.1.1. Statement of identification problem. The parameter identification for mathematical models of antiviral and antibacterial immune responses implies available observation data on corresponding variables of the models. The contemporary level of clinical immunology and medicine does not permit simultaneous measurement of the dynamics of the majority of variables in an organism. Nevertheless, there are numerous data characterizing the development of individual processes described in the models of infectious diseases, in experiments on animals and volunteers, in immunological experiments *in vitro*. This requires the construction of quantitative description of modelled processes, i.e., a generalized picture of a disease, which can be considered as a quantitative representation of a disease characterizing the processes on the average. The generalized picture containing the data on the variables' dynamics of the models of antiviral and antibacterial immune responses was constructed for viral hepatitis B, influenza A, and destructive pneumonia; the details are discussed in Chapters 7, 8. These data were used along with the results of direct measurements for the model parameter identification.

The problem of the quantitative description of data characterizing an object when there is no assumptions about their probabilistic nature is reduced to the problem of fitting the model parameters to the data [281], i.e., to the problem of the search for values of the model parameters minimizing the discrepancy between the mathematical model and the data. The reliability of the description so obtained has to be verified by numerical experiments and by the comparison of simulation results with real experimental data.

Quantitative criterion for the proximity of the model solutions to data. The data provided by the generalized picture of infectious disease possess an uncertainty introduced by the chosen method of their estimates, the measurements' errors, and other factors. We have no reason to connect this uncertainty with the action of some random mechanism with assigned statistical properties and, therefore, to use known statistical criteria of closeness of model predictions to experimental data. In this case usually the weighed sum of squares of deviations of observed data $y^{(i)}_{obs\,j}$ from calculated values $y^{(i)}(t_j, \alpha)$ of model's dependent variables is used, i.e., the function corresponding to the least squares method:

$$\Phi(\alpha) = \sum_{j=1}^{M}\sum_{i=1}^{N} \omega_{ji}(y^{(i)}_{obs\,j} - y^{(i)}(t_j,\alpha))^2, \quad y_{obs\,j} \in \mathrm{R}^N, \quad y(t,\alpha) \in \mathrm{R}^N,$$
$$y(t,\alpha) = [y^{(1)}(t,\alpha)y^{(2)}(t,\alpha), \ \ldots, \ y^{(N)}(t,\alpha)]^T. \tag{5.1.1}$$

The following notation is used here: ω_{ji}, certain weights; $\alpha \in \mathrm{R}^L$, vector of model coefficients; M, the number of experimental observations; N, length of the vector of dependent variables. Though, the functional $\Phi(\alpha)$ possesses different sensitivity, firstly, to deviations of different signs of values of model variables from the observed values $y_{obs\,j}$ ($j = 1, \dots, M$); secondly, to the dispersion of absolute values of observation data for each variable; and thirdly, to the choice of measurement units for the variables of model's state $y(t, \alpha)$. This is why we used a functional in the form:

$$\Phi(\alpha) = \sum_{j=1}^{M} \sum_{i=1}^{N} \left[\left(\frac{y^{(i)}(t_j, \alpha) - y^{(i)}_{obs\,i}}{y^{(i)}_{obs\,j}} \right)^2 + \left(\frac{y^{(i)}_{obs\,j} - y^{(i)}(t_j, \alpha)}{y^{(i)}(t_j, \alpha)} \right)^2 \right]. \quad (5.1.2)$$

One of the ways of fitting the solutions of model to the data may consist of splitting the functional $\Phi(\alpha)$ into the components $\Phi^{(i)}(\alpha_{K_i})$, $\alpha_{K_i} \in \mathrm{R}^{K_i} \subset \mathrm{R}^L$, $1 \le i \le N$, and of solving the sequence of minimization subproblems for separate $\Phi^{(i)}(\alpha_{K_i})$ or for their sums $\sum_{l=1}^{I} \Phi^{(i_l)}\left(\alpha_{K_{i_l}}\right)$, $1 \le i_l \le N$ with respect to the subset of coefficients α_{K_i} that are to be refined, selected from the whole vector α of model's coefficients on the basis of conceptions on physical nature of modelled process.

Thus the identification problem for mathematical models of infectious diseases can be formulated as follows: it is necessary to find

$$\min_{\alpha_K \in \mathrm{R}^K_+} \Phi(\alpha_K),$$

where

$$\Phi(\alpha_K) = F(\mathcal{G}(\alpha_K), Y_{\text{obs}}) = \sum_{l=1}^{I} \Phi^{(i_l)}\left(\alpha_{K_{i_l}}\right),$$

$$1 \le i_l, \quad I \le N, \quad \alpha_{K_{i_l}} \in \mathrm{R}^{K_{i_l}} \subset \mathrm{R}^L, \quad (5.1.3)$$

$$\Phi^{(i)}(\alpha_{K_i}) = \sum_{j=1}^{M} \left[\left(\frac{y^{(i)}(t_j, \alpha_{K_i}) - y^{(i)}_{obs\,j}}{y^{(i)}_{obs\,j}} \right)^2 + \left(\frac{y^{(i)}_{obs\,j} - y^{(i)}(t_j, \alpha_{K_i})}{y^{(i)}(t_j, \alpha_{K_i})} \right)^2 \right],$$

α_K is a certain subset of the set of the model's coefficients α with the values in $\mathrm{R}^K \subset \mathrm{R}^L$, and $\mathcal{G}(\alpha_K) \equiv [y(t_1, \alpha_K), y(t_2, \alpha_K), \dots, y(t_M, \alpha_K)]^T$, $Y_{\text{obs}} \equiv [y_{obs\,1},\ y_{obs\,2},\ \dots,\ y_{obs\,M}]^T$.

5.1.2. Methods of numerical solution of a minimization problem

Smoothness of minimized function. Smoothness properties of the objective function $\Phi(\alpha)$ in the form (5.1.3) are the main characteristics that determine the applicability of methods of the first and the second orders, and, in particular, of quasi-Newton methods for the solution of the minimization problem.

It follows from the substantive statement of the identification problem for the coefficients of models of a fixed structure (4.2.11), (4.4.8), that the definition domain for $\Phi(\alpha)$ represents R_+^L, that is, the set of vectors with positive components. For all $\alpha \in \mathrm{R}_+^L$ there exists a solution of initial value problem for (4.2.11), (4.4.8), which is unique and nonnegative for $t \geq 0$. It can be shown that for $t \in [\delta, T]$, $\delta > 0$, $T < +\infty$ the components of a vector function of the solution of initial value problem are strictly positive. Therefore on the set R_+^L the objective function $\Phi(\alpha) = F(\mathcal{G}(\alpha), Y_{\text{obs}})$, constructed with regard to the properties of strict positiveness of Y_{obs} and $\mathcal{G}(\alpha)$ for $t \in [\delta, T]$, is defined and continuous as a superposition of a finite number of continuous functions.

The first and the second derivatives of $\Phi(\alpha)$ with respect to α which represent respectively the gradient $g(\alpha) = \partial\Phi(\alpha)/\partial\alpha$ and the Hessian matrix $G(\alpha) = \partial^2\Phi(\alpha)/\partial\alpha^2$ of minimized function, are also continuous functions in R_+^L. In fact, we have for $g(\alpha)$ (we restrict ourselves with the case of $L = 1$ for the sake of simplicity):

$$g(\alpha) = \sum_{j=1}^{M} \frac{\partial F(\mathcal{G}(\alpha), Y_{\text{obs}})}{\partial y(t_j, \alpha)} \frac{\partial y(t_j, \alpha)}{\partial \alpha}. \tag{5.1.4}$$

The functions $\partial F(\mathcal{G}(\alpha), Y_{\text{obs}})/\partial y(t_j, \alpha)$ in view of concrete form of (5.1.3) are defined and continuous for the same value of α as for $\Phi(\alpha)$, i.e., in R_+^L. Derivatives of the vector function of solution $\partial y(t, \alpha)/\partial\alpha$ (sensitivity functions) are continuous with respect to α in R_+^L for $t > 0$ by virtue of continuously differentiable right-hand side of system (4.2.11), (4.4.8) $f(y, z_1, \ldots, z_m, \alpha)$ with respect to all arguments and the theorem on differentiability of solutions of initial value problem for the system of delay-differential equations (IVP DDE) with respect to parameters [125]. Therefore the function $g(\alpha)$ is continuous in R^L_+ as a product of two continuous functions. By analogy the Hessian matrix $G(\alpha)$ of the objective function $\Phi(\alpha)$ in the form (5.1.3) is continuous in R^L_+. However, jumps in derivatives of a minimized function can arise when estimating lag function parameters and position of initial point (Baker and Paul [13]).

The values of gradient $g(\alpha)$ and of the Hessian matrix $G(\alpha)$ for the function $\Phi(\alpha)$ are calculated, when solving the minimization problem, by a finite-difference technique. The problem of smoothness of numerical solution $y^h(t, \alpha)$ of IVP DDE, that is used for the estimation of $\Phi(\alpha)$, becomes important as related to the changes in parameters α since it determines the differentiability of $\Phi^h(\alpha)$ and quality of approximation of $g(\alpha)$ and $G(\alpha)$.

It is known that adaptive codes based on varying step and order are the most effective numerical methods to solve the initial value problem for a system of ordinary differential equations (IVP ODE), IVP DDE. But their smoothness properties are not satisfactory, since two slightly different values of some parameter of the initial value problem can generate different sequences of integration step lengths and approximation orders, which results in small jumps (of an order of $O(\varepsilon)$) in the values of a function of solution of initial value problem. As a result the numerical solution $y^h(t, \alpha)$ becomes a piecewise smooth function of the parameter with frequent discontinuities of the first order. The jumps and their size must be taken into account while joining the algorithms of calculation of objective function defined by differential subproblem and the algorithms of minimization.

Gear and Vu notice in [116] that in the case when an integrator is not smooth one should solve an initial value problem with accuracy $\varepsilon \sim O((\Delta\alpha)^s\delta)$ in order to estimate the value of derivative of the solution $\partial^s y(t, \alpha)/\partial\alpha^s$ $(s = 1, 2)$ with accuracy $O(\delta)$ for assigned finite-difference interval $\Delta\alpha$. There are several approaches to the handling of the smoothness problem for numerical solution [116, 118, 212]:

– solution of initial value problem on uniform grid;

– solution of initial value problem on fixed grid;

– construction and solution of initial value problem for the system of sensitivity equations $\partial y(t,\alpha)/\partial\alpha$;

– combined integration of perturbed and unperturbed systems of DDE;

– solution of initial value problem with higher accuracy, etc.

We have chosen the last approach, increasing the accuracy of solution of IVP DDE, taking into account that it needs comparatively small computer time to solve numerically the initial value problem for the model under study. We selected the value $\varepsilon = 10^{-8}$ which proved to be sufficient in the framework of our strategy for using the minimization algorithms.

Algorithms for the minimization of objective function. In order to solve the minimization problem for the function $\Phi(\alpha) : \mathrm{R}_+^L \to \mathrm{R}_+$ we used the algorithms of the MINUIT package of programs: the simplex method and the Davidson–Fletcher–Powell quasi-Newton method [155].

The MINUIT package was designed to solve the minimization problem for nonlinear functions $\Phi(\alpha)$ either with no constraints, or with simple constraints in the form of two-side inequalities:

$$\min\{\Phi(\alpha) : \alpha \in \mathrm{R}^L, \quad a_i \le \alpha^{(i)} \le b_i, -\infty < a_i, \quad b_i < +\infty, \quad i = 1,\dots,L\}. \tag{5.1.5}$$

The system helps analyse the sensitivity of the objective function $\Phi(\alpha)$ to variations of the parameters α in the neighborhood of minimum point α^*. The solution of the minimization problem with linear constraints (5.1.5) is performed by the transition to constrained minimization problem by the substitution of variables in the following form:

$$y^{(i)} = \arcsin\left(2\frac{\alpha^{(i)} - a_i}{b_i - a_i} - 1\right).$$

As was noted above, the identification problem for coefficients of the mathematical models (4.2.11), (4.4.8) can be reduced to a sequence of minimization problems for the function $\Phi(\alpha)$ in the form (5.1.3) with a positive constraint $\alpha \in \mathrm{R}_+^L$, i.e.,

$$\min\{\Phi(\alpha): \ 0 < \delta \leq \alpha^{(i)}, \quad i = 1, 2, \ldots, L\}. \tag{5.1.6}$$

This restriction is a consequence of the requirement for positiveness of parameters. *A priori* information on the domain of admissible values of parameters α allows one to introduce the constraints in the form of inequalities like lower and upper bounds for each parameter and to pass from statement (5.1.6) to statement (5.1.5). One should keep in mind that the constraints in the form of inequalities usually enhance the convergence of the minimization algorithms [118, 294], since there is a possibility of scaling the variables in order to pass from primary measurement units for parameters to new ones which are close in magnitude and consequently are more suitable for computations.

The simplex method is realized according to the procedure proposed by Nelder and Mead [234]. It is the heuristic method of direct search which uses just the information on the values of minimized function. The search for a minimum is made by comparing the values of a function in the test points.

An estimate of a distance to minimum at the kth iteration is provided by the quantity $\delta\Phi_k = \Phi\left(\alpha_H^{(k)}\right) - \Phi\left(\alpha_L^{(k)}\right)$ where $\alpha_H^{(k)}$ is a vertex of the polyhedron characterized by the largest value of the function, and $\alpha_L^{(k)}$ is a vertex with the smallest value. The convergence of $\alpha_L^{(k)}$ to α^* (point of minimum) is achieved when $\delta\Phi_k < \varepsilon$, where ε is a value assigned by a user.

The simplex method is expedient while obtaining "good" first guess for more efficient methods and when insufficient information on the properties of the minimized function or truncation error estimate do not permit us to use the minimization methods of higher order.

The quasi-Newton routine is the realization of Davidson–Fletcher–Powell method suggested by Fletcher [102]. It provides the possibility of using two different formulas for updating the matrix which approximates the inverse Hessian matrix. This method was realized according to the usual scheme for all quasi-Newton methods of unconstrained minimization of smooth functions:

1. Calculation of search direction p_k by the formula $p_k = -H_k g_k$, where $g_k = \partial\Phi(\alpha_k)/\partial\alpha$ is a gradient of the function $\Phi(\alpha)$ at the point α_k, and H_k is an approximation of inverse Hessian matrix constructed on the basis of gradient variations during previous iterations of the descent.

2. Calculation of step length d_k based on approximate one-dimensional minimization with the use of quadratic interpolation.

3. Updating the estimate of solution $\alpha_{k+1} = \alpha_k + d_k p_k$.

4. Calculation of the gradient g_k by finite-differential approximation or by direct evaluation of the derivatives $\partial\Phi(\alpha_k)/\partial\alpha^{(i)}$ $(i = 1, 2, \dots, L)$. Notice, that the choice of the size of finite-differential interval for the approximation of gradient in the MINUIT complex is not correlated with the precision of evaluation of the minimized function, which can account for inefficiency of application of this method. While solving our problem we had to increase the relative accuracy of IVP DDE solution and, therefore, the accuracy of calculation of $\Phi(\alpha)$ up to 10^{-8}.

5. Check if the termination criterion is met. As a criterion of convergence the estimate of distance to the minimum was used:

$$\delta\Phi_k = g_{k+1}^T H_k g_{k+1} < \varepsilon_1,$$

and the quantity δH_k characterizing total change in diagonal elements of matrix H_k

$$\delta H_k = \sum_{i=1}^{L} \left[\frac{(H_k)_{ii} - (H_{k+1})_{ii}}{(H_k)_{ii} + (H_{k+1})_{ii}} \right]^2 < \varepsilon_2,$$

where ε_1 and ε_2 are assigned by user.

6. Updating of the matrix H_k with one of two different formulas which preserve positive-definiteness and symmetry.

The quasi-Newton algorithms, exhibiting the properties of numerical stability and linear (in some cases super-linear) rate of convergence, are considered one of the most effective tools for solving the unconstrained minimization problems for smooth functions provided that the dimensionality of parameter spaces is not too high.

The search for a minimum point α_k for the residual functional $\Phi(\alpha)$ is performed in two stages: the search for the first guess $\bar{\alpha}^*$ to α^* by the

Simplex method, and thereafter the refinement of $\bar{\alpha}^*$ by a quasi-Newton method. The parameters ε, ε_1, and ε_2 that are used in the convergence criteria were chosen as $\varepsilon = 1$, $\varepsilon_1 = 10^{-4}$, $\varepsilon_2 = 10^{-2}$. While forming the lower and the upper bounds for parameters which varied during the search for a minimum, we tried to keep the difference between these bounds within the limits of three orders of magnitude.

5.1.3. Sequential minimization of deviation functional. The complexity and nonlinearity of mathematical models does not allow us to obtain a good fitting of their solutions to the observational data characterizing the modeled process simply by the search for a minimum of the corresponding deviation functional $\Phi(\alpha)$ for the whole time interval $[t_1, t_M]$, where t_1 is the moment of time of the first observation, and t_M corresponds to the last observation. It is difficult to handle the selection problems for identified parameters (i.e., those parameters which will vary during minimization) and to correlate the physical meaning of an identified parameter and properties that minimize the value of the functional.

Using the idea underlying the procedure of sequential estimation of parameters, we have reduced the identification problem for the model coefficients with respect to observation data on the interval $T_{1,M} = [t_1, t_M]$ to the sequence of fitting the problems for the IVP DDE solution to corresponding data on the subintervals $T_{1,2}, T_{1,3}, \ldots, T_{1,M}$ using the deviation functional (5.1.3). In order to preserve the proximity to a local minimum we introduced into the functional sequentially the values of observational data for $t = t_j$ for the model variables. It is possible, using qualitative information on the character of development, interaction, and correlation of virological, clinical and immunological processes, to separate some temporal bounds within the interval $T_{1,M}$ where these processes can be considered as independent. Thus we succeeded in dividing the whole interval $T_{1,M}$ into a sequence of smaller subintervals where the observed kinetics of experimental data is determined by the realization of small number of processes from the whole set of processes described by the model. This splitting makes it possible to solve successfully corresponding approximation problems on the intervals $T_{1,j}$ $(j = 1, 2, \dots, M)$ by physically grounded choice of optimized parameters.

The outlined approach to sequential minimization of deviation functional $\Phi(\alpha)$ using observational data can be formalized as follows.

Construct the sequence of auxiliary minimization problems for $m = 1, 2, \dots, M$ for the set of observation moments $t_1 \leq \dots \leq t_{m-1} < t_{m-1} < t_m < t_{m+1} \leq \dots \leq t_M$,

$$\min_{\alpha^{l_m} \in D} \Phi_m(\alpha^{l_m}), \quad D = \left\{\alpha^{l_m} \in \mathrm{R}^{l_m} \subset \mathrm{R}^L,\ A \leq \alpha^{l_m} \leq B\right\}, \tag{5.1.7}$$

where $\Phi_m(\alpha^{l_m}) \equiv F_m[\mathcal{G}_{[t_1,t_M]}(\alpha^{l_m}), Y_{\text{obs}}]$, $F_m(\cdot,\cdot)$ is a residual functional constructed for observation data on the interval $[t_1, t_M]$, and $\mathcal{G}_{[t_1,t_M]}(\alpha^{l_m}) = [y^{(1)}(t, \alpha^{l_m}), y^{(2)}(t, \alpha^{l_m}), \dots, y^{(N)}(t, \alpha^{l_m})]^T$ is the solution on $t \in [t_1, t_m]$ of the differential subproblem determined by the system of the model's equations

$$\frac{dy}{dt} = f(\alpha, y(t), y(t - \tau)), \quad t_0 \leq t \leq t_m, \tag{5.1.8}$$

$\alpha \in \mathrm{R}^L$ is the vector of parameters, $y(t) = \varphi(t)$, $t_0 - \tau \leq t \leq t_0$ specifies the initial function. Solution for each of these problems is represented by l_m-dimensional vector of parameters $\alpha_*^{l_M} = [\alpha_*^{(j_1)}, \dots, \alpha_*^{j(l_m)}]^T$, and as a whole we obtain a sequence of refined components of the vector of model's parameters $\alpha \subset \mathrm{R}^L: \ \{\alpha_*^{l_m}\}_{m=1}^M$. Thus we have constructed the process of sequential refinement of the vector of the model's parameters $\alpha \in \mathrm{R}^L$:

$$\left\{\alpha_m \in \mathrm{R}^L, \quad \alpha_m = \begin{bmatrix} \alpha_*^{l_m} \\ \alpha_{m-1}^{L-l_m} \end{bmatrix}\right\}_{m=1}^M.$$

The method of approximate solution of the minimization problem for deviation functional $\Phi(\alpha)$ reduces the original problem to a sequence of auxiliary minimization subproblems (5.1.7), (5.1.8) that are less difficult than the original one. In principle this method is similar to the method of local variations for the solution of variational problems.

It should be noted that the fitting of IVP DDE solutions of the model to the data on an interval $T_{1,(j+1)}$ may be followed by the deterioration of approximation quality with respect to some variables attained on preceding intervals $T_{1,j}$, $T_{1,(j-1)}$, etc. It is possible, using structural properties of the model's equations, to fix the obtained quality of fitting on an interval $T_{1,j}$ for some variables, introducing relationships (dependencies) between separate coefficients, i.e., to make *this approximation* independent of variations of corresponding parameters while solving the approximation problems on the next interval $T_{1,(j+1)}$. Further, in Chapters 7, 8, in accordance with the above approach to the solution of the identification problem, sequential stages of its numerical realization will be presented.

5.1.4. Refinement of initial values of parameters by the adjustment of model to spline functions interpolating the observation data. Nonlinear (cubic) terms in the right-hand side of the model equations and the usage of nonmonotone functions when assigning the initial conditions make the parameter identification problem for antiviral and antibacterial immune response computationally expensive and time-consuming. It was manifested in the fitting the model to a generalized picture of influenza A infection data (Chapter 8). It is beneficial, therefore, while solving numerically the problem of search for the point of minimum $\alpha^* = \arg \min\limits_{A \leq \alpha \leq B} \Phi(\alpha)$ where $\Phi(\alpha) \equiv F(\mathcal{G}(\alpha), Y_{\text{obs}})$, to have good initial approximation of the values of model coefficients α_0 to α^*. Here is one of the ways of reaching this goal. Let us have an empirically constructed estimate α_0 for the values of model parameters in the domain of permissible values defined by the inequalities $A \leq \alpha_0 \leq B$; then refine α_0 using the observation data with simplified (short-cut) numerical methods; use this refined estimate $\tilde{\alpha}_0$ in the process of minimization of $\Phi(\alpha)$ with standard algorithms for minimization of nonlinear functions: method of direct search and quasi-Newton method which require high precision estimate of the functional $\Phi(\alpha)$ and, therefore, of model solution $y(t)$.

When solving the inverse problems of chemical kinetics models for differential equations a series of simplified methods for the estimation of parameters was proposed. These methods use spline functions interpolating experimental data rather than solutions of differential problem while calculating $\Phi(\alpha)$ [131, 142, 336]. As a result, the solution procedure of the initial value problem for differential system is not necessary any more, and a simple minimization problem for the functional determined by a system of algebraic equations has to be solved instead of nonlinear programming problem for the functional determined by differential subproblem. These results were used for the construction of a method of the refinement of initial estimates for coefficients of mathematical models of immune response. There are no limitations connected with observability (the presence of observational data) for all model variables and with explicit integrability of equations for unobserved variables, which is a new feature of our approach.

Consider the statement of the identification problem for coefficients of a system of differential equations (for simplicity we shall restrict ourselves to the case of one delay):

$$\min_{A^L \leq \alpha^L \leq B^L} F(\mathcal{G}^N(\alpha^L), Y^N), \quad A^L,\ B^L,\ \alpha^L\ \in \mathrm{R}^L, \tag{5.1.9}$$

$$\begin{gathered} \frac{dy^N(t)}{dt} = f^N(\alpha, y^N(t), y^N(t-\tau)), \\ y^N(t) \in \mathrm{R}^N \quad \text{for each} \quad t \in [t_0, t_0 + T], \\ y^N(t) = \varphi^N(t), \quad t \in [t_0 - \tau, t_0], \end{gathered} \tag{5.1.10}$$

where the indices denote the dimensions of vector variables or vector functions.

Let α_0^L be some initial estimate of the values of model coefficients that is to be refined and let $Y^N = \left\{ \{y^{(i)}_{obs\,j}\}_{j=\overline{J_{i_1}, J_{i_2}}} \right\}_{i=\overline{1,N}}$ be a set of observational data for each of N variables $y^{(i)}(t)$, $(i = \overline{1, N})$ of the model at the moments of time $\{t_j,\ j = \overline{J_{i_1}, J_{i_2}}\}$, $t_0 \leq t_1 \leq t_{J_{i_1}}, < \ldots,$ $< t_{J_{i_2}} \leq t_M \leq t_0 + T$; let $m_i = (J_{i_2} - J_{i_1})$ be the number of observations for ith variable; and let M be the total number of various times of observations.

Let us interpolate the observational data for each of the model's variable $y^{(i)}(t) - \{y^{(i)}_{obs\,j}, t_j\}_{j=\overline{J_{i_1}, J_{i_2}}}$ $(i = 1, 2, \ldots, N)$ with the spline function $s_Y^{(i)}(t)$, solving one of the problems, depending on the character of data: smoothing (m_i are large), the least squares approximation (m_i are large), monotone interpolation of data with piecewise quasi-Hermitean polynomials (m_i are small). As a result we will obtain an

N-dimensional vector of spline functions $S_Y^N(t)$. Considering $S_Y^N(t)$ as a certain prototype of model solution $y^N(t)$ (referred to as a quasi-solution hereafter) corresponding to an unknown set of parameters of differential problem, we will use it for the construction of simplified methods of refinement of the initial estimate α_0^L. When a sufficient amount of observation data is available for the construction of $S_Y^N(t)$ with respect to each of N variables, it is possible, using a direct integral method which is more stable with regard to errors in observational data, to pass from the statement (5.1.9), (5.1.10) to an overdetermined system of algebraic equations determining the vector of parameters, which can be effectively solved by the least squares method. To do so one must pass from differential problem (5.1.10) to equivalent system of integral equations [131]:

$$y^N(t) = y^N(t_0) + \int_{t_0}^{t} f^N(\alpha^L, y^N(\theta), y^N(\theta-\tau))d\theta. \tag{5.1.11}$$

Substituting $S_Y^N(t)$ into (5.1.11) instead of $y^N(\theta)$ and taking (5.1.11) in a set of moments of observations $\{t_k : k = \overline{k_1, k_2},\ t_{k_1} = \max_{i_1}\left(t_{J_{i_1}}\right),\ t_{k_2} = \min_{i_2}\left(t_{J_{i_2}}\right)\}$, we obtain the system of algebraic equations with respect to α^L.

$$\Delta Y^J = \Psi_{S_Y}^J(\alpha^L), \tag{5.1.12}$$

where the vectors ΔY^J and $\Psi_{S_Y}^J(\alpha^L)$ have the following components:

$$(\Delta Y^J)_j = y_{obs\,k}^{(i)} - y_{obs\,k_1}^{(i)},$$

$$\left(\Psi_{S_Y}^J(\alpha^L)\right)_j = \int_{t_{k_1}}^{t_k} \left(f^N(\alpha^L, S_Y^N(\theta), S_Y^N(\theta-\tau))\right)_i d\theta,$$

$$i = 1, 2, \dots, N, \quad k = k_1, k_1+1, \dots, k_2,$$

$$j = (k-1) \times N + i, \quad J = \max\{j\}.$$

If the coefficients α^L appear as linear in the right-hand side of the system of differential equations, equation (5.1.12) is transformed into a linear overdetermined system of algebraic equations with respect to α^L (or with respect to some subset of coefficients α^l, $l < L$, depending on the number of observations and on a value of L) $\Delta Y^m = \Psi_{S_Y}^{m\times L}\alpha^L$, whose solution can be easily obtained using known methods of the solution for linear least squares problems realized in LINPACK [86]:

$$\tilde{\alpha}^L = (\Psi^{m\times L}_{S_Y})^{+}\Delta Y^m, \tag{5.1.13}$$

where $\Psi^{m\times L}_{S_Y}$ is $m \times L$ matrix.

Thus α_0^L can be refined provided that there are observational data which permit us to construct $S_Y^N(t) = [s_Y^{(1)}(t), s_Y^{(2)}(t), \ldots, s_Y^{(N)}(t)]^T$ which is a spline approximation for all the model variables .

When we solve problems of mathematical modelling of infectious diseases, available quantitative characteristics are sufficient for the construction of spline functions just for a part $(n,\ n < N)$ of model's variables $y^n(t)\ :\ \ S_Y^n(t)$. It is possible in this case using a reduced system of differential equations of dimension $(N-n)$ describing the behavior of other $N-n$ model variables $y^{N-n}(t)$,

$$\frac{dy^{N-n}(t)}{dt} = f^{N-n}\left(\alpha^{l_{N-n}}, S_Y^n(t), y^{N-n}(t), S_Y^n(t-\tau), y^{N-n}(t-\tau)\right),$$

$$t \in [t_0^*, t_0^* + T^*], \quad \alpha^{l_{N-n}} \in \mathrm{R}^{l_{N-n}} \subset \mathrm{R}^L, \tag{5.1.14}$$

$$y^{N-n}(t) = \varphi^{N-n}(t), \quad t \in [t_0^* - \tau, t_0^*],$$

which is obtained from (5.1.10) by substitution of $S_Y^n(t)$ instead of $y^n(t)$, to find the numerical solution $y_{S_Y}^{N-n}(t)$. Here the interval $[t_0^*, t_0^*+T^*]$ was chosen in accordance with the set of observation times $\{t_k,\ k = \overline{k_1, k_2}\}$. Having got $y_{S_Y}^{N-n}(t)$ we construct the missing spline functions $S_{\mathcal{G}}^{N-n}(t)$ and obtain, as a result, the quasi-solution $S_{Y,\mathcal{G}}^N(t) = [S_Y^n(t), S_{\mathcal{G}}^{N-n}(t)]^T$. Using $S_{Y,\mathcal{G}}^N(t)$ one can improve the initial guess α_0^L. The applicability of the direct integral method (5.1.11)–(5.1.13) thus extends essentially, since in most of real identification problems for models of complex systems the observational data necessary for the construction of spline functions are available only for a part of variables.

Consider one more approach to the refinement of the initial guess by solving simplified numerical problem as compared to (5.1.9), (5.1.10). Let $S_Y^n(t),\ n \leq N$ be a vector function of quasi-solution obtained from observational data with spline interpolation. Let us denote through $P_n^N\ :\ y^N(\cdot) \to y^n(\cdot)$ a projection operator which maps the N-dimensional vector function $[y^{(1)}(t), y^{(2)}(t), \ldots, y^{(N)}(t)]^T$ onto the vector function of smaller dimension $n \leq N$, i.e., $y^n(t) = [y^{(j_1)}(t), y^{(j_2)}(t), \ldots, y^{(j_n)}(t)]^T$, $1 \leq j_1,\ j_n \leq N$.

The vector function of the quasi-solution $S_Y^n(t)$ or its part of smaller dimension $S_Y^d(t) = P_d^n S_Y^n(t)$ can be used for reducing the dimension of the differential system when solving the minimization problem (5.1.9) for the vector of parameters α^l of dimension $l < L$:

$$\min_{A^l \leq \alpha^l \leq B^l} F\left(\mathcal{G}^{N-d}(\alpha^l), Y^{N-d}\right), \tag{5.1.15}$$

$$\frac{dy^{N-d}(t)}{dt} = f^{N-d}\left(\alpha^l, S_Y^d(t), y^{N-d}(t), S_Y^d(t-\tau), y^{N-d}(t-\tau)\right),$$
$$t \in [t_0, t_0 + T],$$
$$y^{N-d}(t) = \varphi^{N-d}(t), \quad t \in [t_0 - \tau, t_0], \tag{5.1.16}$$

where $y^{N-d}(t) = P_d^N y^N(t)$, $f^{N-d}(\cdot,\cdot,\ldots,\cdot) = P_d^N f^N(\cdot,\cdot,\ldots,\cdot)$, $\mathcal{G}^{N-d}(\alpha^l) = P_d^N \mathcal{G}^N(\alpha^l)$, $Y^{N-d} = P_d^N Y^N$.

All this reduces computer time for calculations of the minimized functional $F(\mathcal{G}, Y)$ and simplifies the problem of refinement of the initial guess α_0^L in correspondence with the observational data Y. Notice that a concrete choice of the solution components $y^N(t)$, which are approximated observation data using spline functions and are eliminated from differential system (5.1.10), is determined by the aims or peculiarities of the identification problem. The usage of quasi-solutions, which implies fixing the components of the vector function $y^N(t)$, speeds up the process of rough fitting the model to data, that is, the process of refined estimation of parameters $\tilde{\alpha}^L$ as compared to α_0^L.

Though being a displaced estimate, $\tilde{\alpha}^L$ allows one to use more effectively precise minimization methods for the functional $\Phi(\alpha)$ for the whole model when we have to solve the problem of parameter identification.

In order to solve numerically the problems of refinement of the initial parameter guess we have constructed a complex of programs using the following algorithms from the programs library IMSL (see IMSL LIbrary Reference Manual 1–3, Houston: IMSL, 1980):

1. IQHSCU, one-dimensional quasi-cubic interpolation on the basis of Hermite polynomials (construction of $S_Y(t)$);

2. DCADRE, calculation of one-dimensional integral (5.1.12);
3. LLBQF, solution of linear least squares problem (5.1.13);
4. ZXMIN, quasi-Newton method minimizing the functions of N variables (5.1.9), (5.1.15).

The identification methods for model parameters discussed in this section were used to fit the mathematical models of antiviral immune response to the data characterizing the course of viral hepatitis B and uncomplicated influenza A, and to fit the model of antibacterial immune response to the data characterizing the course of destructive pneumonia.

5.2. Statistical Estimation of Parameters of the Models of Diseases Based on Experimental Data

5.2.1. Stochastic model for description of observational data. The solution of estimation problems for model parameters based on clinical and laboratory data is of great practical importance. We mean the possibility of estimating the parameters of mathematical models, describing the immune reactions and the course of the disease. Parameters of these models, calculated on the basis of experimental data, are the estimates of parameters of real organism which we try to treat, choosing the therapy which provides a desirable dynamics of the disease. Therefore consider the estimation problem for coefficients of models of diseases based on experimental data; detailed description is represented by Zuev in [355, 356].

Mathematical models presented in this book are the system of nonlinear ordinary delay-differential equations:

$$\frac{dx_t}{dt} = f(x_t, x_{t-\tau}, \alpha), \quad t \in [0, T],$$
$$x_s = g_s, \quad s \in [-\tau, 0], \quad x_t \equiv x(t), \tag{5.2.1}$$

where x_t stands for $x(t)$ and $x_t \in \mathrm{R}^n$ is a solution for the system, $\alpha \in \mathrm{R}^l$ is a vector of coefficients. Such models posses one common property:

all the models can be written so that the vector-function $f(x, y, \alpha)$ will be linear with respect to α. This feature simplifies essentially the problem of estimation. A delay, on the contrary, makes the problem more complicated. Recall, in this connection, physical premises for using the delays.

Take a simple mathematical model of disease. The delay in the right-hand side of this system is caused by the fact that antibody-producing cells C appear, when a certain dose of antigen has penetrated into an organism, with a certain delay. Corresponding immunocompetent cells are to pass through several stages of division before they begin to produce the antibodies F, which takes a time τ. Therefore an equation for C has the form (see Chapter 3)

$$\frac{dC_t}{dt} = \alpha V_{t-\tau} F_{t-\tau} - \mu_c(C_t - C^*), \quad C_0 = C^*.$$

We can model the process of passing through various stages. Let C^i be the number of the cells population in ith stage (i=1,2,...,k) (k corresponds to 6–8 divisions). Then the process of generations of plasma cells can be described by the following system of equations:

$$\begin{aligned} \frac{dC_t^1}{dt} &= hF_tV_t - hC_t^1, && C_0^1 = 0, \\ \frac{dC_t^2}{dt} &= hC_t^1 - hC_t^2, && C_0^2 = 0, \\ &\cdots\cdots\cdots && \cdots \\ \frac{dC_t^k}{dt} &= hC_t^{k-1} - hC_t^k, && C_0^k = 0, \\ \frac{dC_t}{dt} &= \alpha C_t^k - \mu(C_t - C^*), && C_0 = C^*. \end{aligned} \tag{5.2.2}$$

We have constructed a system with no delay describing a simple scheme of disease; its solution, however, should not be considered as an approximation of the solution of a simple mathematical model. These systems represent two methods of modeling the same physical phenomenon and for the same set of coefficients their solutions do not differ qualitatively. We shall assume, therefore, that the model represents the system of ordinary differential equations

$$\frac{dx_t}{dt} = f(x_t, \alpha), \quad x_0 = c = \text{const}, \quad t \in [0, T], \tag{5.2.3}$$

where $x_t \in \mathrm{R}^n$, $\alpha \in \mathrm{R}^l$, and the vector function $f(x, \alpha)$ is linear with respect to α. Later we shall need the existence of the first and the second partial derivatives of the right-hand side with respect to x. Assume therefore, that $f(x, \alpha)$ satisfies this condition. Denote a solution of (5.2.3) by $x_t(\alpha)$, assuming that $t \in [0, T]$. The value of T corresponds to the time of transition process in the modelled system. In other words, the development of immune response and the restoration of healthy state of an organism occurs in the interval $[0, T]$. For instance, $T \sim 40$ days for hepatitis, and $T \sim 30$ days for pneumonia.

Let us have a trajectory of state variables of the model as a result of experiments or clinical observations. It means that we have a set $\theta = \{t_0, t_1, \ldots, t_N\}$ and values of variables $X = [x_{t_0}, x_{t_1}, \ldots, x_{t_N}]$ measured in these instants of time. If an experiment was carried out with a group of m animals, then we obtain a group of trajectories $X_m = \{X^i,\ i = 1, 2, \ldots, m\}$, where $X^i = [x_t^i, t \in \theta]$ corresponds to observations over the ith animal. It may happen that as a result of an experiment we rather the sets of values of variables for $t \in \theta$, rather than the trajectories, i.e., $X'_m = \{x^i,\ i = 1, 2, \ldots, m_t,\ t \in \theta\}$ and all the values x_t^i are statistically independent. This is true when just one measurement is possible for one animal. We assume in any case that animals belong to the same line, so that trajectories obtained can be considered as a result of repeated experiments with the same organism. It is obvious that real trajectories have random behavior and cannot be realized therefore in the framework of model (5.2.3) for certain $\alpha = \tilde{\alpha}$.

Notice at once that the random character of trajectories is conditioned not only by the measurement error in state variables but also by the influence of various factors which are not taken into account in the model (5.2.3). Therefore, while describing real trajectories, it is necessary along with the measurement errors to take into account random oscillations that happen in a system under investigation. The physical meaning of these oscillations is as follows. The mathematical model (5.2.3) describes just the factors which determine the dynamics of the process under study and help to explain the mechanism of the investigated phenomenon. While constructing a model one neglects a lot of unessential connections, and thus the process becomes isolated from the system where it develops. If our conceptions of the mechanisms of a studied phenomenon correspond to reality, then neglected factors are the very factors which do not affect decisively the processes under study, and the model must describe qualitatively its basic features. In the framework of the model (5.2.3) these factors are not controllable, so, while describing real trajectories we consider their influence as random disturbances responsible for short-term deviations of trajectories from some solution $x_i(\bar{\alpha})$.

The problem of the determination of the vector α by X_m or by X'_m consists of finding one of the sets of solutions $\{x_t(\alpha), \alpha \in D \subset \mathrm{R}^l\}$ of problem (5.2.3) that corresponds to a set of sample trajectories $X_m(X'_m)$. In other words, it is necessary to construct the functional $I(x_t, (\alpha), X_m)$ such that

$$\max_{\alpha \in D} I(x_t(\alpha), X_m) = I(x_t(\alpha_m), X_m),$$

where α_m is an estimate of the vector of coefficients.

Let us assume for a while that there is no measurement error. Assume also that the observed trajectory x_t belongs to the set of solutions of system (5.2.3), i.e., for some $\alpha = \bar{\alpha}$

$$x_t(\bar{\alpha}) = x_t, \quad t \in \theta.$$

Then $\bar{\alpha}$ can be found by the methods of perturbation theory along the following scheme.

It is assumed that some (unperturbed) state of the system under study is known, and the vector α_0 corresponds to this state. The observed state, called the perturbed one, is assumed to be described by equation (5.2.3) for $\bar{\alpha} = \alpha_0 + \delta$, where the deviations δ are small compared to α_0. Then we can linearize the problem, determine δ and the parameters corresponding to the perturbed state.

The perturbed state in this case cannot be realized in the framework of system (5.2.3) by the above method, since observed trajectories, due to their random character, do not belong to the set of the model's solutions. Let (5.2.3) describe the unperturbed solution with the set of parameters $\bar{\alpha}$ and let us assume that $\alpha_t = \bar{\alpha} + \delta_t$ corresponds to the perturbed state, where δ_t is a function of time $t \in [0,T]$. Then each sample trajectory X^i has corresponding δ_t^i, $t \in [0,T]$. Since X^i $(i = 1,2,\ldots,m)$ are random in character, it is natural to consider δ_t^i $(i = 1,2,\ldots,m)$ as a realization of some random process $\{\delta_t,\ t \in [0,T]\}$, $E|\delta_t|^2 < \infty$. Since we assume that the differences in trajectories are caused by pure random factors, then $E\delta_t = 0$ for all $t \in [0,T]$ and the existence of random oscillations with respect to trajectory $x_t(\bar{\alpha})$ means that

$$\frac{1}{T}\int_0^T \delta_t dt \approx 0. \tag{5.2.4}$$

Exact equality takes place in the limit when $T \to \infty$ if

$$\operatorname{cov}(\delta_t, \delta_{t+\tau}) \underset{\tau \to \infty}{\longrightarrow} 0.$$

Therefore it is correct for finite T if $\operatorname{cov}(\delta_t, \delta_{t+\tau})$ tends to zero rapidly enough, i.e., for $\tau \ll T$.

In order to write down (5.2.4) correctly, introduce the process ξ_t, such that

$$E|\xi_t|^2 < \infty, \quad E\xi_t = 0, \quad t \in [0,T]_t,$$

$$\operatorname{cov}(\xi_t, \xi_{t+\tau}) \underset{\tau \to \infty}{\longrightarrow} 0,$$

$$\lim_{T\to\infty} \frac{1}{T}\int_0^T\int_0^T E\xi_t^i\xi_s^j dtds = g^{ij}, \quad G = (g^{ij}),$$

$$i = 1,2,\ldots,l, \quad j = 1,2,\ldots,l.$$

Let $\delta_t = \xi_{t/\varepsilon}$, where $\varepsilon > 0$ is a small parameter, and for all $T > 0$, $q > 0$, let

$$\lim_{\varepsilon\to 0} \mathrm{P}\left\{\left|\frac{1}{T}\int_0^{t+T} \xi_{s/\varepsilon} ds\right| > q\right\} = 0$$

uniformly with respect to t. This representation can be explained as follows. Let, for instance, ξ_t and the state variables have a common characteristic time of variation. Then the dynamics of the variable $\delta_t = \xi_{t/\varepsilon}$ is rapid as compared to the dynamics of $x_t(\bar{\alpha})$. Assume that (5.2.4) is satisfied for sufficiently small values of ε. We consider thus the set of trajectories X_m as the narrowing on θ of the set of realizations of random process x_t that satisfies the system of equations

$$\frac{dx_t}{dt} = f(x_t, \alpha_0 + \xi_{t/\varepsilon}), \quad x_0 = c, \quad t \in [0, T]. \tag{5.2.5}$$

As was shown in [264], for any $T > 0$, $\delta > 0$ we can write:

$$\lim_{\varepsilon \to 0} \mathrm{P} \left\{ \sup_{0 \le t \le T} |x_t - x_t(\alpha_0)| > \delta \right\} = 0.$$

Therefore for small ε the model (5.2.5) describes the process of random oscillations of trajectories along general regularity $x_t(\bar{\alpha})$, i.e., a picture observed in reality.

The following stochastic model was suggested in [355, 356] assuming that ε is small:

$$\begin{gathered} \frac{d}{dt} u_t = f_x(\bar{x}_t, \bar{\alpha}) u_t + f_\alpha(\bar{x}_t, \bar{\alpha}) \frac{d}{dt} w_t \\ t \in [0, T], \quad u_0 = 0 \end{gathered} \tag{5.2.6}$$

where $\bar{x}_t = x_t(\bar{\alpha})$; f_x, f_α are the matrices of derivatives of right-hand side of (5.2.3) with respect to x and α; w_t is a Gaussian process with independent increments, zero mathematical expectation, and covariance matrix Γ_t, where $\Gamma = \varepsilon G$. The model is based on the results obtained in [338] for the description of deviations $u_t = x_t - x_t(\alpha)$ of real trajectory x_t from the solution $x_t(\alpha)$ of model (5.2.3).

5.2.2. Calculation of estimates of parameters using observational data. The method proposed in [355, 356] for the calculation of estimates of the vector $\bar{\alpha}$ and matrix Γ, based on properties of solutions (5.2.6), exploits the idea suggested in [195] to use adjoint equations for the solution of inverse problems. The detailed discussion of this method is presented in [356]; skipping it here, consider an iteration process of the calculation of parameter estimates we are interested in.

Let us call adjoint the following systems of equations:

$$\frac{d}{dt}y_{ist} = -f_x^T(\bar{x}_t, \bar{\alpha})y_{ist} + Q^i(t,s),$$
$$i = 1, 2, \ldots, n, \quad s \in \theta, \quad t \in [0,T], \quad y_{ist} = 0. \tag{5.2.7}$$

Take an inner product of (5.2.6) and y_{ist}, and of (5.2.7) and u_t and sum the results to obtain:

$$\langle y_{ist}, \frac{d}{dt}u_t\rangle + \langle u_t, \frac{d}{dt}y_t\rangle = \langle f_\alpha(\bar{x}_t, \bar{\alpha})\frac{d}{dt}w_t, y_{ist}\rangle + \langle Q^i(t,s), u_t\rangle, \tag{5.2.8}$$

where $\langle x, y\rangle$ denotes an inner product in R^n of two vectors x and y. Equation (5.2.8) was obtained due to the equality

$$\langle f_x(\bar{x}, \bar{\alpha})u_t, y_{ist}\rangle - \langle f_x^T(\bar{x}_t, \bar{\alpha})y_t, u_t\rangle = 0.$$

Integrate equality (5.2.8) taking into account that $y_{ist} = 0$ and $u_0 = 0$ to obtain

$$\int_0^T \langle f_\alpha(\bar{x}_t, \bar{\alpha})\frac{d}{dt}w_t, y_{ist}\rangle dt + \int_0^T \langle Q^i(s,t), u_t\rangle dt. \tag{5.2.9}$$

We select the vector function $Q^i(s,t)$ so that all its components, excepting for $q^i(s,t)$, are equal to zero, and $q^i(s,t)$ is a delta-function, i.e.,

$$Q^i(s,t) = (0, 0, \ldots, q^i(s,t), 0, \ldots, 0)$$

$$q^i(s,t) = \delta(t-s)$$

In this case equality (5.2.9) takes the form

$$\int_0^T \langle f_\alpha^T(\bar{x}_t, \bar{\alpha})y_{ist}, dw_t\rangle + u_s^i = 0, \tag{5.2.10}$$

where $u_s^i = x_s^i - x_s^i(\bar{\alpha})$. This equality implies probabilistic properties of deviations u_s^i $(i = 1, 2, \ldots, n, \ s \in \theta)$, namely, there is the Gaussian distribution $Eu_s = 0, \quad \forall s \in [0,T]$ since $E\int\limits_0^T \langle f_\alpha^T(\bar{x}_t, \bar{\alpha}) y_{ist}, dw_t \rangle = 0$ and variance

$$\begin{aligned}\operatorname{var}(u_s^t) &= E\left(\int\limits_0^T \langle f_\alpha^T(\bar{x}_t, \bar{\alpha}) y_{ist}, dw_t \rangle\right)^2 \\ &= \sum_{k,r} \gamma^{kr} \int\limits_0^T \langle f_\alpha^k(\bar{x}_t, \bar{\alpha}), y_{ist} \rangle \langle f_\alpha^r(\bar{x}_t, \bar{\alpha}), y_{ist} \rangle dt,\end{aligned}$$

where f_α^k is the kth column of the matrix f_α. These expressions are determined by properties of the stochastic integral [356].

Let us, for the sake of simplicity, assume that the matrix Γ is diagonal, and consider Γ as l-dimensional vector. Then

$$\operatorname{var}(u_s^t) = \sum_j \gamma^j \int\limits_0^T \left(\langle f_\alpha^j(\bar{x}_t, \bar{\alpha}), y_{ist} \rangle\right)^2 dt.$$

Denote the integral $\int\limits_0^T \left(\langle f_\alpha^i(\bar{x}_t, \bar{\alpha}), y_{ist} \rangle\right)^2 dt$ by $b_s^{ij}(\alpha)$. As a result we can write $\operatorname{var}(u_t^i) = \langle \Gamma, b_t^i(\bar{\alpha}) \rangle$, where $b_t^i(\bar{\alpha})$ is a vector with components

$$b_i^{ij}(\bar{\alpha}), \quad j = 1, 2, \ldots, l.$$

One remark is needed here concerning the observational data. While analyzing the results of immunological experiments we deal with the set X_m'. As has been noted already, it is a set of independent values of phase variables of model, i.e., for $i \neq j, \ s \neq \tau, \ i, j = 1, 2, \ldots, n, \ s, r \in \theta$,

$$\operatorname{cov}(x_s^i, x_\tau^i) = 0,$$

and, therefore,

$$\operatorname{cov}(u_s^i, u_\tau^i) = 0.$$

In other words, u_s^i and u_τ^i belong to different independent realizations

$$\{u_t(\omega_{is}, \ t \in [0,T], \ \omega_{is} \in \Omega\},$$

$$\{u_t(\omega_{j\tau}, \ t \in [0,T], \ \omega_{j\tau} \in \Omega\}.$$

Therefore integrals in (5.2.10) corresponding to these values are independent. It is now possible to write down the maximal likelihood function for the search for estimates of parameters. The joint density of vectors $\{u_s,\ s \in \theta\}$ has the form

$$p(u_s, s \in \theta) = \prod_{t\in\theta} p(u_t).$$

By the conditions of the experiment

$$p(u_t) = \prod_{i=1}^{n} p(u_t^i),$$

where $u_t^i = x_t^i - x_t^i(\alpha)$ and $p(u_t^i)$ is the one-dimensional normal density

$$p(u_t^i) = \frac{1}{\sqrt{2\pi\langle\Gamma, b_t^i(\alpha)\rangle}} \exp\left\{-\frac{(u_t^i)^2}{2\langle\Gamma, b_t^i(\alpha)\rangle}\right\}.$$

In accordance with the principle of maximum likelihood asymptotically unbiased estimates and asymptotically efficient estimates of parameters $\bar{\alpha}$ and $\bar{\Gamma}$ are determined by the condition of maximum for the function $p(u_s, s \in \theta)$. It is equivalent to the minimization of the following function

$$\Phi_m(\alpha, \Gamma) = -\ln p(u_s, s \in \theta),$$

which has the form

$$\Phi_m(\alpha, \Gamma) = \sum_{t\in\theta}\sum_{i=1}^{n}\left[\ln\langle\Gamma, b_t^i(\alpha)\rangle + \frac{\left(x_t^i - x_t^i(\alpha)\right)^2}{\langle\Gamma, b_t^i(\alpha)\rangle}\right]$$

with an accuracy up to a constant, where x_t^i is a value of the ith state variable of the model (5.2.3) observed at the instant t. If at this moment m_t independent measurements of state variables were to be performed, then

$$\Phi_m(\alpha, \Gamma) = \sum_{t\in\theta}\sum_{j=1}^{m_t}\sum_{i=1}^{n}\left[\ln\langle\Gamma, b_t^i(\alpha)\rangle + \frac{\left(x_t^{ij} - x_t^i(\alpha)\right)^2}{\langle\Gamma, b_t^i(\alpha)\rangle}\right]. \qquad (5.2.11)$$

This function depends explicitly on independent parameters comprising the vector Γ, but, this dependence proves to be implicit for the vector α, which creates difficulties for the solution of extremal problem

$$\min_{\alpha,\Gamma} \Phi_m(\alpha, \Gamma) = \Phi_m(\alpha_m, \Gamma_m).$$

Assume, in order to bypass this difficulty, that the initial value α_0 is chosen so close to $\bar{\alpha}$ that it is possible to use the linear approximation

$$x_t^i(\bar{\alpha}) = x_t^i(\alpha_0) + \sum_{k=1}^{l} \frac{\partial}{\partial \alpha^k} x_t^i(\alpha_0)\delta\alpha_0^k, \quad \delta\alpha_0 = \bar{\alpha} - \alpha_0. \tag{5.2.12}$$

It is easy to check that

$$\nabla_\alpha x_s^i(\alpha) = -\int_0^T f_\alpha^T(x_t, \alpha).y_{ist}dt$$

Denote the last integral by the vector

$$a_t^i(\alpha) = \int_0^T f_\alpha^T(x_t(\alpha), \alpha) y_{ist} dt.$$

Then (5.2.12) takes the form

$$x_t^i(\bar{\alpha}) = x_t^i(\alpha_0) - \langle a_t^i(\alpha_0), \delta\alpha_0 \rangle.$$

Substitute this expression into (5.2.11) to obtain

$$\Phi_m(\delta\alpha, \Gamma|\alpha_0)$$

$$= \sum_{t\in\theta} \sum_{j=1}^{m_t} \sum_{i=1}^{n} \left[\ln\langle \Gamma, b_t^i(\alpha)\rangle + \frac{\left(x_t^{ij} - x_t^i(\alpha) + \langle a_t^i(\alpha_0), \delta\alpha_0\rangle\right)^2}{\langle \Gamma, b_t^i(\alpha)\rangle} \right]. \tag{5.2.13}$$

The minimized function now depends explicitly on its arguments Γ and $\delta\alpha$. The extremal problem

$$\min_{\substack{\delta\alpha\in D_1 \\ \Gamma\in D_2}} \Phi_m(\delta\alpha, \Gamma|\alpha_0) = \Phi_m(\delta\alpha_0, \Gamma_0|\alpha_0)$$

is solved quite easily. Knowing $\delta\alpha_0$, find $\alpha_1 = \alpha_0 + \delta\alpha_0$; here $\alpha_1 \neq \bar{\alpha}$, since we use just the linear approximation of solution (5.2.12). Therefore we again choose α_1 as the initial point and repeat the process of calculation of $\delta\alpha_1$, etc. As a result we have arrived at the following iteration process of computing the estimates of parameters:

$$\begin{aligned} &\min_{\substack{\delta\alpha\in D_1 \\ \Gamma\in D_2}} \Phi_m(\delta\alpha, \Gamma|\alpha_k) = \Phi_m(\delta\alpha_k, \Gamma_0|\alpha_k), \\ &\alpha_{k+1} = \alpha_k + \delta\alpha_k, \quad k = 0, 1, 2, \ldots . \end{aligned} \tag{5.2.14}$$

The convergence of this process and properties of estimates so obtained were investigated by Zuev ([285]). Chapter 9 presents some results of applications of this process to the solution of practical problem connected with the analysis of data of immunological experiments.

5.2.3. Goodness-of-fit criteria. There is an important feature of the above method for the solution of estimation problem. As a matter of fact, stochastic models allow one not only to construct the iteration process (5.2.14), but to use known criteria of mathematical statistics for the verification of various hypotheses. Their application is based on the fact that according to (5.2.10) and under conditions of an experiment the random quantities

$$\frac{x_t^{ij} - x_t^i(\bar{\alpha})}{\langle \bar{\Gamma}, b_t^i(\bar{\alpha})\rangle^{1/2}}, \quad i = 1, 2, \ldots, n, \quad j = 1, 2, \ldots, mt, \quad t \in \theta,$$

are independent, have Gaussian distribution, zero mathematical expectation, and unit variance. It is well-known (Cramer [61]) that a sum of squares of such quantities

$$S(X'_m, \bar{\alpha}, \bar{\Gamma}) = \sum_{t\in\theta} \sum_{j=1}^{m_t} \sum_{i=1}^{n} \frac{\left(x_t^{ij} - x_t^i(\bar{\alpha})\right)^2}{\langle \bar{\Gamma}, b_t^i(\bar{\alpha})\rangle} \tag{5.2.15}$$

is governed by the distribution χ^2, with the number of degrees of freedom

$$L = nN \sum_{t\in\theta} m_t.$$

Let α_m, Γ_m be the estimates of parameters obtained on a sample X'_m. Then [356] the quantity $S(X'_m, \alpha_m, \Gamma_m)$ is distributed by the law χ^2 with the number of degrees of freedom $r = L - 2l$, where $2l$ is the number of estimated parameters.

Taking into account these properties, let us verify the hypothesis about correspondence of the model with observational data. Let the value $\chi_p^2(r)$ satisfy the condition

$$P\left(\chi_r^2 > \chi_p^2(r)\right) = p,$$

where p is a small probability (for example, 0.05), and χ_r^2 is a random quantity that has the distribution χ^2 with r degrees of freedom. Then it is unlikely that the event

$$S(X'_m, \alpha_m, \Gamma_m) > \chi^2_p(r)$$

can be explained by pure random factors. Therefore if we have obtained as a result of calculations that

$$S(X'_m, \alpha_m, \Gamma_m) > \chi^2_p(r)$$

then one should conclude that the model hardly corresponds to actual data. On the other hand, the inequality

$$S(X'_m, \alpha, \Gamma) \leq \chi^2_p(r)$$

holds for the values of α and Γ which form a set $Q(X'_m)$, all of whose points can be considered with probability $1-p$ as the values of parameters which do not contradict with actual data. Therefore the set

$$Q(X'_m) = \{\alpha, \Gamma : S(X'_m, \alpha, \Gamma) \leq \chi^2_p(r)\}$$

is a confidence set of the type of $(1-p) \cdot 100\%$ for the parameters. More than that, the criterion χ^2 allows one to check the hypothesis on equality of parameters obtained by the sets of independent trajectories. Let, for example, X'_m be a set of values corresponding to the observations over a group of infected animals who were not given the drugs under study, and let α_m, Γ_m be corresponding estimates of parameters of the model (5.2.3). Then let Y'_m be a result of observations over a group treated with the drugs, and let β_m and R_m be corresponding estimates. Let us verify the zero hypothesis: the drugs did not affect the parameters of the process under study, i.e., $\bar{\alpha} = \bar{\beta}$. But, if the zero hypothesis is true, there must be then no significant difference between α_m and β_m, and, hence

$$\begin{aligned}
&S(X'_m, \alpha_m, \Gamma_m) < \chi^2_p(r), \\
&S(Y'_m, \beta_m, R_m) < \chi^2_p(r), \\
&S(X'_m, \beta_m, \Gamma_m) < \chi^2_p(L-l), \\
&S(Y'_m, \alpha_m, R_m) < \chi^2_p(L-l).
\end{aligned} \tag{5.2.16}$$

If the two first equations hold, the model is not in conflict with observational data. If any of the two last equations does not hold, it means that a difference in the estimations of parameters is caused by nonrandom factors, and the zero hypothesis on the equality $\bar{\alpha} = \bar{\beta}$ must be rejected.

We have outlined the general scheme of data processing for immunological experiments; the reader can find a detailed description in the monograph by Zuev, [356]. We shall present a practical example in Chapter 9 of the solution of the problem of influence of antiviral drugs on the processes in an organism affected by the influenza infection.

The methods of parameter identification for mathematical models set forth in this chapter will be used in what follows when solving various problems of modeling the immune response to viral and bacterial diseases: tuning the models to data, investigation of mechanisms of drugs action, etc. These results are presented in Chapters 7, 8, 9.

5.3. Parameterization in Mathematical Models of Immunophysiological Processes

5.3.1. The conception of parameterization. We have discussed in the preceding section the problem of estimation of model's parameters using observation data from the standpoint of statistical analysis of the data of immunological experiments. If the same problem is considered in view of the prognosis of individual course of disease and its outcome on the basis of mathematical model, then the above statement of the problem proves to be completely unsatisfactory. In fact, the models of diseases contain dozens of unknown parameters. Therefore the number of instants combining the set Θ, must be at least of the same order, and preferably larger. Even if the measurements are performed each day, which is practically impossible, dozens of days will be required to accumulate the necessary volume of information. It is evident that after this time the problem of disease course and outcome will be solved without a mathematical model. It is necessary, therefore, to study the feasibility of the "individualization" of mathematical models in question.

The experience accumulated by the author in mathematical modelling of physical processes enabled him to suggest in the late 70s a hypothesis on the solution of this problem, which can be obtained on the basis of the parameterization of processes under study. The essence of this hypothesis is the assumption about the dependence of model parameters $\alpha = (\alpha^1, \alpha^2, \ldots, \alpha^L)^T$ on some factor H, which describes individual peculiarities of a concrete organism, and known constants $Q = (q^1, q^2, \ldots, q^L)^T$ that are the same for all people.

It is possible to write, according to this hypothesis, that

$$\alpha = \nu(H, Q),$$

where $\nu(H, Q)$ is some continuous function. Then the model (5.2.3) takes the form:

$$\frac{dx_t}{dt} = f(x_t, \nu(H, Q)), \quad x_0 = c, \quad t \in [0, T]. \tag{5.3.1}$$

Since the constants composing the vector Q are known, the model (5.3.1) can be written as

$$\frac{dx_t}{dt} = F(x_t, H), \quad x_0 = c, \quad t \in [0, T], \tag{5.3.2}$$

where there is one unknown parameter H which takes into account the features of the concrete organism and is to be estimated. The transition of model (5.2.3.) into the model (5.3.2) is called in mathematical modelling the *parameterization*. It is clear that one may now think of the prognosis of the course of individual disease on the basis of model (5.3.2), since it contains the unknown parameter H.

We emphasize that it is not obligatory to perform the estimation of the parameter H with the model (5.3.2) that describes the course of disease. It is possible to use some other load tests which are used traditionally in medicine. For example, the well-known glucose tolerance test: the dynamics of glucose concentration in blood is registered when the glucose is received on an empty stomach.

Indeed, let the model that describes the changes in glucose concentration in blood be a system of equations of (5.3.2) type:

$$\frac{dx_t}{dt} = \varphi(x_t, H), \quad x_0 = c, \quad t \in [0, T], \tag{5.3.3}$$

where $x_t = (S_T, G_t, I_t)^T$ is a vector of state, whose components are: S_t, quantity of glucose in stomach; G_t, glucose concentration in blood; I_t, insulin concentration in blood; S_0, known quantity of administered glucose; G_0, I_0, known values of individual homeostatic concentrations of glucose and insulin in blood respectively. It is possible, having the data on the dynamics of glucose concentration in blood $\{G_t,\ t \in \Theta\}$, to estimate personal parameter H in the model (5.3.3) of the organism under study, and to make a prognosis on the course and outcome of disease using the model (5.3.2).

Such are the hypothesis and ideal scheme of prognosis. The main problem of the realization is to establish a form of the function (5.3.1). Pogozhev has devoted his studies to the solution of this problem in early 80s, later Zuev joined him. Their results are presented briefly in the following sections of this chapter.

5.3.2. Microlevel: conceptions and assumptions. The formula (5.3.1) suggests that the search for the dependence of parameters on the factor H that links them should be carried out on the level of interactions between the particles in fluid mediums of organism. In fact, the parameters of models under discussion $\alpha = (\alpha^1, \alpha^2, \dots, \alpha^L)^T$ take into account the probabilities for the particles (viruses, antibodies, lymphocytes, molecules, etc.) to encounter and interact; the changes in their concentrations are described by equations of (5.2.3) type. For example, the coefficient β in simple model of Chapter 2 takes into account the fact of the encounter of viruses with tropic cells of organism where they reproduce; the coefficient γ does the same for viruses and antibodies. Other parameters have a similar sense.

It is necessary therefore to construct a model describing the motion of particles in fluid mediums of organism. This motion is considered as Brownian in the books written by Pogozhev. In order to clarify this conception we consider some particle moving together with lymph in intercellular space. This particle changes its direction many times per second in a random way as a result of collisions with motionless cells of organism and other particles. Its trajectory in space, even observed during one minute, resembles a trajectory of a Brownian particle. Recall that characteristic times of changes of state variables of models of diseases are of an order of a number of hours.

Let us express this considerations in the form of a mathematical model. Let $q(t) \in \mathrm{R}^3$ be the coordinates of a particle at a moment of time t. The increments of the coordinates of a particle $dq(t) = q(t+dt) - q(t)$ for an infinitely small interval of time dt are determined by two components: directed movement of the liquid medium and chaotic motion equivalent to a Brownian one, which is described in mathematical models by the Wiener process $W(t)$. Recall that the Wiener process is a continuous Gaussian stochastic process that possesses the following properties:

1) $W(0) = 0$;
2) $EW(t) = 0, \quad \forall t \geq 0$;
3) $E[W(t)W^T(t)] = It$ where I is the unity 3×3 matrix;
4) the process has independent stationary increments.

Write, according to these properties, for the increments of a particle's coordinates:

$$dq(t) = \Psi(t)dt + BdW(t), \tag{5.3.4}$$

where $\Psi(t)$ is a drift vector describing the motion of a particle together with a liquid medium, $D = BB^T$ is a diffusion matrix accounting for random movements of a particle as related to the moving medium induced by collisions with motionless cells, $W(t)$ is a Wiener process. Let $\bar{q}(t)$ satisfy the equation

$$\frac{d\bar{q}(t)}{dt} = \Psi(t).$$

Then it is convenient to write down the motion of a particle inside a flow in new coordinates $U(t) = q(t) - \bar{q}(t)$ with regard to (5.3.4):

$$dUt = BdW(t), \ . \tag{5.3.5}$$

This very motion leads to the collisions and the following interactions between the particles.

The hypothesis about similarity of live organisms, suggested and formalized mathematically by Pogozhev and Zuev played a key role in the development of these investigations. Summing up the results, one can say that vital functions in all warm-blooded organisms are determined by the interactions of the same particles. Sizes of cells that compose the organisms and intercellular spaces are also the same, and the organisms differ just with the number of cells. Properties of liquid mediums where the particles move are the same too. The main parameter which is different in all alive organisms is the speed of movement of their liquid mediums.

Basing on these results, consider two organisms: an organism under investigation (OUT) and base organism (BO) and introduce the conception of similarity of the organism under investigation to the base one. We define a similarity coefficient H as follows (*the first Pogozhev hypothesis*):

$$H = \frac{V^2\tau}{\underline{V}^2\underline{\tau}}, \tag{5.3.6}$$

where V, $\underline{V}$ are specific velocities of blood circulation for OUT and BO respectively and τ, $\underline{\tau}$ are corresponding durations of a heart-beat. Underlined variables are related hereinafter to the base organism.

Diffusion movement of a particle is considered as Brownian in each organism; therefore, we have for the increments of coordinates, according to (5.3.5) for OUT and BO:

$$U(\Delta t) = Bw_{\Delta t}, \quad \underline{U}(\Delta t) = \underline{B}w_{\Delta t}. \tag{5.3.7}$$

The second Pogozhev hypothesis suggests the validity of the following equality which is to be understood as stochastic equivalence of random processes :

$$U(\Delta t) = \sqrt{H}\underline{U}(\Delta t).$$

In order to clarify the meaning of this hypothesis let us return to the original coordinates. Then with regard to (5.3.8) equation (5.3.4) takes the form:

$$\frac{dq(t)}{dt} = \sqrt{H}\underline{\Psi}(t) + \sqrt{H}\underline{B}\frac{dW(t)}{dt}. \tag{5.3.8}$$

Here

$$\Psi(t) = \sqrt{H}\underline{\Psi}(t), \quad D = H\underline{D}. \tag{5.3.9}$$

It means that the increase in the velocity of motion of the liquid results both in the increase in the velocity of directed drift of particles and in the randomization of their motion, which in its turn increases the probability of their collisions and interactions.

Notice that the parameter H is similar in this respect to the *absolute temperature* of Brownian motion of particles in a liquid, to the *effective temperature* introduced by Myasnikov for multi-component mediums, and to the *fluctuation temperature* suggested by Reggrer.

We can write with regard to (5.3.7) and to the properties of a Wiener's process:

$$U(\Delta t) = \sqrt{H}\underline{U}(\Delta t) = \underline{U}(H\Delta t). \tag{5.3.10}$$

This relationship is a basic one. It is the origin of the similarity relationships for the parameters of an organism under investigation and base organism which are discussed in the next subsection.

5.3.3 Organism level: similarity of parameters. The collisions of particles leading to the interactions are a result of mixing, i.e., of diffusion motion. An intensity of the interactions λ between the particles is defined as an average number of collisions in a unit volume per a unit of time:

$$\lambda = \lim_{\Delta t \to 0} \frac{EZ(\Delta t)}{\Delta t}, \tag{5.3.11}$$

where $Z(\Delta t)$ is the random number of collisions between the particles during the time Δt.

Equalities (5.3.10) permit us to prove the following lemma.

Lemma 1. The intensities of interactions in OUT *and* BO *satisfy the relationship*:

$$\lambda = H\underline{\lambda} \tag{5.3.12}$$

Proof. By definition,

$$EZ(\Delta t) = \Phi(U_\tau,\ 0 \le \tau \le \Delta t),$$

i.e., represents a functional of motion trajectory of a particle. Equalities (5.3.10) open the possibility of changing the time scale:

$$\begin{aligned} EZ(\Delta t) &= \Phi(U(\theta),\ 0 \le \theta \le \Delta t) = \Phi(\underline{U}(\theta),\ 0 \le \theta \le \Delta t) \\ &= \Phi(\underline{U}(\tau),\ 0 \le \tau \le H\Delta t) = E\underline{Z}(H\Delta t). \end{aligned}$$

Hence with regard to (5.3.11) the following equalities are valid

$$\lambda = \lim_{\Delta t \to 0} \frac{EZ(\Delta t)}{\Delta t} = H \lim_{\Delta \tau \to 0} \frac{E\underline{Z}(H\Delta t)}{H\Delta t} = H\underline{\lambda} \qquad \blacksquare$$

Let x_t be a concentration of some free particles in a zone of their interaction at the moment of time t and let X_t be their quantity. By definition

$$X_t = Qx_t,$$

where Q is effective volume of interactions, i.e., a measure of a domain Ω occupied by moving particles. The following lemma results from the model 5.3.3 and from the similarity relationship (5.3.10).

Lemma 2. Effective volumes of interactions Q, $\underline{Q}$ in OUT and BO satisfy the similarity relationship:

$$Q = H^{3/2}\underline{Q}. \tag{5.3.13}$$

Proof. Let $p(t,x) = p(t,x^1,x^2,x^3)$ be the distribution density of particles in the interaction zone such that

$$X_t = \iiint_{\Omega} p(t,x)dx.$$

It follows from the similarity relationship (5.3.10) that

$$p(t,x) = \underline{p}\left(t, \frac{x}{\sqrt{H}}\right). \tag{5.3.14}$$

The amount of particles contained in arbitrary domain Ω^* is determined as follows.

$$\begin{aligned} X_t^{\Omega^*} &= \iiint_{\Omega^*} p(t,u)du = \iiint_{\Omega^*} \underline{p}\left(t, \frac{u}{\sqrt{H}}\right) du \\ &= H^{3/2} \iiint_{\underline{\Omega}^*} \underline{p}(t,u)du = H^{3/2}\underline{X}_t^{\underline{\Omega}^*}, \end{aligned}$$

where $\underline{\Omega}^*$ is a domain that corresponds to Ω^* and depends on H.

The comparison of effective volumes in the systems under consideration has a sense just for the same concentrations, i.e., when $x_t = \underline{x}_t$.

By this equality

$$\frac{\underline{X}_t^{\Omega^*}}{\underline{Q}^*} = \frac{X_t^{\Omega^*}}{Q^*} = \frac{H^{3/2}\underline{X}_t^{\Omega^*}}{Q^*}, \quad \text{hence} \quad Q^* = H^{3/2}\underline{Q}^*,$$

which proves the validity of (5.3.13). ∎

Lemma 3. Stationary concentrations x_∞, $\underline{x}_\infty$ and quantities of particles X_∞, $\underline{X}_\infty$ of OUT and BO are linked by the relationships

$$x_\infty = H^{-1/2}\underline{x}_\infty, \tag{5.3.15}$$

$$X_\infty = H\underline{X}_\infty. \tag{5.3.16}$$

Proof. Let β be an influx velocity of particles coming into the reactor and let γx_∞ be a rate of their absorption due to the collisions and interactions. Then we have for OUT and BO:

$$\gamma x_\infty = \beta, \quad \underline{\gamma}\underline{x}_\infty = \underline{\beta}.$$

According to the above results, $\gamma = H\underline{\gamma}$ and it is possible, by virtue of (5.3.9), to write for the influx of particles provided by the drift $\beta = \underline{\beta}\sqrt{H}$. Then

$$H\underline{\gamma}x_\infty = \sqrt{H}\underline{\beta}, \quad \underline{\gamma}\underline{x}_\infty = \beta,$$

whence the statement (5.3.15) follows. We obtain from the same equalities with regard to the relationship for effective volumes (5.3.13) that

$$H\underline{\gamma} = \frac{X_\infty}{Q} = \sqrt{H}\underline{\beta}, \quad \underline{\gamma}\frac{\underline{X}_\infty}{\underline{Q}} = \underline{\beta},$$

or

$$H\underline{\gamma} = \frac{X_\infty}{\underline{Q}H^{3/2}} = \sqrt{H}\underline{\beta}, \quad \underline{\gamma}\frac{\underline{X}_\infty}{\underline{Q}} = \underline{\beta},$$

which proves the validity of statement (5.3.16). ∎

We assume, according to this line of reasoning, that a reaction of the organism to an external perturbation g takes time T and is described by the following system of ordinary differential equations:

$$\frac{dx_t}{dt} = f(x_t, \alpha, x_\infty), \quad x_0 = g, \quad t \in [0, T], \tag{5.3.17}$$

where $f(\bullet, \bullet, \bullet)$ is a continuous vector function describing the influx of particles, the structure of their interactions, and the utilization of complexes; $\alpha \in \mathrm{R}^L$ is a vector of positive parameters taking into account the interaction between the particles whose components are proportional to the intensity of interaction λ which is defined as a limit (5.3.11); $x_\infty \in \mathrm{R}^N$ is a stationary state corresponding to the state of a healthy organism.

Theorem. If the similarity relationships (5.3.12), (5.3.13), (5.3.15) *are satisfied, a change in the state of an organism under study is described by a model in the form of the following system of differential equations which contains just base parameters and the similarity coefficient H:*

$$\frac{dx_t}{dt} = H^{-3/2} f\left(x_t H^{3/2}, \underline{\alpha} H, \underline{x}_\infty H\right), \quad x_0 = g, \quad t \in [0, T]. \qquad (5.3.18)$$

Proof. Notice first of all that concentration of particles in the models of (5.3.17) type can be defined, generally speaking, in an arbitrary way:

$$x_t = \frac{X_t}{\tilde{Q}},$$

where $\tilde{Q}$ is some volume and X_t is the quantity of interacting particles. In our case of the comparison of two systems the concentrations must be defined strictly, namely as the following relationships:

$$x_t^c = \frac{X_t}{Q}, \quad \underline{x}_t^c = \frac{\underline{X}_t}{\underline{Q}}; \quad \text{then} \quad x_t^c = \frac{X_t}{Q} = x_t \frac{\tilde{Q}}{Q}.$$

Whence, with regard to (5.3.17), we obtain

$$\frac{dx_t^c}{dt} = \frac{\tilde{Q}}{Q} f\left(\frac{Q}{\tilde{Q}} x_t^c, \alpha, \frac{Q}{\tilde{Q}} x_\infty\right), \quad x_0 = g, \quad t \in [0, T].$$

Similarly, for the base system we have

$$\frac{d\underline{x}_t^c}{dt} = \frac{\tilde{Q}}{\underline{Q}} f\left(\frac{\underline{Q}}{\tilde{Q}} \underline{x}_t^c, \underline{\alpha}, \frac{\underline{Q}}{\tilde{Q}} \underline{x}_\infty\right), \quad \underline{x}_0 = g, \quad t \in [0, T].$$

Assign $\tilde{Q} = \underline{Q} = 1$ with regard to $Q = H^{3/2}\underline{Q}$, to obtain

$$\frac{dx_t^c}{dt} = H^{-3/2} f\left(x_t^c H^{3/2}, \alpha, x_\infty H^{3/2}\right), \quad x_0 = g, \quad t \in [0, T],$$

$$\frac{d\underline{x}_t^c}{dt} = (\underline{x}_t^c, \underline{\alpha}, \underline{x}_\infty), \quad \underline{x}_0 = g, \quad t \in [0, T].$$

Substituting the first of these systems into relationships (5.3.12) and (5.3.15) we arrive at system (5.3.18). ■

The value of the similarity relationships we so obtained is determined by their correlation with the observation data. This question deserves special consideration.

5.3.4. The Comparison with observational data. Let us clear up, before we begin to deal with observational data, how the predicted course of disease in a concrete patient depends on a value of the similarity parameter H which will be regarded as known under the assumption, that it is calculated using the data of some test with the corresponding model. To this end, we use the model of Chapter 3, which is investigated thoroughly, and perform the parameterization (5.3.18) for this model according to [280]. We present final dimensionless form of this model in parameterized form

$$\begin{aligned}
\frac{dV}{dt} &= H\left(h_1 V - H^{3/2} h_2 F V\right), && V(0) = V_0,\\
\frac{dC}{dt} &= H\left[h_3 \xi(m) X_k - h_5 (C - 1)\right], && C(0) = 1,\\
\frac{dF}{dt} &= H\left[h_4 (C - F) - H^{3/2} h_8 F V\right], && F(0) = 1,\\
\frac{dm}{dt} &= H\left(H^{3/2} h_6 V - h_7 m\right), && m(0) = 0,\\
\frac{dX_1}{dt} &= H\frac{k}{\tau}\left(H^{3/2} F V - X_1\right), && X_1(0) = 0,\\
&\cdots\cdots\cdots && \cdots\cdots\\
\frac{dX_k}{dt} &= H\frac{k}{\tau}\left(X_{k-1} - X_k\right), && X_k(0) = 0.
\end{aligned} \tag{5.3.19}$$

We intentionally use here the representation of subsection 5.3.2 for the delay term as additional equations, in order to show the parameterization in the systems with a delay. It is seen that the term with a delay $V_{t-\tau}F_{t-\tau}$ turns into $V_{t-\tau/H}F_{t-\tau/H}$ as a result of the parameterization.

We select those values for the parameters as base ones, that provide the interpretation of a solution to this system of equations as subclinical form of disease:

$$\tau = 0.5, \quad h_1 = 8, \quad h_2 = 10, \quad h_3 = 10^4, \quad h_4 = 0.17,$$
$$h_5 = 0.5, \quad h_6 = 10, \quad h_7 = 0.12, \quad h_8 = 8, \quad V(0) = 10^{-6}.$$

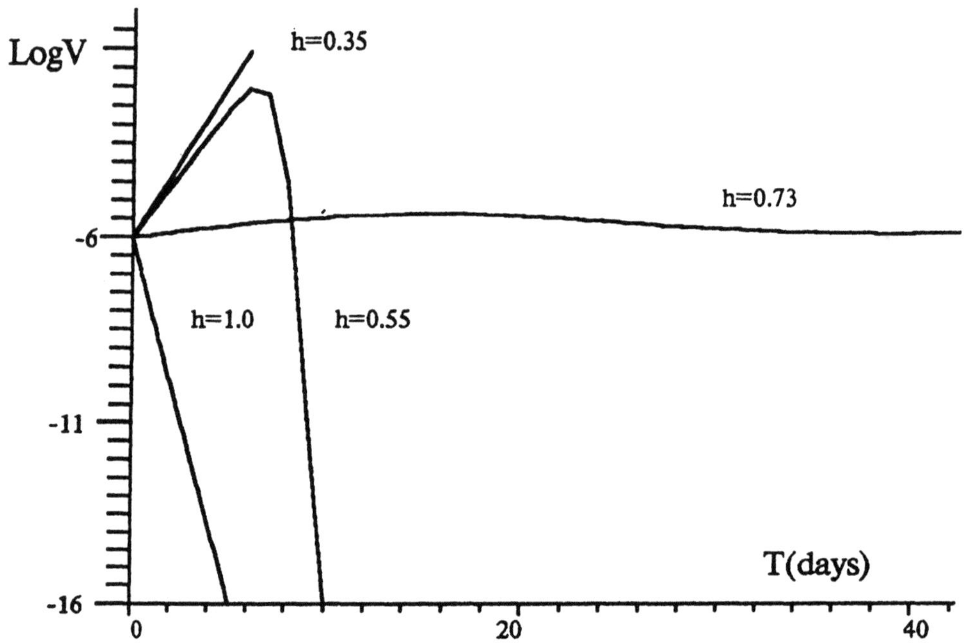

Fig. 34. Solutions of the simple model for the variable $V(t)$ corresponding to various values of the parameter H.

Fig. 34 shows the solutions of the model system (5.3.19) corresponding to various values of individual parameter H for fixed values of the base parameters. The results of modelling demonstrate that firstly, all the diversity of the forms of the course of disease can be explained through the changes of just one personal parameter H; secondly, the decrease of this parameter leads to the increase of the probability of unfavorable course of disease and of lethal outcome; thirdly, the increase or decrease of H during the chronic course of disease leads to the transition of an organism into healthy state. The rise of H is preferable, since such transition goes through subclinical form, whereas the drop implies acute form when lethal outcome is possible.

So, the parameter H obtained somehow for a concrete organism, provides definite information on the predisposition of this organism to a certain form of the course of the disease. This conclusion leads to the statement of a new problem, namely, to the investigation of the possibility of determining this parameter for a concrete person through the data of some physiological test and corresponding mathematical model.

In subsection 5.3.1, while discussing the ideal prognostic scheme, the tolerance test to glucose was mentioned, which appears to be quite "harmless" for an organism under investigation. Here is the essence of this test.

A tested person at morning with empty stomach, when concentrations of glucose and insulin in blood correspond to the homeostasis values, drinks a solution containing 75 g of glucose. This leads to a rise of glucose concentration in blood which is measured every half an hour. The restoration of homeostatic concentrations of glucose and insulin occurs in about 3 hour, and the changes of these variables during this time are described by the following model [280]:

$$
\begin{aligned}
&x(t) = G(t) - \bar{G}, \\
&y(t) = I(t) - \bar{I}, \\
&\frac{dS}{dt} = -\alpha_1 S, && S(0) = g, \\
&\frac{dx}{dt} = \alpha_6 S - \alpha_2 x - \alpha_3 y, && x(0) = 0, \\
&\frac{dy}{dt} = \alpha_4 x - \alpha_5 y, && y(0) = 0,
\end{aligned}
\tag{5.3.20}
$$

where $G(t)$, $I(t)$ are concentrations of glucose and insulin in blood, $\bar{G}$, $\bar{I}$ are their homeostasis values, and g is a quantity of received glucose. Perform the parameterization to obtain

$$x(t) = G(t) - \frac{\underline{\bar{G}}}{\sqrt{H}},$$

$$y(t) = I(t) - \frac{\underline{\bar{I}}}{\sqrt{H}},$$

$$\frac{dS}{dt} = -H\underline{\alpha}_1 S, \qquad S(0) = g, \tag{5.3.21}$$

$$\frac{dx}{dt} = H^{-1/2}\underline{\alpha}_6 S - H(\underline{\alpha}_2 x - \underline{\alpha}_3 y), \quad x(0) = 0,$$

$$\frac{dy}{dt} = H(\alpha_4 x - \alpha_5 y), \qquad y(0) = 0.$$

As in the considerations above, the underlining of a parameter in this model means that its value is fixed and equal to the corresponding value of the base organism.

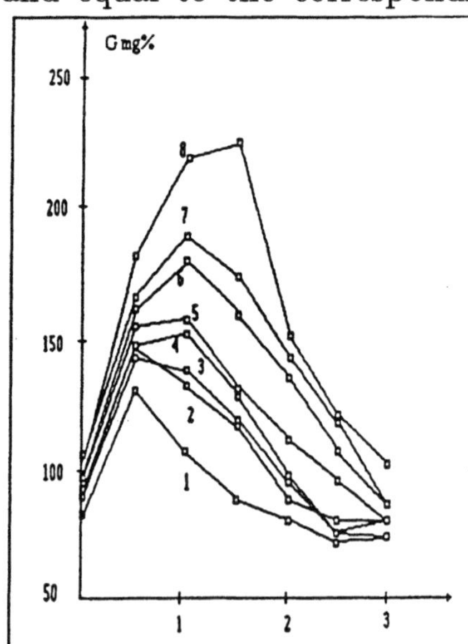

Fig. 35. Sugar concentration in blood of people of different age after glucose load.

Turn now to the observation data. Fig. 35 shows the results of observations while testing various age groups of practically healthy people. Each curve represents the result of averaging over a group of 10 people. Select as basic those values of parameters for which model (5.3.20), i.e., model (5.3.21) for $H = 1$, reproduces the curve corresponding to the age group of 20–30 years. An organism of practically healthy person aged about 25 years will be thus chosen for the BO. In this case $\underline{\bar{G}} = 90$ mg%, $\underline{\bar{I}} = 5$ units.

The following values of base coefficients were obtained as a result of solution of this problem by the least squares method:

$$\underline{\alpha}_1 = 1.57, \quad \underline{\alpha}_2 = 0.50, \quad \underline{\alpha}_3 = 1.97,$$
$$\underline{\alpha}_4 = 1.26, \quad \underline{\alpha}_5 = 1.27, \quad \underline{\alpha}_6 = 219.8.$$

Let us verify now that according to the above reasoning the features of all the rest of curves are taken into account in the model (5.3.21) with just one parameter H. To this end, determine a value of H in the model (5.3.21) for each of the curves by the least squares method too. The results are presented in Fig. 36 which shows that the above hypotheses and the models constructed on their basis do not contradict the data of observations.

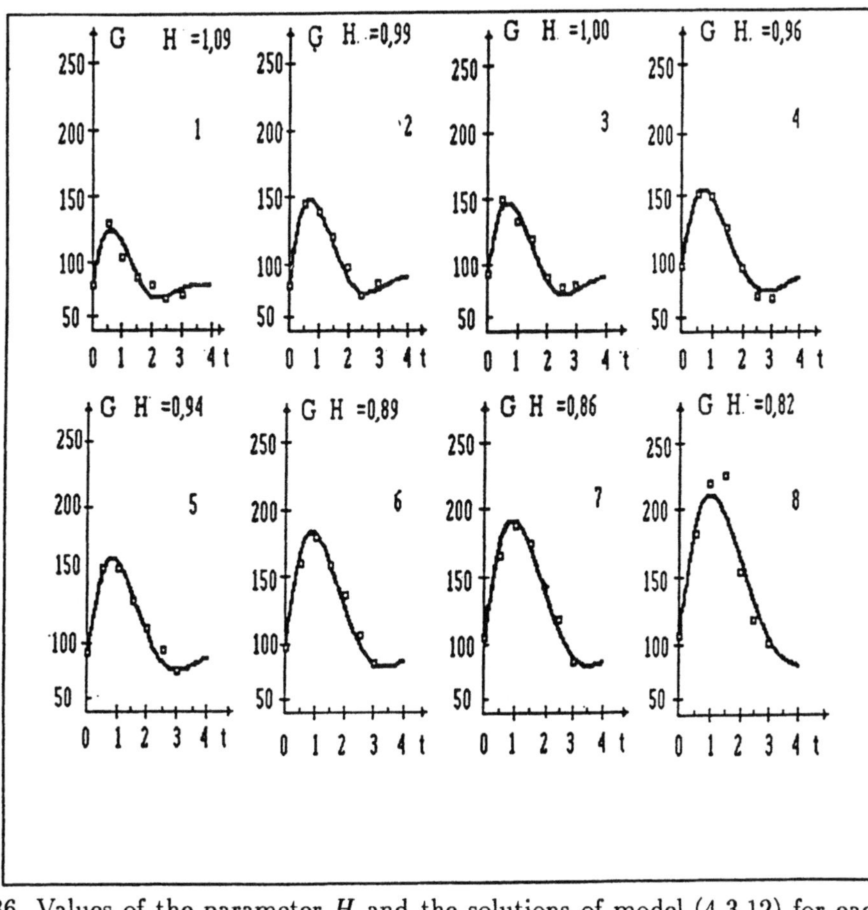

Fig. 36. Values of the parameter H and the solutions of model (4.3.12) for each of the curves presented in Fig. 35.

5.3.5. Level of population: age changes in the mortality index. Notice first of all that as a result of performed correlation of the model with observational data we have estimated the dependence of the parameter H on age, since the curves used to determine this parameter were obtained in eight different age groups. In order to refine this dependence, supplement our data with results of observations over other physiological parameters. So, according to (5.3.15)

$$\bar{G} = \frac{\underline{\bar{G}}}{\sqrt{H}}; \quad \text{hence} \quad H = \left(\frac{\underline{\bar{G}}}{\bar{G}}\right)^2,$$

where $\bar{G}$ is the homeostatic concentration of glucose in blood for a tested person and $\underline{\bar{G}} = 90$ mg% which corresponds to the norm for practically healthy person in the age of 25–30 years.

Let us present one more result obtained by Pogozhev skipping the comments:

$$H = \left(\frac{\underline{M}}{M}\right)^{1/4},$$

where M is the weight of the tested person and $\underline{M}$ is the norm of weight for his stature. Usually they assume $\underline{M}$ = stature − 100 cm.

Such estimates of the parameter H for the people of different age are presented in Fig. 37. The solid curve corresponds to the following dependence of H on the age T:

$$H = \begin{cases} 2.07(1 + 12T)^{1/8}, & T \leq 16, \\ \exp[-0.008(T - 25)], & T > 16. \end{cases} \tag{5.3.22}$$

It is seen, in particular, from this formula, that after 16 years the drop of the parameter H corresponds to the Cliber curve that describes the drop of the basic exchange with age which is measured on the absorption of oxygen by a unit of body mass. Therefore the dimensionless parameter H can be interpreted as an estimate of relative intensity of exchange processes in organism. Its value equals the unit for the age group of 20–30, i.e., for the most healthy individuals in a population.

Notice that the curve (5.3.22) that describes the drop of intensity with age was obtained as a result of the averaging of corresponding data. Therefore it can be considered as a characteristic of the population level. It means that it is natural to perform the comparison of the obtained result with actual data using the observations over some index of this level. We restrict ourselves in this book with the consideration of age changes in the mortality index.

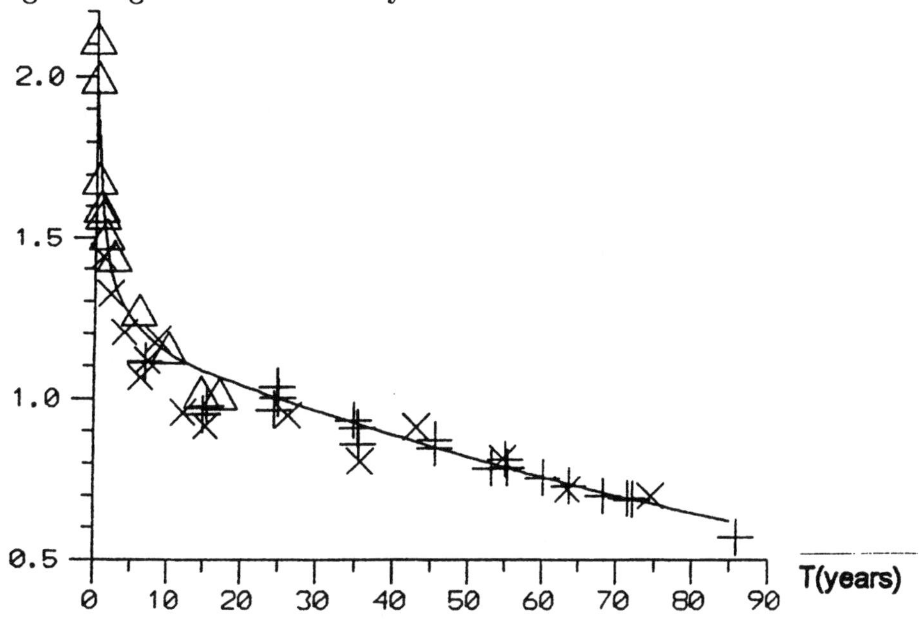

Fig. 37. Estimates of the age function $H(t)$: ×, by homeostatic sugar concentration in blood; +, by the glucose tolerance test; Δ, by mass of a body.

Such a choice was stipulated by the fact that the aging process of an organism is essentially the process of slowing down the intensity of exchange. In its turn, the dependence of intensity of mortality on the age T is well known. In fact, in junior age groups (with the values $H > 1$) the mortality caused by infectious (exogenous) diseases dominates, whereas in older groups the mortality caused by exogenous diseases drops and fast growth of mortality with age is caused by the increase in frequency of cardiac and cancer (endogenous) diseases leading to a lethal outcome.

Assume, keeping this in mind and using the results of [280], that it is possible to write down for the intensity of mortality caused by exogenous diseases μ_α:

$$\ln[\mu_\alpha(T)] = a + b\ln[H(T)],$$

and that for the intensity of mortality caused by endogenous diseases there exists the inverse dependence:

$$\ln[\mu_\alpha(T)] = a - b\ln[H(T)].$$

Then a total index of mortality $\mu(T)$ can be represented by the sum:

$$\mu(T) = \mu_{ex} + \mu_{en} + c = \mu_0[H^b(T) + H^{-b}(T) + q], \tag{5.3.23}$$

where $\mu_0 = e^a$ is the intensity of mortality in the age group 20–30 years which is characterized by $H = 1$, and $c = \mu_0 q$ is the intensity of mortality caused by the events not connected with diseases.

It is worthy of notice now that the dependence (5.3.22) describes averaged over the population decrease with age in the intensity of exchange. This process occurs in every separate individual faster or slower depending on the way and conditions of life. The same may be said about various populations. In order to take this into account we regard the population with the dependence assigned by formula (5.3.22) as the base population assuming that the "aging" curve $h(T)$ for any other population is of the same character but differs of the base curve (5.3.22) with its constant multiplier:

$$h(T) = h_b H(T), \tag{5.3.24}$$

where $H(T)$ is the basic curve (5.3.22), h_b is the parameter of population under study; namely, $h_b = h(25)$. It follows immediately from (5.3.24) and from the fact that $H(25) = 1$. This parameter characterizes thus the relative intensity of exchange processes in the group of the most healthy people in the population under investigation.

As a result of the above reasoning we obtain the following model to describe the age changes in the mortality index of investigated population:

$$\mu(T, \mu_0, h_b) = \mu_0[h^b(T) + h^{-b}(T) + q],$$

where the function of age $h(T)$ is defined by formulas (5.3.24) and (5.3.22). The parameters μ_0, h_b of this model for the population of selected region are found with the least squares method using actual values of the mortality index for various age groups which are published in demographic reference books. Fig. 38 shows the result of the solution of this problem using the data on mortality in Russia for 1958. The diagram demonstrates good correlation between model and actual data, which is confirmed by the statistical criteria too.

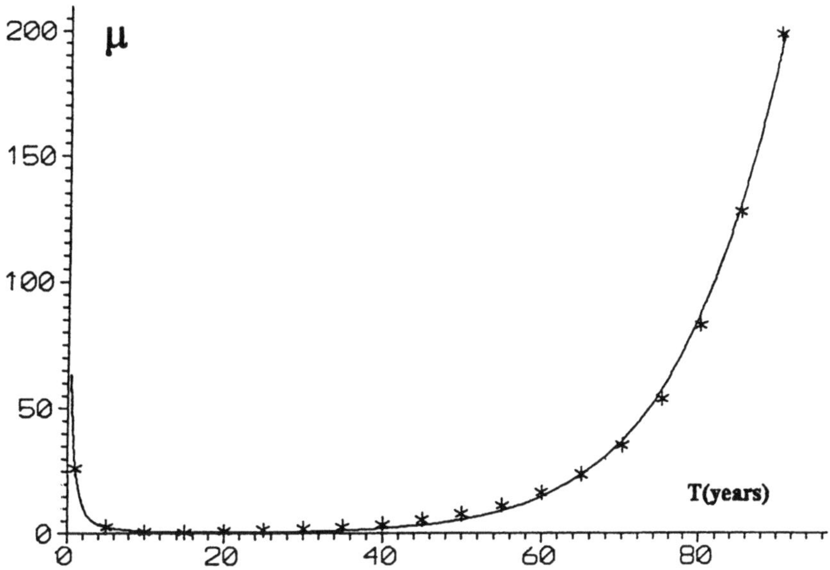

Fig. 38. Observed changes in the mortality index with age (*) and model curve.

It is possible to compare the estimated model parameters with conventional demographic indices for various regions of the world. For example, Fig. 39 shows the dependence of the index of mortality caused by cancer and cardiac diseases on the parameter h_b obtained as a result of such processing the actual data. It is seen on the diagram that the decrease in the values of the parameter h_b corresponds to the growth of mortality caused by these diseases. It correlates reasonably with the results presented in subsection 5.3.4, where it was shown in the framework of the simple model that the decrease of the parameter H leads to the transition of the subclinical form of disease into the lethal one.

Of course, these results require further investigation; though, even now they allow us to make two essential conclusions. First, the processes that occur in an organism on the microlevel determine its state, and, as a consequence, the state of population as a whole. Second, since the environment factors affect the microlevel processes, the parameter h_b can be regarded, on the one hand, as a characteristic of conditions of life for the population, and, on the other hand, as an integral estimate of the state of population's health. These conclusions have a practical value for the solution of medical and ecological problems.

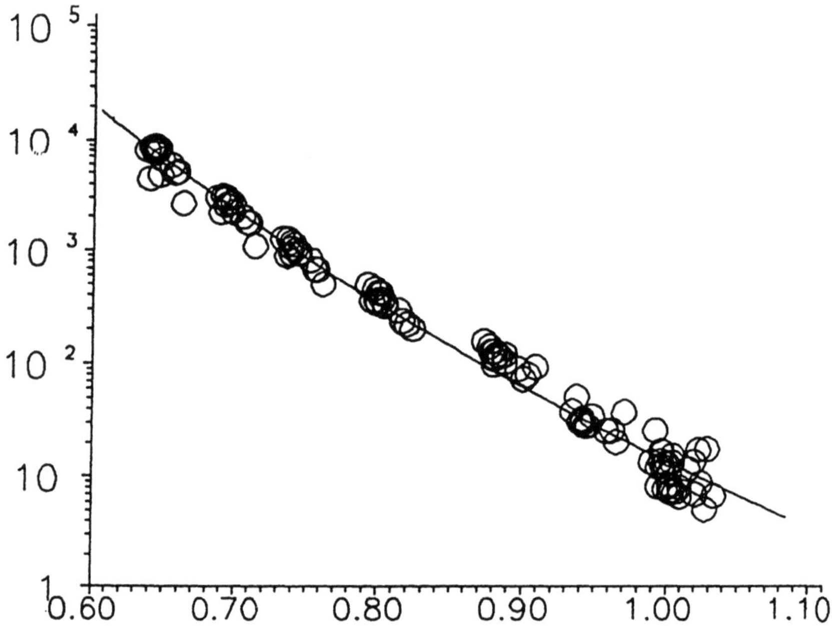

Fig. 39. Dependence of the mortality index on the parameter h_b.

CHAPTER 6

Numerical Realization Algorithms For Mathematical Models

Mathematical models of immune response we have considered are formulated in the form of systems of delay-differential equations. The use of a given class of differential equations is connected with the necessity of description of delay in the formation of cytotoxic T-lymphocytes and plasma cell clones after antigen stimulation. The treatment of the parameter identification and optimal control problem implies the construction of effective methods for the solution of the initial value problem (IVP) with the required accuracy for delay-differential equations (DDE).

The initial value problem for systems of equations of the simple mathematical model of infectious disease and the model of antiviral and antibacterial immune response can be represented as follows:

$$\frac{dy(t)}{dt} = f(y(t), y^{[1]}(t-\tau_1), y^{[2]}(t-\tau_2), \ldots, y^{[m]}(t-\tau_m), \alpha),$$

$$t_0 \leq t \leq t_0 + T,$$

$$y(t) \in \mathrm{R}^N, \quad y^{[i]}(t) \in \mathrm{R}^{N_i}, \quad N_i \leq N, \quad i = 1, 2, \ldots, m,$$

$$y(t_0) = \varphi_0, \quad y^{[i]}(t) = \varphi^{[i]}(t), \quad t \in [t_0 - \tau_i, t_0),$$

$$y \equiv [y^{(1)}, y^{(2)}, \ldots, y^{(N)}]^T, \quad y^{[i]} \equiv [y^{(j_1)}, y^{(j_2)}, \ldots, y^{(j_{N_i})}]^T,$$

where N is the number of state variables in mathematical model, m is the number of delays $\{\tau_i\}_{i=1,m}$, and $\alpha \in \mathrm{R}^L$ is the vector of parameters.

Systems of such type are widely used for the description of processes in electric circuits, population biology, immunology, and physiology. The conventional approach to the construction of numerical methods for solving the initial value problem for delay-differential equations is based essentially on the joint use of methods for the numerical solution of ordinary differential equations and of the approximation of delay variables (the variables with delayed argument).

Algorithms based on the Runge-Kutta methods are discussed in [14, 23, 35, 151, 237, 249–251, 254, 348, 358]; methods that employ the Adams difference schemes are discussed in [14, 38, 144, 145, 237, 347, 348]. These methods proved to be inefficient while solving numerically the systems of delay-differential equations that describe phenomena with very different temporal characteristics of transition processes. In this connection an important problem arises: the adaptation of implicit methods for ordinary differential equations possessing A-, $A(\alpha)$-, L-, or stiff stability properties, as applied to IVP DDE. Results of investigations in this direction are presented in [23, 32, 36, 37, 144, 145, 151, 237, 303, 342].

In order to construct effective numerical algorithms with the required accuracy for differential equations with different constant values of delays we used the Runge–Kutta–Fehlberg methods and two subclasses of linear multistep methods: Adams–Bashforth–Moulton methods (ABM) and the methods that use the Backward Differentiation Formulas (BDF) implemented in the predictor-corrector scheme: $P(EC)^M$, $M = 1, 2, 3$.

We set forth in this chapter the questions connected with the construction of two different algorithms for the numerical realization of mathematical models. Consider the initial value problem for a system of delay-differential equations (we restrict ourselves to one delay case for the sake of simplicity):

$$\begin{gathered}\frac{dy(t)}{dt} = f(t, y(t), y(t-\tau)), \quad t_0 \le t \le t_0 + T, \\ y(s) = \varphi(s), \quad s \in [t_0 - \tau, t_0].\end{gathered} \tag{6.1}$$

Here t is an independent variable; $y(t) \in \mathrm{R}^N$, $\tau = \text{const} > 0$. Let us make the following assumptions concerning problem (6.1):

1. $f(t, y, z) : [t_0, t_0 + T] \times \mathrm{R}^N \times \mathrm{R}^N \to \mathrm{R}^N$ is continuously differentiable l times with respect to all of its arguments, while $l \geq p$ where p is an order of approximation of difference formulas.

2. $f(t, y, z)$ satisfies the Lipschitz condition with respect to y and z with the constants L_1 and L_2 for all $t \in [t_0, t_0 + T]$, $y \in \mathrm{R}^N$, $z \in \mathrm{R}^N$.

3. $\varphi(s) \in C^{l^*}([t_0 - \tau, t_0))$ and the point t_0 is a discontinuity point of the first order for the function $\varphi(s)$. Assume, for simplicity, that $l^* \geq l$.

With the above assumptions, a solution of IVP DDE (6.1) exists, is unique on $[t_0, t_0+T]$, and is continuously differentiable $(l+1)$ times almost everywhere on $[t_0, t_0 + T]$ except for the set of points $\{\theta_i = t_0 + i\tau\}_{i=0}^{l+1}$ where the derivatives up to $(i-1)$th order inclusively are continuous in θ_i [88, 125]).

6.1. Numerical Algorithm for the Initial Value Problem for Delay-Differential Equations Based on the Runge–Kutta–Fehlberg Method

6.1.1. Difference Approximation. Let us construct a method of numerical integration of initial value problem for delay-differential equations of (6.1) type on the intervals of sufficient smoothness of solution $y(t)$, i.e., on the intervals $[t_0 + j\tau, t_0 + (j+1)\tau]$, $j = 0, 1, \ldots, p+1$ and for $t \geq t_0 + (p+2)\tau$ where the solution $y(t)$ is $(p+1)$ times continuously differentiable; the values of kth derivatives of $Y(t)$

$$y^{(k)}(t)|_{t=\theta_j^-}, \quad y^{(k)}(t)|_{t=\theta_j^+}, \quad \left(y^{(k)}(t) \equiv \frac{d^k y(t)}{dt^k}\right)$$

are understood here as the values of left and right limits of $y^{(k)}(t)$ respectively, which exist and are finite due to the assumptions made for the functions $f(t, y, z)$ and $\varphi(s)$.

Present problem (6.1) in the following equivalent form:

$$\begin{gathered} \frac{dy(t)}{dt} = f(t, y(t), z(t)), \quad y,\, z \in \mathrm{R}^N, \quad t_0 \leq t \leq t_0 + T, \\ z(t) = y(t - \tau), \quad y(s) = \varphi(s), \quad s \in [t_0 - \tau, t_0]. \end{gathered} \tag{6.1.1}$$

Problem (6.1.1) can be considered as an initial value problem for nonautonomous ODE in R^N, if $z(t)$ is known function of time. Approximate solution for (6.1.1) can be obtained on the interval $[t_0, t_0 + \tau]$ by some p-order ($p \leq l$) difference scheme for ordinary differential equations, since the function $z(t) = \varphi(t - \tau)$, $t \in [t_0, t_0 + \tau]$ is known. Then the numerical solution of the corresponding difference problem is the set of values $\{y_i\}_{i=0}^{N_1}$, $N_1 \in N^+$ where N^+ is a set of integer positive numbers, that approximate the solution of differential problem on the discrete set of grid points $\{t_i\}_{i=0}^{N_1}$.

In order to obtain an approximate solution of the initial value problem (6.1.1) for $t > t_0 + \tau$ by the same difference scheme for ODE it is necessary to approximate the function $z(t) = y(t - \tau)$ by available values of $\{y_i\}_{i=0}^{N_1}$, $\{f_i\}_{i=0}^{N_1}$, $N \geq N_1$ ($f \equiv dy/dt$), since the use of high-order difference schemes ($p > 3$) or a variable integration step h implies the calculation of the values $z(t) = y(t - \tau)$ at the moments of time $t = t_j + \mu h_j - \tau$, which do not coincide with grid points (here h_j is the size of j-th integration step, $\mu \in [0, 1]$ and depends on the choice of a difference scheme).

It is natural to use the interpolation methods for grid functions in the form of splines, Lagrange polynomials, Hermite polynomials, etc. [14, 151, 249, 254] to approximate delayed variables; then an order of the interpolation should correspond to the order of approximation of difference equation and be at least of the same order [223]. Thus, combining some method of numerical integration for ODEs with a method of grid function interpolation, we construct certain difference approximation for the initial value problem for delay differential equations.

Let us consider one-step Runge–Kutta–Fehlberg method of the order p for the initial value problem for a system of ordinary differential equations (IVP ODE):

$$\frac{dy}{dt} = g(t, y(t)), \quad y(t_0) = y^0,$$

$$y_{n+1} = y_0 + h_n \sum_{m=0}^{r} \gamma_m G_m, \quad r = r(p) \in N^+, \tag{6.1.2}$$

$$G_0 = g(t_n, y_n), \quad G_m = g\left(t_n + \alpha_m h_n, y_n + h_n \sum_{i=0}^{m-1} \beta_{ml} G_l\right),$$

$m = 1, \ldots, r-1$. Concrete values of α_m, γ_m, and β_{ml} are constant and can de found in Forsythe et al. [104]. The function $\psi_p(t_n, y_n, g(t,y)) = \sum_{m=0}^{r} \gamma_m G_m$ is called an increment function, since $y_{n+1} = y_n + h_n \psi_p(t_n, y_n, g(t,y))$, while the approximation order of numerical method (6.1.2) is equal to p.

In order to approximate $y(t)$ by available values $\{y_i\}_{i=0}^{n}$, $\{f_i\}_{i=0}^{n}$ on the grid $\{t_i\}_{i=0}^{n}$, $0 \le n \le N$ we use the Hermite interpolation polynomials π_q of an order q defined as follows: let $t \in [t_{k^*}, t_{k^*+\Delta_k}]$, $0 \le k^* \le n$, $0 \le k^* + \Delta k \le n$, $\Delta k = \max(\Delta k_1, \Delta k_2)$, $\Delta k_1, \Delta k_2 \in N^+$, $\Delta k_1 + \Delta k_2 = q+1$; then

$$y(t) = \pi_q(t) + \varepsilon(t), \quad \pi_q(t) = \sum_{j=k^*}^{k^*+\Delta k_1} y_j c_j(t) + \sum_{j=k^*}^{k^*+\Delta k_2} f_j \tilde{c}_j(t),$$

where

$$c_j(t) = [1 - 2(t - t_j)\dot{U}_j(t_j)]U_j^2(t), \quad j = 1, 2, \ldots, \Delta k_1,$$

$$\tilde{c}_j(t) = (t - t_j)U_j^2(t), \quad j = 1, 2, \ldots, \Delta k_2,$$

and $U_j(t)$ are the Lagrange multipliers [292]. Take the case of $\Delta k_1 = \Delta k_2 = \Delta k$ $(2\Delta k = q+1)$; then the expression for an error of the interpolation method has the form:

$$\varepsilon(t) = \prod_{j=k^*}^{k^*+\Delta k} (t - t_j)^2 \frac{y^{(2\Delta k)}(\xi)}{(2\Delta k)!}, \quad \xi \in [t_{k^*}, t_{k^*+\Delta k}],$$

i.e., $\varepsilon(t) = O(h^{2\Delta k}) = (h^{q+1})$ where $h = \max_{k^* \le n \le k^*+\Delta k} h_n$.

6.1.2. Analysis of convergence of the difference method. We have approximated the initial value problem for DDE (6.1.1) with the difference scheme

$$y_{n+1} = y_n + h_n \psi_p(t_n, y_n, f(t, y, z_n(t))), \quad n = 0, 1, 2, \ldots, N,$$

$$t_{n+1} = t_n + h_n, \tag{6.1.3}$$

$$z_n(t) = \begin{cases} \varphi(t - \tau), & t \in [t_0, t_0 + \tau], \\ \pi_{q,n}(t - \tau) \triangleq \pi_q(\tilde{t}), & t > t_0 + \tau, \\ t \in [t_0, t_{n+1}], & \tilde{t} = t - \tau. \end{cases}$$

which we will the call one-step method $(\psi_p(t_n, y_n, z_n)\pi_q)$, where ψ_p denotes the method (6.1.2). Introduce the notation $\psi(t_n, y_n, z_n) \equiv \psi_p(t_n, y_n, f(t, y, z_n))$. We shall consider the case of variable step of integration h_N. Assume, therefore, that there exists the function $\theta(t)$ such that $0 < \Delta \le \theta(t) \le 1$, $t \ge t_0$, $h_n = h\theta(t_n)$ where $h = \max\limits_{0 \le n \le N} h_n$, $\Delta = \text{const} > 0$. The following statement is valid on the convergence of the solution y_n of the difference problem to the solution $y(t_n)$ of the differential problem for $h \to 0$ for the difference scheme (6.1.3) and for the case of variable step of integration.

Statement 1. Let $p \ge 1$ be an order of the one-step Runge–Kutta–Fehlberg method ψ_p, and let q be an order of the Hermite interpolation polynomial π_q. Let the assumption on smoothness of $f(t, y, z)$ and $\varphi(s)$ be valid and let the Lipschitz condition for $f(t, y, z)$ in y and z be satisfied. Then the one-step method (ψ_p, π_q) (6.1.13) converges and the following estimate of convergence rate is valid:

$$\|y(t_n) - y_n\| = 0(h^{\min(p,q+1)}), \quad \forall t_n \in [t_0, t_0 + T], \quad 0 \le n \le N,$$

where $\|\cdot\|$ is any norm in R^N.

Corollary. The difference scheme (6.1.13), using the one-step Runge–Kutta–Fehlberg method of the 5th order and the Hermite interpolation polynomials of the 5th order, is of the 5th order of approximation.

6.1.3. Realization of the RKF45–DDE algorithm. The Runge-Kutta-Fehlberg methods ψ_p, designed for ordinary differential equations, provide an estimate of the method's local error $\|y(t_n) - y_n\|$ at each integration step. The adaptation of the ψ_p method, which is the RKF45 algorithm [104], for numerical solution of delay-differential equations implying the interpolation of delayed variables, has to preserve its essential characteristics: the possibility of estimating local error and to choose a step size. Existence of asymptotic expansion of global error of the method (ψ_p, π_q) with respect to powers of h is the necessary condition for that. The following statement is true.

Statement 2. Let exact solution $y(t)$ of initial value problem for delay-differential equations and the function $f(t, y(t), y(t-\tau))$ be sufficiently smooth. Then, if local approximation error of the method ψ_p

is of an order $p+1$, and an error of interpolation of delayed variables is of an order $q+1$, and $q+1 \geq p+1 \geq 2$, then

$$\|y(t_n) - y_n\| = h^p \|\delta(t_n)\| + O(h^{p+1}), \tag{6.1.4}$$

where $\delta(t)$ satisfies linear nonhomogeneous equation in R^N:

$$\frac{d\delta(t)}{dt} = (f_y(t, y, y_\tau) + f_{y_\tau}(t, y, y_\tau))\delta(t) + d(t),$$

$$t \geq t_0, \quad \delta(s) = 0, \quad t_0 - \tau \leq s \leq t_0.$$

Here $\delta(t)$ is the principal error function of the method ψ_p, and f_y, f_{y_τ} are the Frechét derivatives along the solution $y(t)$.

The conditions of Statement 2 do not hold in the general case for $t \geq t_0$, since there are the discontinuities of derivatives of the IVP DDE solution at the moments $\theta_i = t_0 + i\tau$, $i \in N^+$.

The above requirements of smoothness of $y(t)$, $y(t-\tau)$ at each step of integration $t_{j+1} = t_j + h_j$ lead to the necessity to organize the numerical process so that

$$\theta_i \notin (t_j, t_{j+1}), \quad \theta_i \notin (t_{k^*}, t_{k^*+\Delta k})$$

where $[t_{k^*}, t_{k^*+\Delta k}] \ni (t_j + \alpha_m h_j - \tau)$, $\{\alpha_m\}_{m=0}^r$ are coefficients from (6.1.2) and $t_{k^*}, t_{k^*+1}, \ldots, t_{k^*+\Delta k}$ are the grid points for the construction of the interpolation polynomial. Thus the discontinuity points $\{\theta_i\}_{i=1}^M$, $M \in N^+$ of solution's derivatives should belong to the set of grid points, and must not be internal points of an interval where the interpolation polynomial is constructed. Notice that the use of one-step Runge–Kutta–Fehlberg methods of an order p for solving DDEs on $[t_i, t_{i+1}]$ requires only the first $(p+1)$ derivatives of $y(t)$ to be continuous on $[t_i, t_{i+1}]$. Therefore a control over the coincidence of $\{\theta_i\}_{i=1}^M$ with the set of grid points $\{t_j\}_{j=0}^N$ is required only on the interval $[t_0, t_0 + (p+1)\tau]$, taking into account that a solution of IVP DDE is smoothed, and for $t > t_0 + (p+1)\tau$ all the derivatives $y^{(i)}$, $i = 1, 2, \ldots, p$ are continuous.

Consider in conclusion the main characteristics of the numerical algorithm RKF45–DDE for initial value problem for differential equations with various constant delays.

Integration method. The code RKF45 [104] was chosen as an integrator for initial value problem (6.1), designed for the solution with assigned accuracy of ordinary differential equations on the basis of the Runge–Kutta–Fehlberg methods of the 4th and 5th orders.

Interpolation method. Hermite interpolation polynomials of the 5th order were used to approximate delayed variables $y(t-\tau_m)$, $m = 1, 2, \ldots, k$ by calculated values $\{y_i\}_{i=0}^n$, $\{f_i\}_{i=0}^n$, $\{t_i\}_{i=0}^n$, $0 \leq n \leq N$. If t_j is a current moment of integration and $t_\tau = (t_j + h_j - \tau) \in [t_{i-1}, t_{i+1}]$ and, besides, $t_\tau \in [t_i, t_{i+2}]$, then of these two sets of grid points for the interpolation is chosen the set where t_τ is closer to the central grid point.

Discontinuity points of the first derivatives of solution. Since the Runge–Kutta–Fehlberg methods of the 4th and 5th orders are used, one must take into account the existence of discontinuity points just for the first seven derivatives of solution, i.e., the points $\{t_0 + i\tau\}_{i=1}^7$ must belong to the set of integration grid points. In the case of k different constant delays it is necessary that the points $\{t_0+i\tau_1\}_{i=1}^7, \ldots,$ $\{t_0+i\tau_k\}_{i=1}^7$ belong to the same set. Approaching the point $\theta_1 = t_0 + \tau$ from the left during the integration process, so that $t_{n+1} = t_n + h_n = \tau$, it is necessary to use the value of left-side limit $f(\tau, y(\tau), y(0^-))$ while calculating $f(t, y(t), y(t-\tau))$ which is connected with peculiarities of numerical integration of piecewise continuous functions.

Choice of integration step. The algorithm that chooses the integration step size remained unchanged.

It is necessary, in order to construct the Hermite interpolation polynomials of the 5th order, to have at least three grid points inside each of intervals $[t_0 + n, t_0 + (n+1)\tau)$, $n \in N^+$; for this reason the maximal integration step equals $h_{\max} = \dfrac{1}{3} \min\limits_{1 \leq i \leq k} (\tau_i)$.

The results of test calculations show that the algorithm RKF45–DDE is efficient while solving non-stiff systems of delay-differential equations under low requirements for accuracy (relative error $\leq 10^{-6}$).

6.2. Numerical Algorithm for Initial Value Problem for Delay-Differential Equations Based on Linear Multistep Methods

We discuss in this section the questions arising in the process of adaptation of the algorithm DIFSUB [114], solving initial value problem for stiff systems of ordinary differential equations, to the case of delay-differential equations.

6.2.1. Difference approximation of delay-differential equations on the basis of linear multistep methods. Let us construct a numerical method for problem (6.1) on the basis of the linear k-step method of pth order on the intervals of sufficient smoothness of solution $y(t)$, i.e., on the intervals $[t_0 + j\tau, t_0 + (j+1)\tau]$, $j = 0, 1, 2, \ldots, p+1$, and for $t \geq t_0 + (p+2)\tau$. It is evident that for $t \in [t_0, t_0 + (p+2)\tau]$ the maximal integration step h must not exceed τ, whereas h may for $t \geq t_0 + (p+2)\tau$ be of arbitrary size.

Let us use linear the k-step formula [112, 177] to obtain difference equation for y_{n+1} (for the sake of simplicity we consider just the scalar case $y \in \mathrm{R}^N$, $N = 1$):

$$\sum_{i=0}^{k} \alpha_{n,i} y_{n+1-i} = \sum_{i=0}^{k} \beta_{n,i} h_{k+1-i} f(t_{n+1-i}, y_{n+1-i}, y(t_{n+1-i} - \tau)) \qquad (6.2.1)$$

This requires us to approximate the values $z(t_{n+1-i}) \equiv y(t_{n+1-i} - \tau)$ for $t_{n+1-k} > t_0 + \tau$ by values of $\{y_i, f_i\}_{i=0}^{n}$ calculated before at the points t, $t = t_{n+1-i} - \tau$, $0 \leq i \leq k$ which in the general case may not coincide with the set of grid points $\{t_i\}_{i=0}^{n}$. One can use Lagrange or Hermite interpolation polynomials to approximate delayed variables; the order of interpolation polynomials should correspond to an order of approximation of difference formula for ordinary differential equations. Notice, that in case of $h_{n+1} \geq \tau$ an interpolation formula can use $\{y_{n+i}, f_{n+i}\}$, i.e., be implicit.

Consider the interpolation polynomial $\pi_q\left(Y_{\sigma_1,\sigma_{\nu_1}}, Y'_{\tilde{\sigma}_1,\tilde{\sigma}_{\nu_2}}, t\right)$ of qth order of approximation which uses the vectors $Y_{\sigma_1,\sigma_{\nu_1}}$ and $Y'_{\tilde{\sigma}_1,\tilde{\sigma}_{\nu_2}}$ of the following form:

$$Y_{\sigma_1,\sigma_{\nu_1}} = \left[y_{\sigma_1}, y_{\sigma_2}, \ldots, y_{\sigma_{\nu_1}}\right]^T, \quad Y'_{\tilde{\sigma}_1,\tilde{\sigma}_{\nu_2}} = \left[y'_{\tilde{\sigma}_1}, y'_{\tilde{\sigma}_2}, \ldots, y'_{\tilde{\sigma}_{\nu_2}}\right]^T, \quad (6.2.2)$$

where $\nu_1 + \nu_2 = q + 1$, $t \in \left[t_{\min(\sigma_1,\tilde{\sigma}_1)}, t_{\max(\sigma_{\nu_1},\tilde{\sigma}_{\nu_2})}\right]$, $y' \equiv dy/dt$.

Combining (6.2.1) and (6.2.2), we arrive at the following difference approximation of initial value problem for the system of delay-differential equations on the basis of linear k-step formula:

$$\sum_{i=0}^{k} \alpha_{n,i} y_{n+1-i} = \sum_{i=0}^{k} \beta_{n,i} h_{n+1-i} f(t_{n+1-i}, y_{n+1-i}, z_{n+1-i}),$$

$$z_{n+1-i} = \begin{cases} \pi_{q,n+1}\left(Y_{\sigma_1^{(i)},\sigma_{\nu_1}^{(i)}}, Y'_{\tilde{\sigma}_1^{(i)},\tilde{\sigma}_{\nu_2}^{(i)}}, t_{n+1-i}-\tau\right), t_{n+1-i} > t_0+\tau, \\ \varphi(t_{n+1-i}-\tau), \qquad t_{n+1-i} \le t_0+\tau. \end{cases} \tag{6.2.3}$$

Notice, that $\sigma_{\nu_1}^{(i)}$ and $\tilde{\sigma}_{\nu_2}^{(i)}$ in (6.2.3) satisfy the inequality .

$$\sigma_{\nu_1}^{(i)},\ \tilde{\sigma}_{\nu_2}^{(i)} \le n+1, \quad 0 \le i \le k$$

It is possible to represent difference equations approximating initial value problem for DDE (6.1) on the basis of predictor-corrector methods that use the Adams–Bashforth–Moulton formula (ABM) or backward differentiation formulas (BDF), and interpolation polynomials $\pi_q\left(Y_{\sigma_1,\sigma_{\nu_1}}, Y'_{\tilde{\sigma}_1,\tilde{\sigma}_{\nu_2}}, t\right)$ in a vector form ($\bar{y}_n$ is the vector of solution of difference equations; A_n is a matrix and l_n is a vector determined by linear k-step formulas) [112, 113, 177]:

1) ABM-$P(EC)^M$ method:

$$\bar{y}_{n+1}^{(0)} = A_n \bar{y}_n,$$

$$\bar{y}_{n+1}^{(m+1)} = \bar{y}_{n+1}^{(m)} + \bar{l}_n F_n(\bar{y}_{n+1}^{(m)}), z_{n+1}), \tag{6.2.4}$$

$$\bar{y}_n \in \mathrm{R}^{k+1}, \quad \bar{y}_n^{(m)} \in \mathrm{R}^{k+1}, \quad m = 0, 1, \ldots, M-1.$$

2) BDF-$P(EC)^M$ method:

$$\bar{y}_{n+1}^{(0)} = A_n \bar{y}_n,$$

$$\bar{y}_{n+1}^{(m+1)} = \bar{y}_{n+1}^{(m)} + \bar{l}_n \left[-\frac{\partial F_n(\bar{y}_{n+1}^{(m)}, z_{n+1})}{\partial \bar{y}} \bar{l}_n \right]^{-1} F_n(\bar{y}_{n+1}^{(m)}, z_{n+1}). \tag{6.2.5}$$

$$\bar{y}_n \in \mathrm{R}^{k+1}, \quad \bar{y}_n^{(m)} \in \mathrm{R}^{k+1}, \quad m = 0, 1, \ldots, M-1.$$

Notice that the function $F_n(\cdot,\cdot)$ in (6.2.4), (6.2.5) depends on delayed variable denoted by $z_{n+1} \equiv y(t_{n+1} - \tau)$. In addition,

$$F_n(\bar{y}^{(m)}_{n+1}, z_{n+1}) \equiv h_{n+1}[f(t_{n+1}, y^{(m)}_{n+1}, z_{n+1}) - y'^{(m)}_{n+1}],$$

$$y'^{(m)}_{n+1} = f(t_{n+1}, y^{(m-1)}_{n+1}, z_{n+1}), \quad m = 1, 2, \ldots, M-1,$$

$$z_{n+1-i} = \begin{cases} \pi_{q,n+1}\left(Y_{\sigma_1,\sigma_{\nu_1}}, Y'_{\tilde{\sigma}_1,\tilde{\sigma}_{\nu_2}}, t_{n+1} - \tau\right), & t_{n+1} > t_0 + \tau, \\ \varphi(t_{n+1} - \tau), & t_{n+1} \le t_0 + \tau, \end{cases} \tag{6.2.6}$$

$$\sigma_{\nu_1},\ \tilde{\sigma}_{\nu_2} \le n+1, \quad 0 \le n \le N-1.$$

If, while constructing the interpolation polynomial $\pi_{q,n+1}(\cdot,\cdot,t)$, unknown values $[y_{n+1}, y'_{n+1}]$ are used in the grid point t_{n+1} (for example, for $\tau < h_{n+1}$), then $z_{n+1} = z_{n+1}(y^{(m)}_{n+1})$ and, therefore, z_{n+1} are changed during the iterative process of correction. In this case the Jacobian $\dfrac{\partial F_n(\bar{y}^{(n)}_{n+1}, z_{n+1}(y^{(m)}_{n+1})}{\partial \bar{y}}$ changes too.

The transition to the representation of the solution vector $\bar{y}_n = \mathrm{R}^{k+1}$ of the difference problem in the Nordsieck form does not affect the form of equations (6.2.5), (6.2.6), but provides a convenient approximation of delayed variables. In fact the vector $\bar{y}_n$ which may be written [112, 177] as

$$\bar{y}_n = \left[y_n, h_n y'_n, \ldots, \frac{h^p_n y^{(p)}_n}{p!} \right]^T \in \mathrm{R}^{p+1} \tag{6.2.7}$$

determines in a neighborhood of the point t_n a polynomial of pth degree that takes the values $\{y_{n-i}\}^k_{i=0}$ and/or $\{y'_{n-i}\}^k_{i=0}$ at the grid points $\{t_{n-i}\}^k_{i=0}$. This polynomial coincides locally, i.e., on $[t_{n-k}, t_n]$, with the polynomial approximating a solution of differential problem by linear k-step difference formula of pth order. Therefore, while it is applied for the approximation of the delayed variables on $[t_{n-k}, t_n]$ it provides the method of interpolation consistent with employed linear k-step formula.

Let, for instance, $(t-\tau) \in [t_{n-1}, t_n]$. Then the polynomial

$$\pi_{q,n}\left(Y_{\sigma_1,\sigma_{\nu_1}}, Y'_{\tilde{\sigma}_1,\tilde{\sigma}_{\nu_2}}, t-\tau\right) \equiv \pi_{q,n}(\bar{y}_n, t-\tau) = C\left(\frac{t-\tau-t_n}{h_n}\right)\bar{y}_n, \tag{6.2.8}$$

where $C(\alpha) = [1, \alpha, \dots, \alpha^p]$, and $\bar{y}_n = \bar{y}_n(t_n)$ approximates $y(t - \tau)$ with accuracy $O(h^{p+1})$, $h = (t - \tau - t_n) = (h_n - \mu)$, $0 \le \mu \le h_n$. Notice that though the polynomial

$$\pi_{q,n}(\bar{y}_{n-1}, t - \tau) = C\left(\frac{t - \tau - t_{n-1}}{h_{n-1}}\right)\bar{y}_{n-1}$$

has the same order of approximation p as (6.2.8) does, however, theoretically its use leads to larger error, since the approximation is of an extrapolation character.

Taking into account that $\bar{y}_n$ is the Nordsieck vector [112, 115, 314], equations (6.2.5), (6.2.6) can be presented as a recursion relation:

$$\begin{gathered}
\bar{y}_{n+1} = SC_{n+1}\bar{y}_n + h_{n+1}\bar{\psi}(t_n, C_{n+1}\bar{y}_n, z_{n+1}, h_{n+1}), \\
C_{n+1} \equiv C(h_{n+1}/h_n), \\
t_{n+1} = t_n + h_{n+1}, \quad h_{n+1} = \theta(t_{n+1}, h)h, \\
0 < \Delta \le \theta < 1, \quad \forall h > 0, \quad t_0 \le t \le t_0 + T, \qquad (6.2.9) \\
z_{n+1} = \begin{cases} \pi_{q,n+1}(Y_{\sigma_1,\sigma_\nu}, t_{n+1} - \tau), & t_{n+1} > t_0 + \tau, \quad \sigma_\nu \le n+1, \\ \varphi(t_{n+1} - \tau), & t_{n+1} \le t_0 + \tau, \end{cases} \\
Y_{\sigma_1,\sigma} = \left[\bar{y}_{\sigma_1}^T, \bar{y}_{\sigma_2}^T, \dots, \bar{y}_{\sigma_\nu}^T\right]^T, \quad t_{n+1} - \tau \in [t_{\sigma_1}, t_{\sigma_\nu}].
\end{gathered}$$

The propagation matrix (step operator) S and the connection of increment function ψ with f are the same as in the case of ordinary differential equations, but now ψ depends on the delayed variable $\psi(t, \bar{y}, y(t - \tau), h)$. In (6.2.9) $\theta(t, h)$ is a function describing changes of integration step or the irregularity of a grid.

We have constructed the system of difference equations (6.2.9) for the initial value problem (6.2.1); we call this system the linear k-step method for delay-differential equations which uses the representation of the solution vector of difference equations in the Nordsieck form. Denote it by (ψ_p^S, π_q), where ψ_p^S means the linear k-step method of pth order for ordinary differential equations, and π_q stands for an interpolation polynomial of qth order approximating the delayed variables.

The condition $p = k$ is satisfied for the methods ψ_p^S (in the case of methods ABM and BDF).

It is convenient, in order to analyse the stability, approximation, and convergence of the (ψ_p^S, π_q) method, to write down (6.2.9) in the form of "one-step" recursion:

$$\bar{y}_{n+1} = SC_{n+1}\bar{y}_n + h_{n+1}\bar{\psi}(t_n, C_{n+1}\bar{y}_n, \pi_{q,n+1}(Y_{\sigma_1,\sigma_\nu,t_{n+1}-\tau}), h_{n+1}). \quad (6.2.10)$$

6.2.2. Analysis of convergence for difference method. The method (ψ_p^S, π_q) was constructed on the basis of the convergent method [115, 117] for ordinary differential equations and interpolation method defined by the polynomial $\pi_{q,n}(Y_{\sigma_1,\sigma_\nu}, \mu)$, $\sigma_\nu \le n+1$, $\mu \in [t_{\sigma_1}, t_{\sigma_\nu}]$ and can be written as

$$\bar{y}_{n+1} = SC_{n+1}\bar{y}_n + h_{n+1}\bar{l}\psi(t_n, C_{n+1}\bar{y}_n, \pi_{q,n+1}(Y_{\sigma_1,\sigma_\nu,t_{n+1}-\tau}), h_{n+1}). \quad (6.2.11)$$

where $\bar{l} \in \mathrm{R}^{p+1}$ is the same vector of coefficients as in (6.2.4), (6.2.5). The investigation of convergence for this method is connected with the analysis of the behavior of global error $\bar{\varepsilon} = \bar{y}_n - \bar{y}(t_n)$ for $h \to 0$. Here $\bar{y}(t_n)$ is a vector-function corresponding to the vector-function of solution $\bar{y}_n$ of difference equations (6.2.11), but constructed on the basis of the exact solution of the initial value problem (6.1). Let us formulate a number of assumptions.

1. Assume, that the method ψ_p^S, which is characterized by the function $\theta(t, h)$ of step change, has an order of approximation $p \ge 1$ and converges with an order p, i.e.,

$$\begin{aligned} \bar{d}_n &= SC_{n+1}\bar{y}(t_n) + h_{n+1}\bar{l}\psi(t_n, C_{n+1}\bar{y}(t_n), h_{n+1}) - \bar{y}(t_{n+1}) \\ &= O(h_{n+1}^{p+1}) = O(h^{p+1}), \end{aligned}$$

$$\|\bar{y}(t_n) - \bar{y}_n\|_{\mathrm{R}^{p+1}} = O(h^p), \quad 0 \le n \le N.$$

By virtue of the assumption on Lipschitz-continuity of the function $f(t, y, z)$ with respect to y and z, we have that for sufficiently small h the increment functions $\psi(t, \bar{y}, z, h)$, corresponding to the schemes for the realization of linear multistep methods in the DIFSUB algorithm, also satisfy the Lipschitz condition with respect to $\bar{y}$ and z with constants L_y, L_z, i.e.,

$$|\psi(t,\bar{y},z,h_n) - \psi(t,\bar{y}^*,z^*,h_n)| \leq L_y\|\bar{y}-\bar{y}^*\|_{\mathrm{R}^{p+1}} + L_z|z-z^*|$$

$$\text{for} \quad \forall \bar{y},\bar{y}^* \in \mathrm{R}^{p+1}, \quad \forall z,\bar{z}^* \in \mathrm{R}, \quad \forall t \in [t_0, t_0+T],$$

$$\forall h_n \in (0,h], \quad 0 \leq n \leq N.$$

2. Let interpolation method defined by the polynomial $\pi_{q,n}(Y_{\sigma_1,\sigma_\nu},\mu)$, have an order $q \geq 0$ and satisfy the Lipschitz condition in $[t_0, t_0+T]$ for the step change function $\theta(t,h)$

$$|\pi_{q,n}(Y_{\sigma_1,\sigma_\nu},t) - \pi_{q,n},(\bar{X}_{\sigma_1,\sigma_\nu},t)| \leq L \max_{0\leq\sigma_1\leq j\leq\sigma_\nu\leq n} \|\bar{y}_j - \bar{x}_j\|_{\mathrm{R}^{p+1}}. \quad (6.2.12)$$

where $t \in [t_{\sigma_1}, t_{\sigma_\nu}]$, $0 \leq n \leq N$. In this case an error e_{int} of interpolation of the function $y(t)$ by the polynomial $\pi_q(\cdot,\cdot)$ is such that

$$e_{\text{int}} = \max_{0\leq\sigma_1\leq\sigma_\nu\leq N}\left(\max_{t\in[t_0,t_0+T]}|\pi_q(\Delta_{-\nu}^{\sigma_\nu}\bar{y},t) - y(t)|\right) = O(h^{q+1}),$$

$$t \in [t_{\sigma_1}, t_{\sigma_\nu}].$$

where $\Delta_{-\nu}^{\sigma_\nu}$ is a digitization operator: $\Delta_{-\nu}^{\sigma_\nu}\bar{y}(t) \equiv \left[\bar{y}^T(t_{\sigma_1}),\ldots,\bar{y}^T(t_{\sigma_\nu})\right]^T$ and $\Delta_{-\nu}^{\sigma_\nu}\bar{y}(t)$ corresponds to Y_{σ_1,σ_ν}, but on the exact solution $y(t)$ of IVP DDE: hence, the following statements hold [36, 37].

Statement 1. If the assumptions 1–2 are true, the method (ψ_p^S, π_q) is convergent, i.e., $\|\bar{\varepsilon}_n\| \to 0$ for $h \to 0$, $\forall t_n \in [t_0, t_0+T]$.

Statement 2. If $p \geq 1$ is an approximation order of the method ψ_p^S and $q \geq 0$ is an order of interpolation polynomial approximating the delayed variables, then for $h \to 0$

$$\|\bar{\epsilon}_n\| = O(h^{\min(p,q+1)}) \quad \forall t_n \in [t_0, t_0+T].$$

Thus, $q = p-1$ is sufficient to preserve the order p for the convergence of the method (ψ_p^S, π_q).

Asymptotic expansion of global error of the method (ψ_p^S, π_q) in powers of integration step. The algorithm DIFSUB, as was noted above, is based on the methods of ABMs and BDFs of variable step and variable order, which allows one to obtain a solution with assigned accuracy (an estimate of error is meant) using maximal possible step of integration. The adaptation of the DIFSUB algorithm for delay-differential equations, connected with the introduction of interpolation of delayed variables, may lead, generally speaking, to the necessity of modifying the procedure of error control choice the step-size value, since the global error $\varepsilon_n = y(t_n) - y_n$, $0 \leq n \leq N$ is determined both by the error of linear k-step method for ordinary differential equations and by the error of interpolation of delayed variables. Let us define sufficient conditions for the order of interpolation polynomial approximating the delayed variables such that a leading term in the expansion of global error of the method (ψ_p^S, π_q) in powers of h is defined asymptotically by the principal error function of the linear k-step method for ordinary differential equations. In this case the procedure of error control and choice of integration step (the weakest spot while modifying the algorithm) can be left unchanged, and this conclusion is correct asymptotically, i.e., for $h \to 0$.

It is known from the theory of numerical methods that there exists an asymptotic expansion in series in powers of the step size of integration h for global error of solution of ordinary differential equation obtained with linear multistep methods [112, 113, 177, 320]. Consider the difference methods based on strongly stable linear k-step formulas realized by the scheme $P(EC)^M$ or $P(EC)^M E$; here an order of predictor formula $p^* \geq p$, where p is an order of corrector formula. It is shown in [58, 320], that in the case when an error of initial conditions is of order $(p+1)$, i.e., $O(h^{p+1})$, then global error of the method, that uses the step-size variation scheme $\theta(t,h) \in C^2[t_0, t_0+T]$, permits the expansion

$$\varepsilon_n = y(t_n) - y_n = h^p e(t_n) + O(h^{p+1}), \quad 0 \leq n \leq N, \tag{6.2.13}$$

where the function $e(t)$ is a solution of differential problem

$$\frac{de(t)}{dt} = \frac{\partial f(t, y(t))}{\partial y} e(t) + C_{p+1}[\theta(t,h)]^p y^{(p+1)}(t), \quad e(t_0) = 0. \tag{6.2.14}$$

In addition, local approximation error value d_n in the point t_n has the form

$$\begin{aligned} d_n &= h^{p+1}[\theta(t,h)]^{p+1}C_{p+1}y^{(p+1)}(t_n)+O(h^{p+2}) \\ &\equiv h^{p+1}\Phi_{p+1}(t_n)+O(h^{p+2}). \end{aligned} \tag{6.2.15}$$

A conventional way of controlling the global error ε_n in algorithms consists of estimation of the leading term in the asymptotic expansion (6.2.15) for d_n, which is used for choosing the values of step and order. This approach is based on the connection between d_n and ε_n assigned by relations (6.2.13)–(6.2.15).

The transition to multi-valued methods, which use, for example, the equivalent representation of solution $\bar{y}_n$ in the Nordsieck form, can change the character of the connection between $\bar{\varepsilon}_n$ and $\bar{d}_n$: an order of convergence may be by a unit higher than an order of approximation. This phenomenon is connected both with spectral properties of propagation matrix S and with location of vectors $\bar{\Phi}_p(t)$ and $\bar{\Phi}_{p+1}(t)$ in R^{k+1}, where $\bar{\Phi}_p$ and $\bar{\Phi}_{p+1}$ are respectively the first (leading) and the second terms in asymptotic expansion of $\bar{d}_n$ in powers of h: $d_n = \bar{\Phi}_p(t_n)h^p + \bar{\Phi}_{p+1}(t_n)h^{p+1} + O(h^{p+2})$. This phenomenon was studied in [5, 314–319]. We assume hereafter that the method ψ_p^S satisfies a strong root condition, i.e., $\lambda = 1$ is a simple eigenvalue of the matrix S, and all the rest of eigenvalues λ are inside the unit circle, and consider the case when

$$\bar{d}_n = h^{p+1}\bar{\Phi}_{p+1}(t_n) + O(h^{p+2}) \tag{6.2.16}$$

(notice, that these assumptions are correct in case of methods realized in DIFSUB).

The asymptotic behavior of global error $\bar{\varepsilon}_n$ ($h = \text{const}$) for the methods using the Nordsieck representation was studied in [314]. It was shown, in particular, in this work, that if initial conditions are exact and there exists a continuous derivative

$$\frac{\partial\bar{\psi}(t,\bar{u},h)}{\partial\bar{u}}, \quad t \in [t_0, t_0+T], \quad \bar{u} \in \mathrm{R}^{k+1}, \quad h \in (0,\tilde{h}),$$

and $\bar{d}_n$ has the form (6.2.16), then the following expansion is true

$$\bar{y}_n - \bar{y}(t_n) \equiv \bar{\varepsilon}_n = h^p\bar{e}(t_n) + O(h^{p+1}), \quad O \le n \le N, \tag{6.2.17}$$

where $\bar{e}(t)$ is a solution of the initial value problem:

$$\frac{d\bar{e}(t)}{dt} = \Lambda f_y(t, y(t)) M^T \bar{e}(t) + E\bar{\Phi}_{p+1}(t), \quad \bar{e}(0) = 0.$$

Let us explain the meaning of Λ, M^T, and E. It is possible, using the spectral expansion of matrices, to represent S as $S = E + T$, where E is an accompanying matrix corresponding to the eigenvalue $\lambda = 1$. Here $E = \Lambda M^T$, $S\Lambda = \Lambda$, where Λ and M^T are the right and the left eigenvectors of S corresponding to $\lambda = 1$, and $\bar{\psi}(t, \Lambda y, 0) = \Lambda f(t, y)$ [314–316].

Let us analyse the value of the global error for the method (ψ_p^S, π_q) for $t = t_n$: $\bar{\varepsilon}_n = \bar{y}_n - \bar{y}(t_n)$, $0 \le n \le N$. Part of the global error $\bar{\varepsilon}_n$ which is due just to the method ψ_p^S is equal to $h^p \bar{e}(t)$, where $\bar{e}(t)$ is a solution of the initial value problem similar to (6.2.17), except for f which depends both on $y(t)$ and $y(t - \tau)$. It is convenient, in order to obtain an equation for $\bar{\varepsilon}_n$, to consider the function $f(t, y(t), y(t - \tau))$ as a superposition of the operator

$$\begin{aligned} F(y, y_\tau) \;&\equiv\; f(t, y(t), y(t - \tau)) : C[t_0, t_0 + T] C[t_0 - \tau, t_0 + T - \tau] \\ &\to C[t_0, t_0 + T] \end{aligned}$$

and the shift operator

$$y_\tau \equiv D_\tau y(t) \equiv y(t - \tau) : C[t_0, t_0 + T] \to C[t_0 - \tau, t_0 + T - \tau].$$

Taking this into account, the derivative of $f(t, y(t), y(t - \tau))$ with respect to $y(t)$ in (6.2.18) is the Frechét derivative of the operator $F(y, y_\tau)$ in the point y and has the form

$$F'(y, y_\tau) = \frac{\partial f(t, y, y_\tau)}{\partial y} + \frac{\partial f(t, y, y_\tau)}{\partial y_\tau} D_\tau.$$

Therefore the behavior of $\bar{\varepsilon}(t)$ is described by the initial value problem for DDE

$$\frac{d\bar{e}(t)}{dt} = \Lambda \left\{ \frac{\partial f}{\partial y} M^T \bar{e}(t) + \frac{\partial f}{\partial y} M^T \bar{e}(t - \tau) \right\} + E\bar{\Phi}_{p+1}(t),$$

$$\bar{e}(\theta) = 0, \quad \theta \in [t_0 - \tau, t_0].$$

Using these results one can prove the following statement [37].

Statement 3. Assume that:

1) the method ψ_p^S, satisfying the strict root condition, converges with an order $p \geq 1$;

2) the interpolation method, defined by the polynomial $\pi_q(Y_{\sigma_1,\sigma_\nu}, t)$ of an order $(q \geq 1)$, satisfies the condition 2 of item 6.2.2, i.e., $e_{\text{int}} = O(h^{q+1})$,

$$|\pi_{q,n},(Y_{\sigma_1,\sigma_\nu}, t) - |\pi_{q,n},(X_{\sigma_1,\sigma_\nu}, t)| \leq L_\pi \max_{0 \leq \sigma_1 \leq j \leq \sigma_\nu \leq n} \|\bar{y}_j - \bar{x}_j\|_{\mathrm{R}^{p+1}},$$

$$t \in [t_{\sigma_1}, t_{\sigma_\nu}] \subset [t_0, t_0 + T], \quad 0 \leq n \leq N;$$

3) the function $f(t, y, z)$ satisfies the smoothness and Lipschitz continuity conditions 1–2 formulated in Section 6.1; in addition, $f(t, y, z)$ is twice continuously differentiable in y and z;

4) we consider the methods ψ_p^S in the form (6.2.4), (6.2.5) and, therefore, the increment function $\bar{\psi}(t, y, z, h)$ possesses the same properties of smoothness and Lipschitz-continuity with respect to $\bar{y} \in \mathrm{R}^{p+1}$, $z \in \mathrm{R}$ as $f(t, y, z)$ does uniformly with respect to $t \in [t_0, t_0 + T]$, $h \in (0, \tilde{h})$. If $q \geq p \geq 1$ then the global error of the method (ψ_p^S, π_q) for $h \to 0$ has the form

$$\bar{y}_n - \bar{y}(t_n) = h^p \bar{e}(t_n) + \bar{w}_n,$$

where $\bar{w}_n = \bar{O}(h^{\min(p+1,q+1)}) = \bar{O}(h^{p+1})$, $0 \leq n \leq N$, i.e., the leading term in asymptotic expansion of global error of the method ψ_p^S for ODE exists and is determined by properties of the method (ψ_p^S, π_q), i.e., by concrete difference formulas underlying the method.

The investigation of convergence conditions for the method (ψ_p^S, π_q) on various nonuniform grids which are used in algorithms for ordinary differential equations based on the methods ψ_p^S (ABM and BDF), made it possible to obtain the following sufficient condition (for the order of interpolation polynomial) of convergence for the method (ψ_p^S, π_q) with pth order: $p: \ q \geq p - 1$. The analysis of asymptotic expansion of global error of the method (ψ_p^S, π_q) in powers of the integration step have shown that the preservation of the form of the first term in this expansion independent of the interpolation polynomial for delayed variables requires stronger condition for the value of q: $q \geq p$.

6.2.3. Realization of the algorithm DIFSUB–DDE [33, 34]

Integration method. The algorithm DIFSUB-DDE contains two subclasses of linear multistep methods: the Adams–Bashforth–Moulton methods (ABMs) of variable step and variable order p ($p = 1, 2, \ldots, 7$) and the Gear methods (BDFs) of variable step and variable order p ($p = 1, 2, \ldots, 6$), which are realized in the predictor-corrector mode $P(EC)^M$, $M \leq 3$. In order to solve an implicit equation of the form

$$y_{n+1} = h_{n+1}\beta_0 f(t_{n+1}, y_{n+1}) + \psi_n$$

in the case of ABM methods, simple iterations are used, and in the case of BDF methods an iteration process based on the Newton-Rafson method is used.

Approximation of delayed variables. To approximate the delayed variables the components of Nordsieck vector $\bar{y}_n$ are used that determines the pth order polynomial of in the neighborhood of the point t_n so that

$$y(t-\tau) = C(\alpha)\bar{y}_n + O((\alpha h_n)^{p+1}) + C(\alpha)\bar{\varepsilon}_n, \quad y \in \mathrm{R},$$

where

$$\bar{y}_n = \left[y_n, h_n, y_n', \ldots, \frac{h_n^{p_n} y_n^{(p_n)}}{p_n!} \right]^T, \quad C(\alpha) = [1, \alpha, \alpha^2, \cdots, \alpha^{p_n}],$$

$$(t-\tau) \in (t_{n-1}, t_n], \quad \alpha = \frac{(t-\tau-t_n)}{t_n - t_{n-1}}, \quad |\alpha| < 1,$$

and the quantity $C(\alpha)\bar{\varepsilon}_n$ describes an influence of global error of approximation of the vector-function $\bar{y}(t_n)$ with the help of $\bar{y}_n$: $\varepsilon \equiv \bar{y}(t_n) - \bar{y}_n$. Nordsieck polynomial $\pi_{p_n}(\bar{y}_n, t-\tau) = C(\alpha)\bar{y}(t_n)$ has the order p equal to the order of linear multistep method $\psi_{p_n}^S$, employed to obtain the vector $\bar{y}_n$ in the point t_n, and, therefore, its use is correct.

Discontinuity points of derivatives of solution of differential problem. In general case the interval $[t_0, t_0 + T]$ where a solution $y(t)$ of initial value problem is to be found with p-order method, can be divided into two intervals: the interval $I_1 = [t_0, t_0 + (p+1)\tau]$ of insufficient smoothness of solution and the interval $I_2 = [t_0 + (p+1)\tau, T]$ where the first $(p+1)$ derivatives of solution $y(t)$ are continuous. The solution $y(t)$ must satisfy the requirements of smoothness on each step $t_{n+1} = t_n + h_{n+1}$. Therefore, it is necessary to organize a numerical process on I_1 in such a way that the set of discontinuity points $\{t_0 + l\tau\}_{l=0}^{p+1}$ of solution derivatives must belong to the set of grid points $\{t_j\}_{j=0}^{n}$. In case of m different delays $\{\tau_i\}_{i=1}^{m}$ the points $\{\{t_0 + l\tau_i\}_{l=0}^{p+1}\}_{i=1}^{m}$ must belong to the set of grid points.

Algorithm DIFSUB-DDE uses the methods of variable order, where maximal permissible order $p_{\max}$ is assigned by user. Taking this into account, in the case of different delays, I_1 and I_2 have the forms:

$$I_1 = \left[t_0, t_0 + (p_{\max} + M_0) \max_{1 \le i \le m} \{\tau_i\}\right],$$

$$I_2 = \left[t_0 + (p_{\max} + M_0) \max_{1 \le i \le m} \{\tau_i\}, t_0 + T\right],$$

where $M_0 = 2$. Maximal value of the step $h_{\max}$ is limited in the first interval of integration of initial value problem:

$$h_{\max} \le \max_j \{\xi_j - \xi_{j-1}\}, \quad 1 \le j \le p_{\max} m,$$

where ξ_j and ξ_{j-1} are the neighboring discontinuity points of derivatives $y^{(i)}(t)$, $i \le p_{\max} + M_0$. The maximal value of the step $h_{\max}$ is not limited on the interval I_2. Since only the first $(j-1)$ derivatives of solution $y(t)$ are continuous in the points $\theta_j^{(i)} = j\tau_i$, $j = 1, 2, \ldots$, $i = 1, 2, \ldots, m$ when $t_n = \theta_j^{(i)}$ is reached, the adjustment of an order of linear multistep method and of an order of solution smoothness is performed.

Features of the algorithm DIFSUB–DDE. The DIFSUB–DDE algorithm uses, as DIFSUB does, two subclasses of linear multistep methods: the Adams–Bashforth–Moulton methods of variable step and variable order p ($p = 1, 2, \ldots, 7$) and Gear methods of variable step and variable order p ($p = 1, 2, \ldots, 6$) which are realized with the scheme $P(EC)^M$ ($M = 1, 2, 3$). The Jacobian $\partial f / \partial y$ which is employed for the solution of a system of nonlinear equations generated by BDF methods, is calculated with a help of finite-difference approximation.

Interpolation corrector polynomials formed on the basis of Nordsieck vector are used for the approximation of delayed variables $y(t - \tau)$. While solving an initial value problem, discontinuity points of the first eight derivatives of solution $y(t)$ are matched automatically in the process of solution with the grid points of integration.

The coordination between the order of interpolation polynomial approximating the delayed variables and the order of approximation of a linear multistep method used on current size of integration step permits to use the scheme of control of solution accuracy which is realized in DIFSUB and based on the adjustment of assigned value of error ε to an estimate of local error value, which is contributed by ABM or BDF methods. As a result the algorithm DIFSUB–DDE provides a numerical solution of initial value problem with prescribed relative accuracy using maximal possible step of integration; more than that, its value in case of sufficient smoothness of solution $y(t)$ of differential problem is not limited by the values of delays.

The analysis of values of coefficients and characteristic turn-over times of processes described by the system of differential equations of the immune response models allows one to infer, that these systems are stiff, i.e., describe the processes with essentially different times of dumping. The BDF methods of DIFSUB–DDE algorithm are efficient tools for solving the initial value problem for a system of differential equations describing the processes with rapidly and slowly decaying components. These methods, being adapted to the solution of delay-differential equations, preserve qualitatively the properties of unboundedness of region of absolute stability for test equations similar to the properties of A_0-, A-, and $A(\alpha)$-stability in case of ordinary differential equations [32, 303, 342].

Notice, in conclusion, that the BDF method of the DIFSUB–DDE algorithm of variable step and variable order was the basic method of repeated numerical solution of initial value problem for the identification of the model's coefficients. The relative accuracy of the numerical solution was kept to the prescribed level ε. The reliability of the BDF method was checked periodically with ABM methods of DIFSUB–DDE algorithm and with RKF45–DDE method. It should be noted that ABM/DIFSUB–DDE and RKF45–DDE methods proved to be less effective by 1–2 orders, than BDF/DIFSUB–DDE methods, with respect to the number of integration steps required for the solution of initial value problem in the region of coefficients values, we are interested in and for $\varepsilon = 10^{-8}$.

The average computer time required for the solution of initial value problem for the model of antiviral immune response was about 1 minute for a computer with 100,000 flops performance and 8-byte word length.

PART II

Models of Viral and Bacterial Infections

Viral Hepatitis B

Viral and Bacterial Infections of Respiratory Organs

Model of Influenza Infection

Adjoint Equations and Sensitivity Study for Mathematical Models of Infectious Diseases

CHAPTER 7

Viral Hepatitis B

This chapter deals with mathematical modeling of viral hepatitis B. Viral hepatitis B is a world-spread infectious disease which is accompanied usually by primary damage to liver and metabolism disfuncion [33, 34]).

The inducer (aetiological agent) of this disease, the hepatitis B virus HBV, can be transferred by the transfusion of blood, plasma or serum from a donor (virus carrier). The hepatitis B virus infects the liver cells (hepatocytes) and reproduces there. It is an important feature of HBV that this virus does not destroy an infected cell, moreover, the hepatitis B virus is characterized by a low rate of reproduction.

Four major clinical variants of disease outcome after hepatitis B virus has penetrated into an organism can be isolated: the acute form of disease, which ends with complete recovery and formation of immunity; the subclinical form of disease (with minimal or no clinical and biochemical indicators of disease); the chronic form of disease when the virus reproduces during a long period (up to dozens of years) with clinical and biochemical indicators of disease; the acute form of disease with serious destruction of the liver which leads to the death of the organism (malignant form). Chronic and hypertoxic (malignant) forms of viral hepatitis B are most dangerous to human life [33, 34, 240].

We consider mathematical modeling of the acute course of viral hepatitis B as a starting point in modeling the mechanisms of development of chronic and hypertoxic forms of viral hepatitis B and their treatment.

To this end we use the mathematical model of antiviral immune response constructed in Chapter 4. We will analyse the processes described by the model as applied to the processes that occur in an organism during the development of viral hepatitis B.

It was assumed, while constructing the mathematical model of antiviral immune response, that the immune reaction to the reproduction of virus inside the target cells is a decisive factor in the clinical course of viral disease. This reaction leads to the destruction of infected cells by effector cells of the immune system and to neutralization of free viral particles by antibodies. When describing the reaction of the immune system, we use two main mechanisms of control over proliferation and differentiation of immune system cells: the principle of MHC restricted antigen recognition by cells [351, 352] and the principle of clonal selection [49]. The model takes into account the mechanism of nonspecific immunodepression caused by damage to the cells of target organ and, therefore, by general decrease in functional reserves of an organism. Recently, on the basis of mathematical models, questions were investigated connected with the damage and regeneration of a target organ [16], and with the reproduction of a virus inside target cells [209].

We have not considered the role of specific mechanisms of immunosuppression, since their contribution to the development of acute viral hepatitis infection remains unclear. Distribution and dynamical resettling of antibodies population with regard to affinity is not considered in the model either. This simplification is based on the fact that during acute viral disease the phase of mass production and neutralization of a virus and its antibodies is of short duration, and, therefore, there is no sense to account for these phenomena in the model. The generation and neutralization of only one viral antigen HBsAg was considered while constructing the model's equations. This is a convenient and natural simplification, in our opinion, since while solving the problem of modelling a disease it makes sense to consider those antigens of the whole set of antigens connected with a given virus that are the most immunogenic and mass in quantity and to whom the immune response neutralizes the viruses and possesses the defence effect. In this case, naturally, the model takes into account that B- and T-cellular response is induced by different antigenic determinants. The description of the process of virus multiplication in a cell and processes of destruction and regeneration of a target organ is rather of principal character and must be refined if required.

7.1. Parameter Identification for the Model of Antiviral Immune Response

7.1.1. Physical meaning of model variables. It is necessary, while constructing the mathematical model of viral hepatitis B, to make the physical meaning of the variables in mathematical model of antiviral immune response more concrete, taking into account the localization of various processes in different tissues and organs. It is known that the processes of immune response occur mainly in the lymph node system surrounding a target organ; the processes of interaction between viral particles and antibodies occur in lymph, blood, and target organs; the processes of infection, reproduction of viruses, and destruction of infected cells of an organ are localized in a target organ. Thus our model describes the processes that occur in three systems of organism: in lymph nodes, in target organs (liver), and in blood. The following estimates are possible in the case of viral hepatitis B: volume of lymph nodes involved in immune reaction to viral hepatitis B of average severity is about 50 ml; blood volume $\sim$ 1000 ml; extracellular space volume for liver $\sim$ 100 ml; total volume of all liver cells $\sim$ 300 ml.

Let us explain the physical meaning of model variables for viral hepatitis B, which we will use hereafter:

$V_f(t)$, the number of free viral particles (HBV) and 22 nm particles HBsAg in 1 ml of each compartment . Concentration is measured in the units: 1 particle/ml or 1 mol/ml; 1 mol/ml $= 6.02 \times 10^{23}$ particles/ml. The concentrations ratio HBV/HBsAg is assumed to be constant and equal to $10^{-2} - 10^{-3}$;

$M_V(t)$, the number of stimulated (antigen-presenting and producing the interleukin 1 (IL-1)) macrophages in 1 ml of the lymph nodes compartment (cell/ml);

$H_E(t)$, the number of activated (producing the interleukin 2 (IL-2)) helper T-lymphocytes providing the proliferation of effector T-lymphocytes in 1 ml of the lymph nodes compartment (cell/ml);

$H_B(t)$, the number of activated helper T-lymphocytes providing the proliferation of B-lymphocytes in 1 ml of the lymph nodes compartment (cell/ml);

$E(t)$, the number of effector T-lymphocytes in 1 ml of the lymph nodes compartment (cell/ml);

$B(t)$, the number of B-lymphocytes in 1 ml of the lymph nodes compartment (cell/ml);

$P(t)$, the number of plasma cells in 1 ml of the lymph nodes compartment (cell/ml);

$F(t)$, the number of antibodies IgG molecules in 1 ml of each compartment (molecule/ml);

$C_V(t)$, the number of cells infected by viruses in 1 ml of liver compartment, 900 ml volume (cell/ml);

$m(t)$, the number of affected (destroyed) cells in 1 ml liver compartment, 900 ml volume (cell/ml).

Having defined the variable m as the number of damaged organ cells which is manifested in the reduction of liver function, we can assume that m is connected linearly with the level of enzyme activity and bilirubin concentration in blood serum. Since the biochemical index (φ_B) in case of average severity form of disease is a function close to linear function of these parameters [199], we can assume that there exists linear relation $m = \alpha\varphi_S$, where α is unknown constant.

Using the concentration values of above model variables and having estimates of block volumes for lymph nodes, liver, and blood, one can easily obtain the estimates of values of corresponding characteristics of viral disease in the whole organism.

7.1.2. Generalized picture of acute course of viral hepatitis B of average severity. Results of clinical and laboratory observations, data of tests on animals, clinical and experimental data on viral hepatitis B, make it possible to construct a generalized picture of acute form of viral hepatitis B of average severity.

By the generalized picture of disease we mean quantitative representation of virological, immunological and pathological processes that occur in an organism during the acute form of viral hepatitis B of average severity. The generalized picture is obtained as a result of combination of three types of data:

1) virological data on incubation period and on period of disease;

2) immunological data describing the development of immune response in the system of lymph nodes that drain liver;

3) data on liver describing the dynamics of the number of infected and destroyed hepatocytes.

We have constructed a generalized picture of acute course of viral hepatitis B of average severity [202] for the analysis and interpretation of various data. The following concepts [202] were used:

a) anatomical localization of processes of immune response;

b) similarity between immune response processes in case of hepatitis B and experimental antigen stimulation of peripheral lymph node;

c) conditions that determine a start of immune response;

d) clinical evidence of the start of immune response;

e) possibility of the construction of generalized picture of a disease by joining the data obtained both in experiments and in clinical observations.

Fig. 40 shows the data on the dynamics of indices, corresponding to variables of mathematical model of antiviral immune response, which characterize on the interval of 0–200 days the processes of acute course of viral hepatitis B.

It follows from the analysis performed in [202] that the data on the acute course of viral hepatitis B can be divided into two groups from the certainty standpoint. The first group consists of the data that are results of measurements of model variables or of characteristics related linearly to them; the second group consists of the data that are results of indirect but realistic estimations. In particular, data on the dynamics of V_f (quantity of HBsAg in 1 ml of blood), C_V, and that of m (the number of infected and damaged hepatocytes in a unit of liver volume) belong to the first group; the data on the dynamics of M_V, H_E, H_B, E, B, P in a unit of lymph node volume belong to the second group.

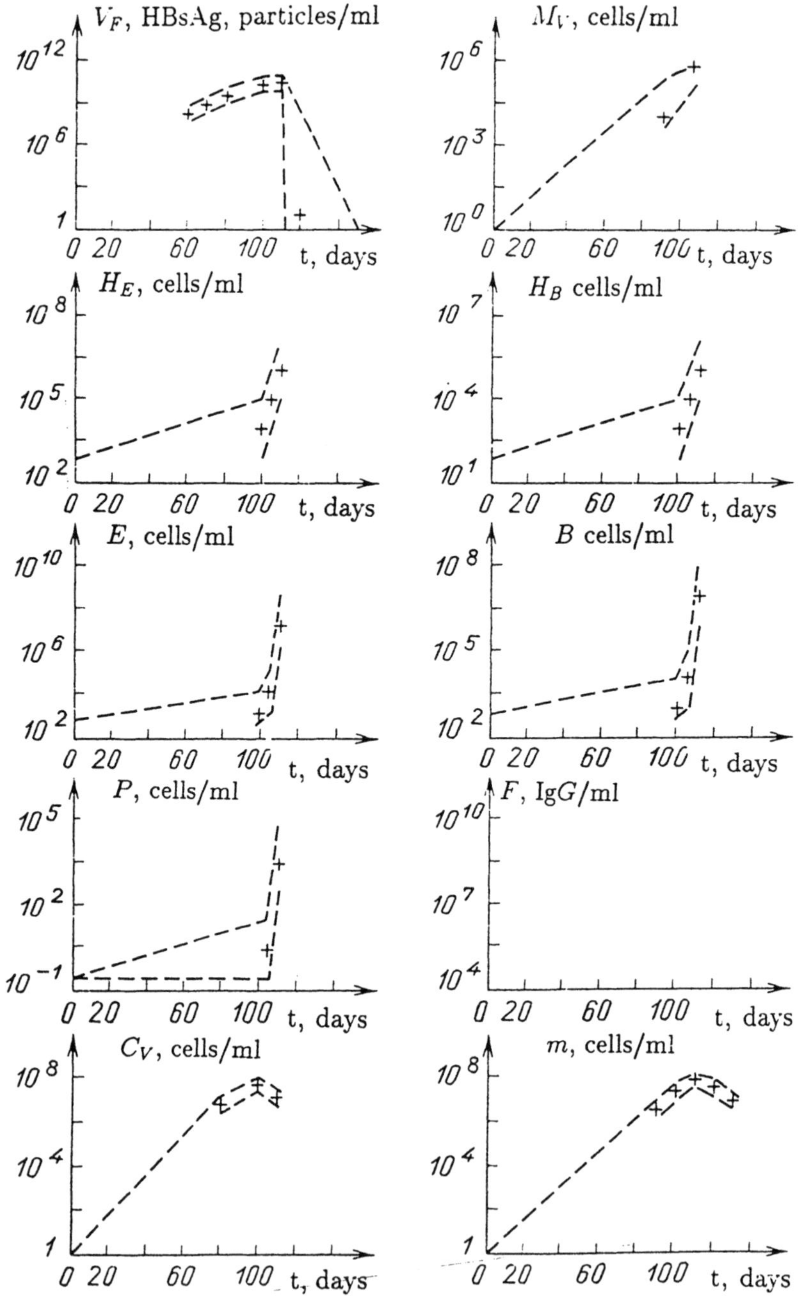

Fig. 40. Generalized picture of acute course of viral hepatitis B: pluses mark the estimates of values of model variables; dashed curves mark the boundaries for permissible values of model variables.

We believe that the essential distinction between the degrees of their certainty manifests itself in the fact that the permissible degree of deviation of upper bounds of observed values of C_V and m from lower ones in the case of the acute course of viral hepatitis B is not more than one order of magnitude. At the same time, permissible dispersion of estimates for the number of M_V, H_E, H_B, E, B, P in a lymph node can be one–three orders. Moreover, the variables V_f, C_V, and m can be measured during the acute course of HBV infection, i.e., in the phase of the development of disease and in the phase of recovery, whereas the quantities of M_V, H_E, H_B, E, B, P can be estimated only for the phase of development of antiviral immune response.

7.1.3. Statement of parameter identification problem for the model of antiviral immune response. The identification of coefficients of the mathematical model of antiviral immune response by experimental, clinical, and laboratory data, that characterize the acute course of viral hepatitis B, is an important stage in tuning the mathematical model of antiviral immune response to concrete viral disease, that is, viral hepatitis.

Taking into account the character of generalized picture, we have formulated the aim of parameter identification problem for mathematical model of antiviral immune response by the data on the acute course of viral hepatitis B as follows: to determine the values of model's coefficients which provide minimal uniform relative deviation (with respect to chosen quantitative criterion) of model solutions from the data on the dynamics of V_f, C_V, and m and to check, if the rest of variables belong to permissible range of values in the neighborhoods of conditional trajectories.

The identification of coefficients for the mathematical model of antiviral immune response by the data, which characterize the generalized picture of acute course of viral hepatitis B, was performed by solving numerically a sequence of minimization problems of the type of $\min_{\alpha_K \in R_+^K} \Phi(\alpha_K)$ according to the approach set forth in Section 4.1.

Calculation of values of the functional $\Phi(\alpha_K)$ implies the solution of initial value problem for the system of model equations.

Initial value problem for model's equations. The m athematical model of antiviral immune response is stated as a system of delay-differential equations describing the changes with time in ten variables of state:

$$\frac{dV_f}{dt} = \nu C_V + nb_{CE}C_V E - \gamma_{VF}FV_f - \gamma_{VM}MV_f - \gamma_{VC}(C^* - C_V - m)V_f,$$

$$\frac{dM_V}{dt} = \gamma_{VM}M^*V_f - \alpha_M M_V, \tag{7.1.1}$$

$$\frac{dH_E}{dt} = b_H^{(E)}\left[\xi(m)\rho_H^{(E)}M_V(t-\tau_H^{(E)})H_E(t-\tau_H^{(E)}) - M_V H_E\right] - b_p^{(H_E)}M_V H_E E + \alpha_H^{(E)}(H_E^* - H_E),$$

$$\frac{dH_B}{dt} = b_H^{(B)}\left[\xi(m)\rho_H^{(B)}M_V(t-\tau_H^{(B)})H_B(t-\tau_H^{(B)}) - M_V H_B\right] - b_p^{(H_B)}M_V H_B B + \alpha_H^{(B)}(H_B^* - H_B),$$

$$\frac{dE}{dt} = b_p^{(E)}\left[\xi(m)\rho_E M_V(t-\tau_E)H_E(t-\tau_E)E(t-\tau_E) - M_V H_E E\right] - b_{EC}C_V E + \alpha_E(E^* - E),$$

$$\frac{dB}{dt} = b_p^{(B)}\left[\xi(m)\rho_B M_V(t-\tau_B)H_B(t-\tau_B)B(t-\tau_B) - M_V H_B B\right] + \alpha_B(B^* - B),$$

$$\frac{dP}{dt} = b_p^{(P)}\xi(m)\rho_P M_V(t-\tau_P)H_B(t-\tau_P)B(t-\tau_P) + \alpha_P(P^* - P),$$

$$\frac{dF}{dt} = \rho_F P - \gamma_{FV}V_f F - \alpha_F F,$$

$$\frac{dC_V}{dt} = \sigma V_f(C^* - C_V - m) - b_{CE}C_V E - b_m C_V,$$

$$\frac{dm}{dt} = b_{CE}C_V E + b_m C_V - \alpha_m m, \qquad \xi(m) = 1 - \frac{m}{C^*}.$$

The physical meaning of the model's variables $V_f(t)$, $M_V(t)$, $H_E(t)$, $H_B(t)$, $E(t)$, $B(t)$, $P(t)$, $F(t)$, $C_V(t)$, and $m(t)$, which is of principal importance while identifying the parameters, was discussed in item 7.1.1.

The process of the infection of a healthy organism can be described by the following system of initial conditions at the instant $t = t_0$ (one can assign $t_0 = 0$ since the system of model equations of antiviral immune response is autonomous):

$$\begin{gathered} V_f(0) = V_f^0, \quad M_V(0) = M_V^0, \quad H_E(0) = H_E^*, \quad H_B(0) = H_B^*, \\ E(0) = E^*, \quad B(0) = B^*, \quad P(0) = P^*, \quad F(0) = \rho_F P^*/\alpha_F, \\ C_V(0) = 0, \quad m(0) = 0, \qquad (7.1.2) \\ M_V(s)H_E(s) = 0, \quad M_V(s)H_B(s) = 0, \\ M_V(s)H_E(s)E(s) = 0, \quad M_V(s)H_B(s)B(s) = 0, \\ -\tau \le s \le 0, \quad \tau = \max\{\tau_H^{(E)}, \tau_H^{(B)}, \tau_E, \tau_B, \tau_P\}. \end{gathered}$$

We denote the set of the model's coefficients in the right-hand side of (7.1.1) by α: $\alpha \in \mathrm{R}^L$. Notice that the right-hand side $f(\cdot, \cdot, \ldots, \cdot, \alpha)$ of the system (7.1.1) depends linearly on all components of the vector-parameter α.

In order to solve the problem of quantitative fitting of the model (7.1.1), (7.1.2) to observational data on the dynamics of immune processes for viral hepatitis B, the analysis of coefficients α of the model of antiviral immune response was performed and permissible regions were constructed for values of each model parameter. Results of estimations of permissible ranges for model parameters are presented in Table 2. Values of some model coefficients must be refined in the process of the solution of the identification problem by means of fitting the solutions of initial value problem for the system of model equations (7.1.1), (7.1.2) to the data $(t_1, y_{obs\,1}), (t_2, y_{obs\,2}), \ldots, (t_M, y_{obs\,M})$ of generalized picture.

To this end, variables of the model of antiviral immune response were rescaled as follows

$$\begin{gathered} \bar{V}_f = V_f/V_f^*, \quad \bar{M}_V = M_V/M^*, \quad \bar{H}_E = H_E/V_E^*, \quad \bar{H}_B = H_B/H_B^*, \\ \bar{E} = E/E^*, \quad \bar{B} = B/B^*, \quad \bar{P} = P/P^*, \quad \bar{F} = F/F^*, \\ \bar{C}_V = C_V/C^*, \quad \bar{m} = m/C^*, \end{gathered}$$

Table 2

Results of the parameter estimation

Parameter, unit	Biological meaning of parameter	Permissible range	Initial value
M^*, mol/ml	Concentration of MHC class I or II macrophages in a lymph node	$(0.5-3)10^{-18}$	10^{-18}
H_E^*, mol/ml	Concentration of specific T-helpers for CTLs in a lymph node	$(0.1-1)10^{-20}$	10^{-21}
H_B^*, mol/ml	Concentration of specific T-helpers for B-cells in a lymph node	$(0.1-1)10^{-21}$	10^{-22}
E^*, mol/ml	Concentration of specific precursors of CTLs in a lymph node	$(0.1-1)10^{-20}$	10^{-21}
B^*, mol/ml	Concentration of specific B-lymphocytes in a lymph node	$(0.5-5)10^{-21}$	10^{-21}
P^*, mol/ml	Concentration of specific plasma cells in a lymph node	$(0.1-1.7)10^{-24}$	$4.3\cdot10^{-25}$
F^*, mol/ml	Concentration of specific antibodies in blood, lymph node, and liver	$(1.7-17)10^{-17}$	$8.5\cdot10^{-17}$
C^*, mol/ml	Concentration of hepatocytes in liver	$(0.5-5)10^{-16}$	$0.5\cdot10^{-15}$
α_M, 1/day	Rate constant of loss of stimulated state by a macrophage	1.0 – 1.5	1.2
$\alpha_H^{(E)}$, 1/day	Rate constant of loss of stimulated state by T-helpers for CTLs	0.8 – 1.2	1
$\alpha_H^{(B)}$, 1/day	Rate constant of loss of stimulated state by T-helpers for B-cells	0.8 – 1.2	1
α_E, 1/day	Rate constant of natural death of CTLs	0.33 – 0.50	0.4
α_B, 1/day	Rate constant of natural death of B-lymphocytes	0.05 – 0.10	0.1
α_P, 1/day	Rate constant of natural death of plasma cells	0.33 – 0.50	0.4
α_F, 1/day	Rate constant of natural death of antibodies	0.043	0.043
$\tau_H^{(E)}$, day	Duration of the division cycle for Th1 cells	0.4 – 0.8	0.6
$\tau_H^{(B)}$, day	Duration of the division cycle for Th2 cells	0.4 – 0.8	0.6

See the continuation

Table 2 (continuation)

Results of the parameter estimation

Parameter, unit	Biological meaning of parameter	Permissible range	Initial value
τ_E, day	Duration of the division cycle for CTLs	2–3	2
τ_B, day	Duration of the division cycle for B-lymphocytes	2–3	2
τ_P, day	Duration of division and differentiation for B-cells resulting in the appearance of plasma cells	3–4	3
$\rho_H^{(E)}$	Number of Th1 cells created during single division cycle	2	2
$\rho_H^{(B)}$	Number of Th2 cells created during single division cycle	2	2
ρ_E	Number of CTLs in a clone created during the division(s series)	10–18	16
ρ_B	Number of B-cells in a clone created during the division(s series)	10–18	16
ρ_P	Number of plasma cells in a clone created during the division(s series)	2–5	3
$b_H^{(E)}$, $\frac{\text{ml}}{\text{mol}\cdot\text{day}}$	Rate constant of stimulation for Th1 cells	$(0.4-3.0)10^{18}$	10^{18}
$b_H^{(B)}$, $\frac{\text{ml}}{\text{mol}\cdot\text{day}}$	Rate constant of stimulation for Th2 cells	$(0.4-3.0)10^{18}$	10^{18}
$b_p^{(E)}$, $\frac{\text{ml}^2}{\text{mol}^2\cdot day}$	Rate constant of stimulation for CTLs	$(0.5-70)10^{38}$	10^{38}
$b_p^{(B)}$, $\frac{\text{ml}^2}{\text{mol}^2\cdot day}$	Rate constant of stimulation for B-cells for the number of B-lymphocytes	$(1-100)10^{38}$	10^{38}
$b_p^{(P)}$, $\frac{\text{ml}}{\text{mol}\cdot\text{day}}$	Rate constant of stimulation for B-cells resulting in the proliferation and differentiation into plasma cells	$(1-100)10^{38}$	10^{38}
$b_p^{(H_E)}$, $\frac{\text{ml}^2}{\text{mol}^2\cdot day}$	Expenditure rate of Th1 for the stimulation of CTLs; $b_p^{(H_E)} = \delta_{H_E} b_p^{(E)}$	$\delta_{H_E} = 10^{-4}$	10^{34}
$b_p^{(H_B)}$, $\frac{\text{mol}^2}{\text{ml}^2\cdot day}$	Expenditure rate of Th2 for the stimulation of CTLs; $b_p^{(H_B)} = \delta_{H_B} b_p^{(B)}$	$\delta_{H_B} = 10^{-4}$	10^{34}
γ_{MV}, $\frac{\text{mol}}{\text{ml}\cdot\text{day}}$	Rate constant for the antigenic stimulation of macrophages in l. n.	$10^{11}-4\cdot10^{16}$	10^{12}

See the continuation

Table 2 (continuation)

Results of the parameter estimation

Parameter, unit	Biological meaning of parameter	Permissible range	Initial value
γ_{VM}, 1/day	Rate constant for the removal of antigen particles by macrophages	$0.1-0.7$	0.4
γ_{FV}, $\frac{\text{ml}}{\text{mol·day}}$	Rate constant for the binding of one IgG molecule with 22 nm HBsAg particle	$(0.86-86)10^{14}$	$8.6\cdot 10^{14}$
γ_{VF}, $\frac{\text{ml}}{\text{mol·day}}$	Rate constant for the binding of 22 nm HBsAg particles with 10 IgG molecules	$(0.1-86)10^{14}$	$3\cdot 10^{14}$
ρ_F^*, 1/day	Rate constant for the synthesis of IgG molecules by one plasma cell	$(8.5-17)10^7$	$1.7\cdot 10^8$
σ, $\frac{\text{ml}}{\text{mol·day}}$	Rate constant for the infection of hepatocytes b hepatitis B viruses	$(2-20)10^{11}$	$2.5\cdot 10^{11}$
b_{CE}, $\frac{\text{mol}}{\text{mol·day}}$	Rate constant for the destruction of hepatocytes by cytotoxic T-cells	$6\cdot 10^{15}-4\cdot 10^{20}$	$1.6\cdot 10^{18}$
b_{EC}, $\frac{\text{mol}}{\text{mol·day}}$	Rate constant for the destruction of cytotoxic T-cells due to lytic interaction with infected cells	$6\cdot 10^{14}-4\cdot 10^{19}$	$1.6\cdot 10^{17}$
b_m, 1/day	Rate constant for the destruction of infected hepatocytes due to the cytopaticity of viruses	$0.005-0.1$	0.01
α_m, 1/day	Rate constant for the regeneration of hepatocytes	0.12	0.12
ν, 1/day	Rate constant for the secretion of 22 nm HBsAg particles by one infected hepatocyte	10^2-10^5	$6\cdot 10^3$
n	Number of 22 nm HBsAg particles released by the destruction of of one hepatocyte by T-effector	10^3-10^6	$2\cdot 10^4$
γ_{VC}, $\frac{\text{mol}}{\text{mol·day}}$	Rate constant for the adsorption of viruses by uninfected hepatocytes	$3\cdot 10^9-2\cdot 10^{12}$	$2.5\cdot 10^{10}$

where V_f^* is the mole concentration corresponding to 1 particle of HBsAg per ml, and M^*, H_E^*, H_B^*, E^*, B^*, P^*, F^*, C^* are the homeostasis concentrations (mole/ml) respectively for: MHC class I or II molecules bearing macrophages in LN; specific T-helpers for T-effectors and B-lymphocytes; specific precursors of T-effectors; B-lymphocytes; plasma cells; antibodies; hepatocytes in 1 ml of liver volume. Initial problem (7.1.1), (7.1.2) was solved numerically on corresponding fitting interval

$[0, t_j]$ to get the values of the vector function of solution $y(t, \alpha)$ of the model's equations at the moments $(t_1, t_2, \ldots, t_j)$, $j \leq M$, corresponding to the moments of experimental observations. We shall discuss below characteristics of the fitting interval with respect to dominating processes, justification of selection of optimized parameters, and results of the solution of minimization problems for each interval.

7.1.4. Sequential parameter identification for the model of antiviral immune response. Four phases can be separated in the course of acute viral hepatitis B; they differ by processes and by their contribution into observed dynamics of data in generalized picture of acute viral hepatitis B of average severity: incubation period, period of activation and development of immune response, peak of disease, and period of recovery.

Fitting the model to the data on incubation period. The incubation period of acute viral hepatitis B is characterized by the development of processes of infection and replication of viral particles in infected cells. Processes of specific activation of immune system cells are absent. Duration of this period is determined by the time which is necessary to reach HBsAg concentration in blood level comparable with a sensitivity threshold starting the reaction of immune system. For the concrete set of data of generalized picture of acute course of viral hepatitis B (Fig. 36) the incubation period lasts for 90 days from the moment of infection.

There are data on dynamics of concentrations of HBsAg, stimulated macrophages, infected and destructed cells for the observation interval of 60–90 days. The fitting of corresponding components of the vector function of model solutions $V_f(t)$, $C_V(t)$, $m(t)$, and $M_V(t)$ to these data was carried out stage by stage on a sequence of expanding time intervals.

These stages are shown in Table 3 (stages 1–3). About one thousand calculations of residual functional $\Phi(\alpha)$, constructed for the components V_f, M_V, C_V, and m were performed to solve numerically the set of minimization problems (see Section 5.1).

Table 3

Relative numbers of stages	Times of observations used at given stage of data fitting (days)	Model variables that are fitted to observational data	Parameters varied while minimizing the functional
1	60, 70	V_f	$\nu,\ V_f^0$
2	60, 70, 80	$V_f,\ C_V$	$\nu,\ \sigma,\ V_f^0$
3	60, 70, 80, 90	$V_f,\ C_V,\ m$	$\nu,\ \sigma,\ b_m,\ V_f^0$
4	60, 70, 80, 90, 100	$V_f,\ C_V,\ m$	$\nu,\ \sigma,\ b_m,\ V_f^0$
5	90, 105	M_V	γ_{MV}
6	100	$H_E,\ H_B$	$b_H^{(E)},\ b_H^{(B)}$
7	100, 105	$H_E,\ H_B$	$b_H^{(E)},\ b_H^{(B)}$
8	100	$E,\ B$	$b_p^{(E)},\ b_p^{(B)}$
9	100, 105	$E,\ B$	$b_p^{(E)},\ b_p^{(B)}$
10	105	P	$b_p^{(P)}$
11	60, 70, 80, 90, 100, 105	$V_f,\ M_V,\ H_E,\ H_B,$ $E,\ B,\ P,\ C_V,\ m$	$\nu,\ V_f^0,\ \gamma_{MV},\ b_H^{(E)},$ $b_H^{(B)},\ b_p^{(E)},\ b_p^{(B)},$ $b_p^{(P)},\ \sigma,\ b_m$

Fitting the model to the data on period of activation and development of immune response. Period of activation and development of antiviral immune response is characterized both by the development of processes of infection and replication of viruses in infected cells and by the development of processes of specific stimulation, proliferation and differentiation of T- and B-lymphocytes. In addition, contribution of specific effector T-lymphocytes into the destruction of infected hepatocytes increases. This period is about 15 days for the above data on acute viral hepatitis B. There are data on the dynamics of V_f, M_V, H_E, H_B, E, B, P, C_V, and m on the observation interval of 90–105 days. It is necessary to fit corresponding components of the vector function of model solutions to these data, providing minimal deviation of variables V_f, C_V, and m in terms of the functional $\Phi(\alpha)$ and the belonging of variables M_V, H_E, H_B, E, B, P to the permissible range of data of the generalized picture. A constructive approach to the solution of this problem was found in step by step solution of a sequence of more simple fitting problems, described in Table 3 (stages 4–11), on sequentially expanding intervals of observations. It took about 300 calculations of the functional to fit the model to the data on period of activation and development of immune response.

Fitting the model to the data on illness peak period of acute viral hepatitis B. The peak of acute viral hepatitis B is characterized by the development of processes of proliferation and differentiation of T- and B-lymphocytes and by intensive destruction of infected hepatocytes by specific T-effectors as a result of considerable increase in their number during the development of specific immune response. So, within the framework of the model, it is the interaction between cytotoxic T-lymphocytes and antibodies on the one hand, and between infected cells and viral particles on the other hand, that determines the dynamics of the data on V_f, C_V, and m for the peak of the disease, and that, in turn, influences the choice of coefficients nb_{CE}, b_{CE} as varied parameters for fitting V_f, C_V, and m. The observation interval of 105–110 days corresponds to this period for the data set of generalized picture of acute viral hepatitis B. It is necessary to fit all but $F(t)$ components of the vector function of the model's solution. The splitting of this problem into a sequence of more simple problems, as we explained above, permits us to solve the fitting problem by 300 calculations of $\Phi(\alpha)$. The procedure of tuning the parameters is to be clarified. In order to fit simultaneously the components C_V and m to the data on the interval of 60–110 days, the model coefficient b_{CE}

was varied. In addition, in order to preserve the quality of fitting of variables C_V and m on the interval of 60–110 days, which is mostly determined by the sum $b_m + b_{CE}E(t)$, every change in b_{CE} during the numerical fitting was accompanied by a change in b_m according to the relation $b_m = b^* - b_{CE}E^*$, where b^* is a quantity, identified while fitting V_f and C_V to the data on the interval of 60–80 days.

Fitting the model to the data on recovery period. The recovery period of acute viral hepatitis B is characterized by complete elimination of HBsAg and viral particles HBV from an organism and by the regeneration of destroyed hepatocytes. This period covers the time interval of 110–130 days for the data set on acute viral hepatitis B.

There are data on the dynamics of V_f and m for this interval. Fitting of the model to the data on the number of destroyed hepatocyte cells was done by the minimization of the functional $\Phi(\alpha)$, constructed for the components V_f, C_V, and m, and by varying the parameter α_m, which is the rate of regeneration of liver. Since identified values of coefficients of the model ensured that an interval of trajectory $V_f(t)$, $t \in [110, 130]$ is within permissible time bounds for the elimination of viruses from blood, and the value of virus concentration by 120th day is known just approximately (up to several orders of magnitude), we have not tried to make V_f to pass through this point.

Fig. 41 shows the solution that corresponds to resulting set of model parameters. Values of the model's coefficients obtained as a result of identification are presented in Table 4. We consider the quality of fitting the model to the data thus acieved as sufficient to regard the stated identification problem for the model of antiviral immune response by the data on acute viral hepatitis B as being solved.

It should be noticed that, strictly speaking, the set of values for teh model's coefficients, obtained as a result of identification, is not unique. It just permits us to reproduce quantitatively characteristic features of modelled disease in accordance with the region of permissible values.

It is possible to verify the biological correctness of this set by deeper analysis of results of modeling the experimental situations and by the comparison of results with observational data.

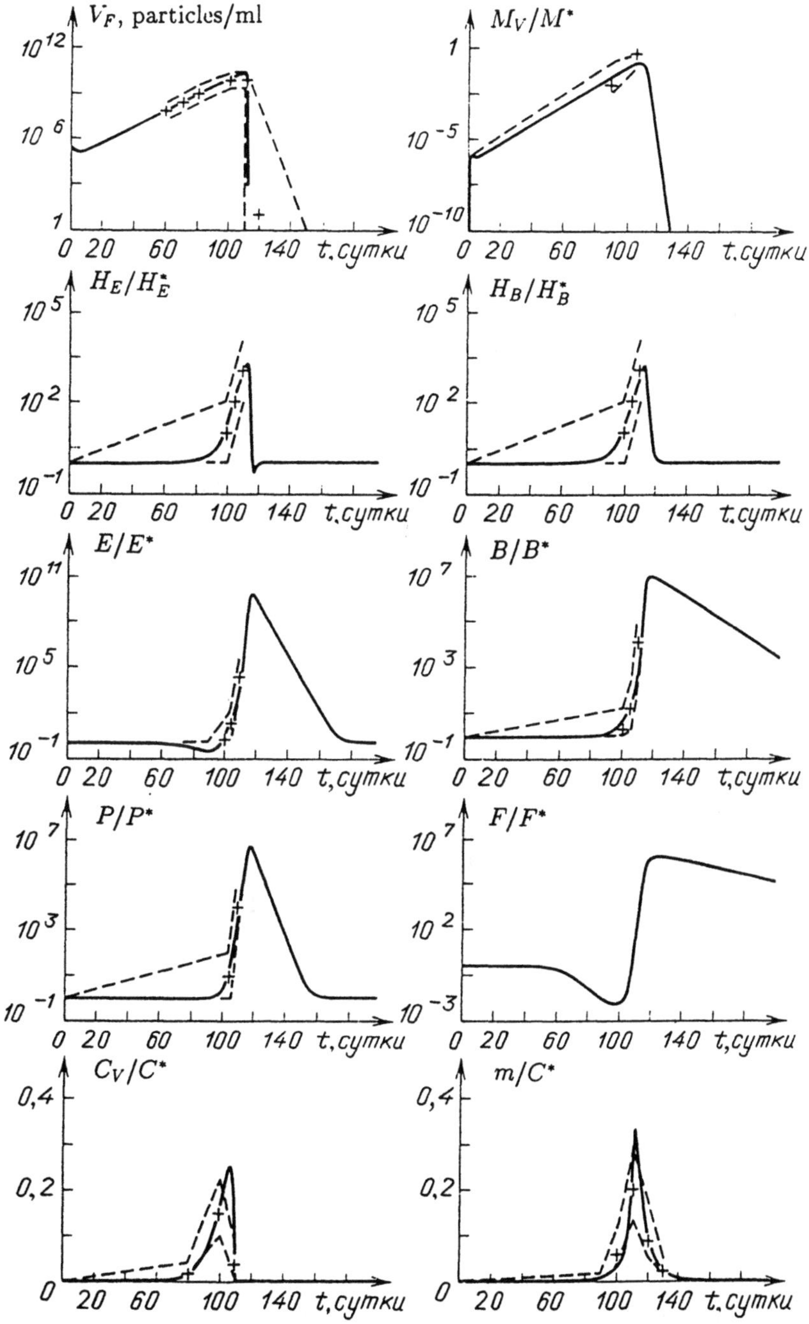

Fig. 41. Model solution obtained as a result of the parameter identification by the data (+) of generalized picture of acute viral hepatitis B.

Table 4

Results of the parameter identification

Parameter	Parameter's value	Parameter	Parameter's value	Initial guess
M^*	10^{-18} mol/ml	ρ_E	16	
H_E^*	10^{-21} mol/ml	ρ_B	16	
H_B^*	10^{-22} mol/ml	ρ_P	3	
E^*	10^{-21} mol/ml	ρ_F	$1.7 \cdot 10^8$ molec/(day·cell)	
B^*	10^{-21} mol/ml	$b_H^{(E)}$	$2.7 \cdot 10^{19}$ ml/(mol·day)	10^{18}
P^*	$4.3 \cdot 10^{-25}$ mol/ml	$b_H^{(B)}$	$2.7 \cdot 10^{19}$ ml^2/(mol^2·day)	10^{18}
F^*	$8.5 \cdot 10^{-17}$ mol/ml	$b_p^{(E)}$	$5.3 \cdot 10^{39}$ ml^2/(mol^2·day)	10^{38}
C^*	$0.5 \cdot 10^{-15}$ mol/ml	$b_p^{(B)}$	$8.0 \cdot 10^{38}$ ml^2/(mol^2·day)	10^{38}
α_M	1.2 day^{-1}	$b_p^{(P)}$	$1.7 \cdot 10^{36}$ ml^2/(mol^2·day)	10^{38}
α_E	0.4 day^{-1}	σ	$2.3 \cdot 10^{12}$ ml/(mol·day)	$2.5 \cdot 10^{11}$
α_B	0.1 day^{-1}	b_{CE}	$6.6 \cdot 10^{17}$ ml/(mol·day)	$1.6 \cdot 10^{18}$
α_P	0.4 day^{-1}	b_m	0.052 day^{-1}	0.01
α_F	0.043 day^{-1}	α_m	0.15 day^{-1}	0.12
V_F^0	$1.7 \cdot 10^5$ particle/ml	ν	83 day^{-1}	$6 \cdot 10^3$
	$2.9 \cdot 10^{-19}$ mol/ml	γ_{VC}	$2.5 \cdot 10^{10}$ ml/(mol·day)	
$\tau_H^{(E)}$	0.6 day	γ_{VM}	0.4 day^{-1}	
$\tau_H^{(B)}$	0.6 day	γ_{VF}	$3 \cdot 10^{14}$ ml/(mol·day)	
$\alpha_H^{(E)}$	1 day^{-1}	γ_{MV}	$9.4 \cdot 10^{12}$ ml/(mol·day)	10^{12}
$\alpha_H^{(B)}$	1 day^{-1}	γ_{FV}	$8.6 \cdot 10^{14}$ ml/(mol·day)	
τ_E	2 day	$b_p^{(H_E)}$	$5.3 \cdot 10^{33}$ ml^2/(mol^2·day)	10^{37}
τ_B	2 day	$b_p^{(H_B)}$	$8.0 \cdot 10^{34}$ ml^2/(mol^2·day)	10^{37}
τ_P	3 day	b_{EC}	$1.6 \cdot 10^{17}$ ml/(mol·day)	
$\rho_H^{(E)}$	2	nb_{CE}	$3.4 \cdot 10^{18}$ ml/(mol·day)	$3.2 \cdot 10^{22}$
$\rho_H^{(B)}$	2	n	5	

7.2. Modeling the Elements of Immunotherapy and Processes of Pathogenesis in Case of Viral Hepatitis BFri 03-15-1996

We suggest using the set of parameters for the model of viral hepatitis B, obtained as a result of the construction of zero approximation and numerical identification, for the description of a wider range of phenomena than the data on generalized picture of viral hepatitis B.

Therefore it appears to be interesting to analyze whether the model with a given set of parameters can describe clinical and experimental situations that differ from acute course of the disease. We shall consider the following situations:

– development of infection (relationship between concentration of specific antibodies, concentration of viral particles, and the development of disease);

– vaccination at the moment of infection or shortly after that moment (influence of vaccination on clinical course of the disease);

– infection of the organism by various doses of virus (influence of quantity $V_f(t_0)$ on duration of incubation period and on severity of disease).

In addition, we shall consider in details the data on the destruction of infected hepatocytes destruction and discuss the sensitivity study for the model.

7.2.1. Modeling the infection of organism by hepatitis B viruses. Let us introduce a definition: the *infection* means creating at an instant t_0 a nonzero concentration of viruses $V_t(t_0) \geq \delta > 0$ in the organism. Assume that as a result of infection an organism can be in one of two situations:

The infection has happened, i.e., the viruses entered sensitive cells and started to multiply there, and further relationships between viruses and the organism are determined by reactions of immune and other defence systems. Formally, the infection has happened if, for $t_1 > t_0$, $C_V(t_1) \geq \bar{C}_V > 0$, where $\bar{C}_V$ is some threshold concentration.

The infection has not happened, i.e., the viruses have been neutralized by the action of preexisting antibodies and nonspecific defence, and sensitive cells were left intact.

According to data from the work [29], 1–10 viruses of hepatitis B are sufficient to infect a chimpanzee ($CID_{50} \sim$ 1–10 particles of HBV). Therefore, one may assume that the infection has occurred if 1–10 (5 on average) of infected hepatocytes have appeared in teh organism, and $\bar{C}_V = 5\,\text{cell/l} = 5 \cdot 10^{-3}\text{cell/ml} \simeq 10^{-26}\text{mol/ml}$. Results of experiments show that the period of time necessary for the infection of these cells is less than 2 days [139]. An estimate of the coefficient describing the rate of the infection of target cells by viruses was obtained in here as well as an estimate of the rate of nonspecific elimination of HBsAg and HBV particles.

Let us define what is a nonimmune organism to a given antigen. A nonimmune organism is characterized qualitatively by the fact that it has a very low concentration of antibodies against a given antigen. Concentration of all immunocompetent cells specific to given antigen is very low too. This state can be described quantitatively by a conditional threshold concentration separating nonimmune organisms from immune ones.

This approach is used in clinical investigations: when the test for antibodies against HBsAg yields a negative result, the organism is qualified as nonimmune [139, 140].

It is possible, using the data of the work [154] on limiting dilution analysis of monoclonal antibodies against HBsAg with known content of protein in solution, to estimate this threshold concentration. It is about 10^7 IgG molecules/ml. The sensitivity threshold in the radioimmunoassay method (RIA) is also connected with a conventional unit for the measurement of antibodies concentration; in international units (IU) it seems to be about 1 mIU/ml$\approx 10^7$ molecules [287, 288]. It is known that in case of infecting a chimpanzee by the dose containing 1–10 viral particles the possibility of the disease to develop is determined by the concentration of specific antibodies in blood plasma of animal. If this level equals or is lower than the threshold level (1mIU/ml), the disease develops in about 50% of tests; if the level of antibodies is about 100mIU/ml, the disease does not develop and the organism is regarded as immune protected [73].

Fig. 42 presents the results of numerical simulation of these cases ($F_1^* = 5\,\mathrm{m}IU/\mathrm{ml}$, $F_2^* = 500\,\mathrm{m}IU/\mathrm{ml}$, $V_F^0 = 5\,\mathrm{virus/l}$). It is seen that for $F^* = F_1^*$ infection does occur, which is manifested in the increase in the number of infected cells and in the increase in concentration of virus particles beginning from the fourth day after infection. There is no infection in the case of $F^* = F_2^*$. The concentration of viral particles decreases in two days lower than one viral particle in organism.

Concentration of infected cells decreases slower owing to the low value of E^* and large lifetime of infected cells. Results of modelling the case $F_2^* = 100\,\mathrm{m}IU/\mathrm{ml}$ imply that the model describes adequately the case of passive immune defence, when an organism is infected by 600 IU of antibodies against HBsAg [73]. If we assume that these antibodies are distributed in a volume of 3–6 liters, then concentration of antibodies in blood should be equal to 200–100 $\mathrm{m}IU/\mathrm{ml}$.

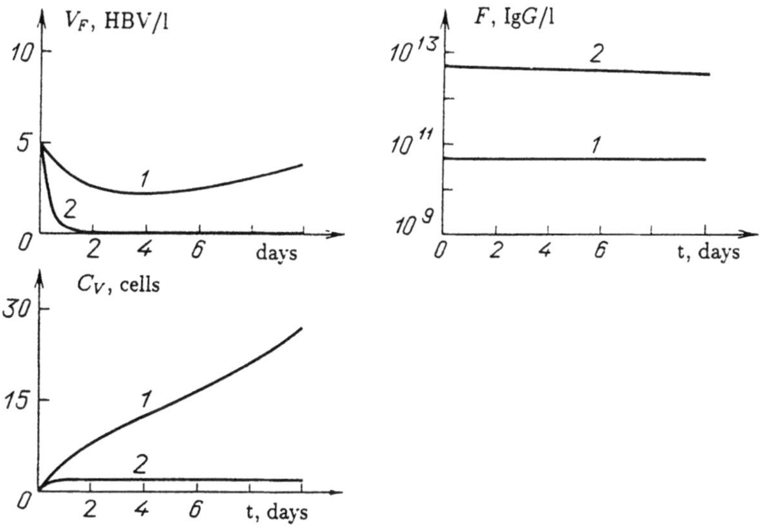

Fig. 42. Modeling the defence action of antibodies in case of infection by hepatitis B viruses. Initial concentration of antibodies: (1) $5 \cdot 10^{10}$ IgG molecule/l, (2) $5 \cdot 10^{12}$ IgG molecule/l.

7.2.2. Infection and vaccination. Consider the case of active immune defence, that is, vaccination. The usual practice of vaccination by HBsAg is intramuscular or subcutaneous injection of vaccine dose containing from 20 to 50 μg HBsAg. The processes of antigenic stimulation and of division of immunocompetent cells are localized in this case in lymphoid tissue draining the site of vaccine injection. The situation described in [287] was chosen to verify the results of simulation. Five chimpanzees were injected intravenously with 500 μg of HBsAg. The injection of large amount of HBsAg into blood provoked strong immune response in three of five animals. It is evident that the immune response to the vaccine would influence strongly the course of disease in case of simultaneous infection by HBV. For example, 100 doses of HBsAg, injected simultaneously with 10–100 viruses, provided the immune defense of organism and prevented the development of disease [325]. Analogously, in [290] the injection of vaccine 48 hours after the infection by a dose containing 3500–3500 viruses protected three of four animals.

Consider the following problem: as a result of infection the concentration of viral particles (HBV) $V_f^0 = 50$ particle/ml was formed in an organism; simultaneously, as a result of vaccine injection the concentration $V_{\text{HBs}}^0 = 1.3 \cdot 10^{12}$ particles/ml of HBsAg is created in blood. To simulate the vaccination, we add in (7.1.1) the equation describing the changes in concentration of vaccine:

$$\frac{dV_{\text{HBs}}}{dt} = -\gamma_{VM} V_{\text{HBs}} - \gamma_{VF} F V_{\text{HBs}},$$

and modify the equations for M_V and F as follows:

$$\frac{dM_V}{dt} = \gamma_{MV} M^* (V_f + V_{\text{HBs}}) - \alpha_M M_V,$$

$$\frac{dF}{dt} = \rho_F C - \gamma_{FV}(V_f + V_{\text{HBs}})F - \alpha_F F.$$

Fig. 43 shows the results of simulation for concentrations of viral particles (solid line), vaccine HBsAg particles (dashed line), and infected cells. Antibodies and effector T-cells that appear in two weeks after vaccination, neutralize viral particles, HBsAg particles, and destroy infected cells. The interval of two weeks corresponds to the time when large quantities of antibodies appear after vaccination [140].

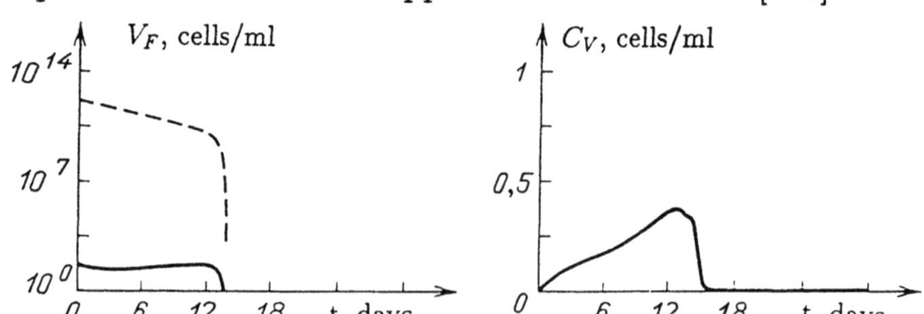

Fig. 43. Simulation of vaccination effect in case of infection by hepatitis B viruses (solid line, dynamics of viruses; dashed line, dynamics of HBsAg vaccine).

7.2.3. Dependence of incubation period duration and severity of disease on initial dose of viruses. In [325] the influence of inoculum size on the duration of incubation period and severity of disease was studied. The duration of incubation period was estimated by positive results of radioimmunoassay (RIA) on HBsAg in blood plasma, and severity of disease was estimated by the maximal level of ALT enzyme. It was established that the duration of incubation period is inversely related to the initial dose of viruses, and that the severity of disease does not depend on the dose of infection. We studied this dependence using the model of viral hepatitis B. It was assumed, while simulating, that incubation period is over when $t = t^* : V_f(t^*) = \max_{0 \le t \le 200} V_f(t)$. The severity was estimated by maximal values of $m(t)$, which is a part of damaged cells of a target organ.

Fig. 44 shows the dependence of duration of incubation period on the initial dose of viruses. Experimental data from [325] for subtype of HBsAg are marked by triangles, results of simulations are shown by solid line. Simulations have confirmed, that there is no influence of inoculum size on severity of disease.

7.2.4. Imitation of data on hepatocyte cytolysis. The clinical phase of acute viral hepatitis is characterized by the rise in the level of activity of a number of enzymes (AST, ALT) in blood. This rise is caused by the disruption of integrity of cellular membrane of hepatocytes, which leads to the release of intracellular enzymes into blood. Being released into blood, where they can be observed, the enzyme molecules lose their activity.

For example, the half life for ALT is $t_{1/2} = 47 \pm 10$ hours. The investigation of teh dynamics of these enzymes is an important component in the diagnosis, prognosis, and estimation of effectiveness of treatment in case of viral hepatitis. That is why we have formulated the problem of simulation of ALT activity dynamics in blood. We have made a number of assumptions.

– the ejection of enzymes into blood is connected directly with destruction of hepatocytes (as noted in literature, the integrity of a membrane can be disrupt in case of reversible damage of hepatocytes too;

direct connection between the processes of reversible and irreversible damage to hepatocytes has also been established);

— for $V_f = 0$ and $C_V = 0$ level of enzyme activity is equal to A^* and does not depend on the destruction of hepatocytes (it is connected with the fact that in the model of viral hepatitis B a target organ is not destroyed in a norm).

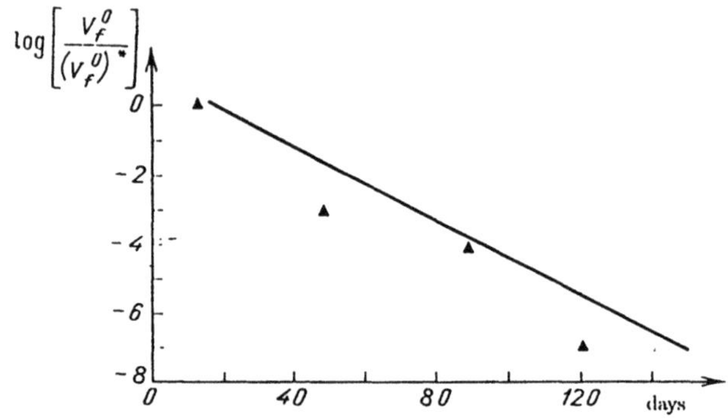

Fig. 44. Dependence of duration of incubation period of acute viral hepatitis B on the dose of infection. Δ, experimental data; solid line, results of simulation. Relative value of infection dose is plotted on the ordinate axis.

Thus, the equation for ALT activity has the form:

$$\frac{dA}{dt} = \sigma_A(\rho C_V E + kC_V) - \alpha(A - A^*).$$

The value of the coefficient σ was chosen so that the ratio $\max\limits_{0 \le t \le 200} \dfrac{A(t)}{A^*} \simeq 20$ corresponds to average severity of modelled form of viral hepatitis.

Results of numerical simulation are shown in Fig. 45. The maximal value of ALT precedes by 1–2 days the maximum of m. The variable m is interpreted as a characteristic of damage to the organ and, therefore, its maximal value corresponds to the peak of the disease. This relationship is in agreement with clinical observations. Detailed analysis of correlation between various model solutions and data on ALT dynamics is a task for further investigations.

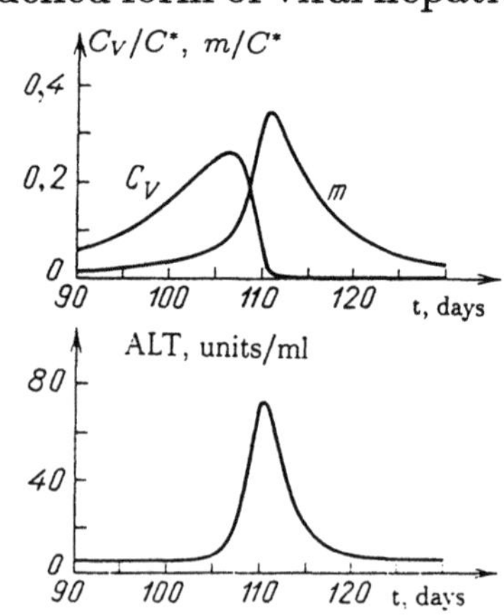

Fig. 45. Imitation of data on the cytolysis of hepatocytes.

7.2.5. Evaluation of contribution of separate processes to modelled disease. The set of parameters presented in Table 4 can be analysed not only from the standpoint of correspondence between clinical and experimental data on the dynamics of model variables, but from the viewpoint of integral characteristics of the modeled phenomenon too. For instance, adding the equations for auxiliary variables to the model system of equations of antiviral immune response, we can estimate: the total number of viral particles secreted by infected hepatocytes; the part of antibodies spent on the neutralization of viral particles, as related to teh total number of antibodies produced, etc. Unfortunately, experimental data on these characteristics are meagre, so the conclusions will be of an approximate character.

Let us introduce the following auxiliary characteristics of the modelled phenomenon (per milliliter of volume):

$V^C = \int_0^T \nu C_V dt$, total amount of V_f secreted by infected hepatocytes with no participation of effector cells in the interval of time $[0, T]$;

$V^E = \int_0^T n b_{CE} C_V E dt$, total amount of V_f released during the destruction of infected hepatocytes by cytotoxic T-cells in the interval of time $[0, T]$;

$V^M = \int_0^T \gamma_{VM} V_f dt$, total amount of V_f neutralized by macrophages in the interval of time $[0, T]$;

$V^F = \int_0^T \gamma_{VF} F V_f dt$, total amount of V_f neutralized by antibodies in the interval of time $[0, T]$;

$F^P = \int_0^T \rho_F P dt$, total amount of F produced by plasma cells in the interval of time $[0, T]$;

$F^V = \int_0^T \gamma_{FV} F V_f dt$, total amount of antibodies F bound with V_f in the interval of time $[0, T]$;

$C^V = \int_0^T \sigma V_f(C^* - C_V - m)dt$, total amount of hepatocytes infected in the interval of time $[0, T]$;

$C^E = \int_0^T b_{CE} C_V E dt$, total amount of infected hepatocytes destroyed by cytotoxic T-cells in the interval of time $[0, T]$;

$H_E^D = \int_0^T b_P^{(E)} \left(M_V(t)H_E(t) - M_V(t - \tau_H^E)H_E(t - \tau_H^E)\right) dt$, amount of helper T-cells that are in the division cycle at the moment of time T.

We shall use these auxiliary characteristics as follows. Several qualitatively different stages can be separated in the course of a disease: infection, incubation period, period of development of immune response (which is the beginning of disease), peak of disease, and recovery period. Defence immune mechanisms work actively just after incubation period and untill teh recovery period.

We will consider the solution describing the form of average severity of acute viral hepatitis (Fig. 37). Peak of the disease (maximum of $m(t)$) falls on 110th day. Processes connected with immune response manifest themselves most obviously in the interval of 90–120 days. Therefore Table 5 shows the relative characteristics for the following intervals of time: $\{0, 120\,\text{days}\}$, the whole period of disease; $\{110, 120\,\text{days}\}$, period of immune response.

The following conclusions can be drawn when analysing the data of Table 5:

– the part of V_f released from infected hepatocytes destroyed by CTLs becomes significant only in the period of immune response (110–120th days after the moment of infection);

– during the whole period of disease $\{0, 120\,\text{days}\}$ viral particles are mostly neutralized by non-specific mechanisms, but in the period of immune response the contribution of humoral immunity becomes decisive;

– cytotoxic T-cells are the main mechanism that destroys infected cells in the whole course of disease, but in the period of immune response their contribution into destruction of infected hepatocytes becomes dominant;

– part of antibodies bound to viral particles is very small during the whole period of disease $\{0, 120\,\text{days}\}$, since most antibodies are produced after 110th day and accumulated in excess; estimates of the quantity F^V/F^P for the period $\{0, 110\,\text{days}\}$ show that the share of antibodies bound to viral particles reaches 91%;

– one can see, comparing proportion of cells participating in the cell cycle on 110th day and on 120th day after infection, that during the period of development of immune response the main part of cells is in the cell cycle, while the 120th day marks complete cessation of cell division.

Table 5

Relative contribution of separate processes into modelled mechanisms of antiviral immune response

Characterized process	Notation	0–120 days	110–120 days
Relation of the number of viral particles freed during the destruction of C_V to the number of particles secreted by infected hepatocytes C_V	V^E/V^C	0.007	0.23
Part of viral particles neutralized by antibodies	$V^F/(V^C + V^E)$	0.1	0.75
Part of viral particles cleared by macrophages	$V^M/(V^C + V^E)$	0.9	0.25
Part of infected hepatocytes destroyed by cytotoxic T-cells	C^E/C^V	0.68	0.99
Part of antibodies bound with viral antigens	F^V/F^P	$6 \cdot 10^{-4}$	$4 \cdot 10^{-4}$
Part of helper T-cells H_E in the cell cycle at moment T	$H_E^D / \left(H_E^D + H_E(T)\right)$	$T = 110$ 0.7	$T = 120$ $4 \cdot 10^{-5}$

The estimates of contributions shown in Table 5 seem to be in agreement with experimental and literature data we are aware of. Meanwhile, it was found while analysing the auxiliary characteristics that one effector-cell is spent for the destruction of 100 target cells. The estimate is obviously not quite consistent with experimental data, and possibly requires the correction of a number of model coefficients. However, one cannot rule out that some neglected by the model processes are disguised in this way.

7.2.6. Stochastic sensitivity analysis for severity characteristic of disease. Let $\bar{\alpha}_i$ $(i = 1, \dots, 45)$ be the basic values of parameters presented in Table 4. In order to estimate the sensitivity of solution to initial value of ith parameter, a random value ξ_{ij} $(j = 1, \dots, N)$ is added, which is uniformly distributed, has mathematical expectation $E(\xi_{ij}) = 0$, and variance $D(\xi_{ij}) = \varepsilon\bar{\alpha}_i$, $\varepsilon = 0.1$. As a result we obtain the perturbed jth value of ith parameter:

$$\alpha_{ij} = \bar{\alpha}_i + \xi_{ij}.$$

All 45 parameters of the model were perturbed by such a randomization mechanism. The initial value problem is solved with these new values of parameters in the interval of $\{0, 120\,\text{days}\}$ and for the solution so obtained some chosen characteristic $m_{\max} = \max_{0 \le t \le 200} m(t)$ is estimated, which is interpreted as the severity of disease. Then new random perturbations $\alpha_{ij} = \bar{\alpha}_i + \xi_{ij+1}$ are added to initial set of parameters, and the value $m_{\max}$ is estimated again. This operation is repeated $N = 100$ times.

The result can be presented as a matrix with 46 columns (45 columns are for the values of independent parameters and 46th column is for the dependent variable (characteristic of severity) and 100 lines. Regression analysis is one of the conventional methods for the estimation of the influence of independent variables on the dependent variable. The degree of influence can be represented quantitatively either by the correlation coefficient between the actual value of dependence of the variable and its predicted value, or by the square of correlation coefficient which is equal to the part of dispersion of dependent variable described by the regression formula obtained.

Results of calculations are presented partially in Table 6. One can see that the characteristics of virus interaction with target cell ν and σ influence the severity of disease to the largest extent; the characteristics of virus interaction with the macrophagal system γ_{VM}, γ_{MV}, and α_M are less important, and the characteristics of the immune reaction $\rho_H^{(E)}$, $\rho_H^{(B)}$, and ρ_E are still less important. A model of multiple regression for the set of only four parameters σ, ν, γ_{MV}, and γ_{VM} allows us to describe 64% of variation of dependent variable $m_{\max}$. This quantity is about 50% for the parameters of immune system (25 parameters). Both these results and the data presented in Table 6 allows for the conclusion that the severity of disease is determined largely by parameters of virus-macrophage interaction rather than by immune system characteristics.

Table 6

Estimation of the influence of model's parameters on the characteristics of severity of disease

Parameter	Description of the parameter	ρ^2
σ	Rate of the infection of target cells by a virus	50 %
ν	Rate of the secretion of viral and 22 nm HBsAg particles by an infected cell	12 %
γ_{MV}	Rate of the activation of macrophages by viral antigens	10 %
α_M	Rate constant for the loss of activated state by a macrophage	8.4 %
γ_{VM}	Rate of the absorption by macrophages of viral and 22 nm particles of HBsAg	7 %
ρ_E	Number of the CTL-daughter cells as a result of one division cycle	5 %
$\rho_H^{(E)}$	Number of the Th1 helper cells formed as a result of one division cycle	5 %

It should be noted that for small perturbations (up to 10%) from the basic set of model's parameters all solutions may be interpreted as an acute form of disease. Such solutions are characterized by the possibility of separating the period of intensive destruction of infected cells and neutralization of viruses, which results in their full elimination within an interval of 110–140 days after the infection. Though, when perturbing the model parameters to a greater extent, the solutions arise where viral particles and infected cells are not eliminated completely in the interval of 0–300 days. To study these situations, a new characteristic of solution was introduced

$$V_f^{\min} = \min_{t_1 \leq t \leq 300} V_f(t),$$

where t_1 is the moment when the first maximum of variable $V_f(t)$ is attained. Values of parameters in this case were varied in a regular way. The model solution where $V_f^{\min} \geq 1$ particle/l may be interpreted as corresponding to chronic viral hepatitis. Therefore we call the value $V_f^{\min}$ the characteristic of outcome of disease.

7.2.7. Influence of the variation of model parameters on indices of severity and outcome of disease. We have investigated the relationship between some parameters and the values $m_{\max}$, $V_f^{\min}$ for the model of antiviral immune response. Fig. 46 shows the dependence of these characteristics on the parameters ν, σ, γ_{MV}, $b_p^{(E)}$. The values of parameters were varied within the limits of their permissible ranges.

The analysis of curves presented in Fig. 43 shows that the increase in the parameters ν and σ leads to significant increase in the characteristic of severity $m_{\max}$, but does not influence the outcome characteristic ($V_f^{\min} \leq 1$ particle/liter), which may be interpreted as hypertoxic form of disease.

The variation of the values of the parameters γ_{MV} and $b_p^{(E)}$ (rates of stimulation of macrophages and cytotoxic T-cells, respectively) results in the decrease of severity characteristics both in domain of high and low values of parameters in comparison with initial set of parameters, but unlike parameters ν and σ, these solutions are characterized by the inequality $V_f^{\min} > 1$ particle/liter. These solutions, as was noticed above, may be interpreted as the transition to chronic viral hepatitis.

In our opinion the results presented do not contradict clinical and experimental data and hypotheses suggested on the basis of these data for the pathogenesis of hypertoxic and chronic forms of the course of viral hepatitis B [87, 141, 284].

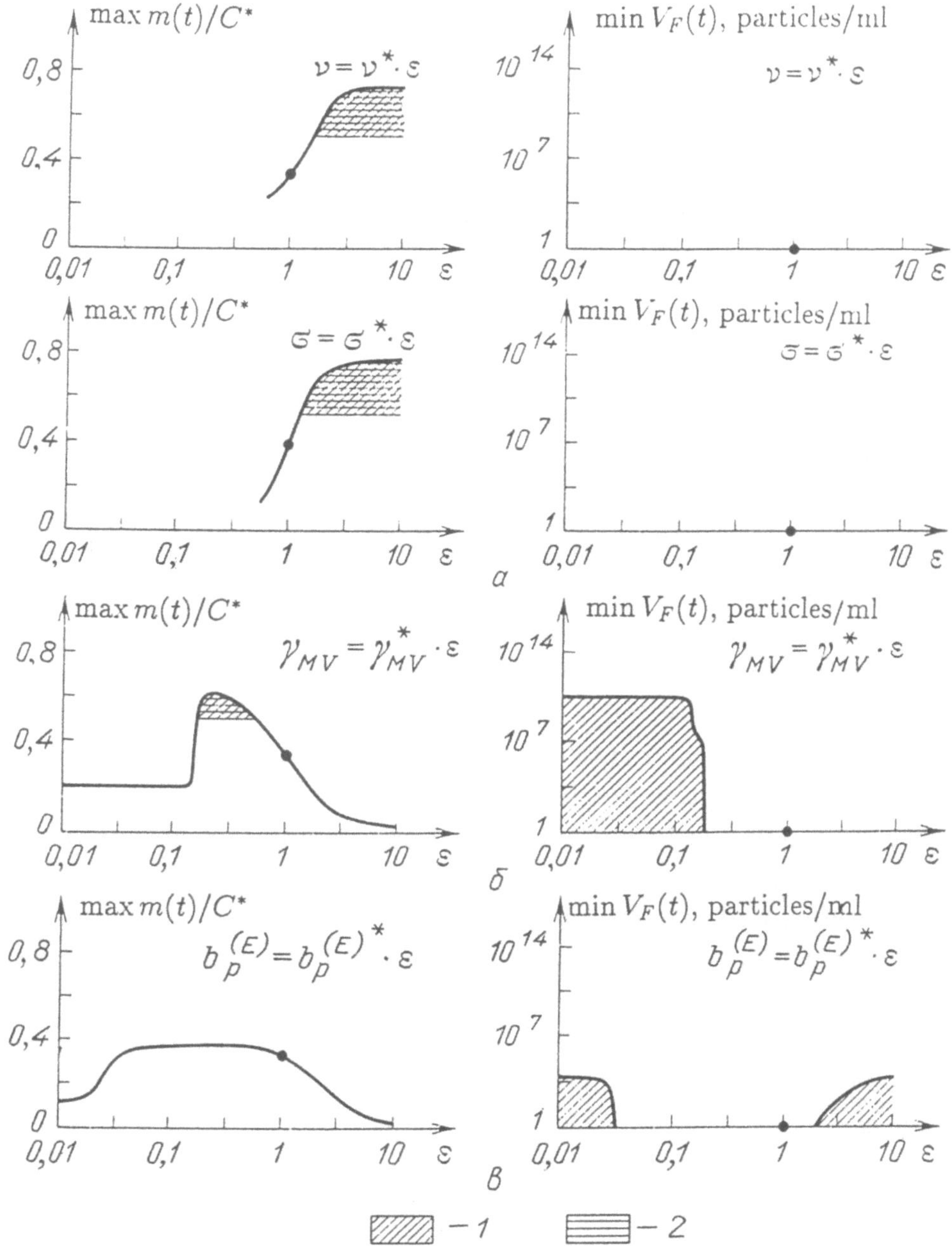

Fig. 46. Investigation of dependence of characteristics of severity and outcome of disease $m_{\max}$ and $V_f^{\min}$ on model parameters: a, rate of replication (ν) and infection rate (σ); b, macrophages activation rate (γ_{MV}); c, cytotoxic T-lymphocytes stimulation rate (b_p^E); •, values of characteristics for identified set of parameters ν^*, σ^*, γ_{MV}^*, $b_p^{(E)*}$; 1, domain of chronization; 2, domain of hypertoxic form.

So, the mathematical model of antiviral immune response with the basic set of parameters corresponding to acute viral hepatitis B infection describes quantitatively:

– dependence of incubation period duration on inoculum size;

– the development of disease in relation to concentration of specific antibodies (preexisting or injected);

– effect of vaccination within the period of infection on the clinical course of disease.

The results of sensitivity analysis for the model's solutions to variations of parameters are in agreement with known hypotheses on the mechanisms of pathogenesis of malignant and chronic forms of viral hepatitis B.

7.3. Modeling of Interleukin-2 Treatment During Viral Hepatitis B on the Basis of Mathematical Model of Antiviral Immune Response with Refined Description of T-lymphocyte Proliferation

The modeling of the control over the process of immune response requires the description in the model equations the mechanisms of the action of drugs that are used. Nowadays a number of immunomodulators are used in clinical practice and experiments, such as interferons, interleukins, and corticosteroids.

The mechanism of the action of some of them works by regulating the rate of separate phases of immune response. Interleukins 1 and 2 (IL-1, IL-2) are the obligatory components of immune response, since immunocompetent cells interact between themselves through these substances.

Results of experimental and clinical studies show that IL-2 is a promising drug for the stimulation of immune response in patients with primary and secondary immunodeficiency.

Results of modeling the IL-2 effect on immune response to viral hepatitis B will be presented in this Section. Notice that in clinical practice IL-2 was used mostly for the treatment of patients with cancer, AIDS, and some other diseases.

7.3.1. Cell cycle of a lymphocyte. Multiplication processes of lymphocytes under the influence of an antigenic signal consists of several phases preparing a cell for its division. All these phases together are called the cell cycle.

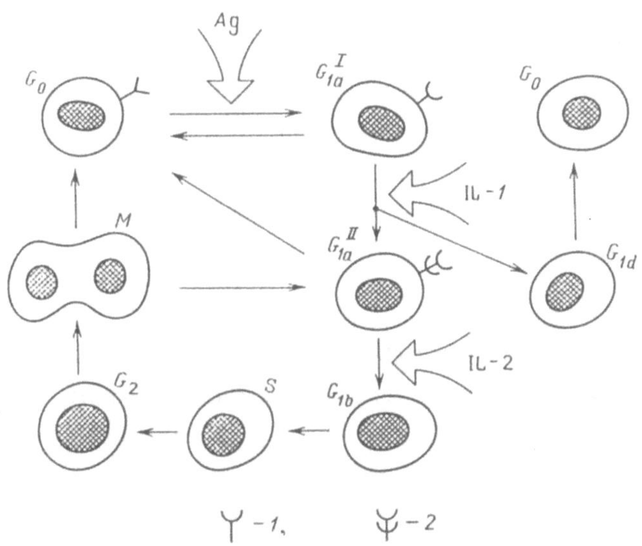

Fig. 47. Cell cycle of a lymphocyte. G_0, G_{1a}^{I}, $G_{1a}^{\mathrm{II}^\bullet}$, G_{1b}, G_{1d}, S, G_2, M are the phases of cell cycle. $\longrightarrow$, phase-to-phase transition; $\Longrightarrow$, signal for the phase-to-phase transition; Y, receptors to an antigen; 1, receptors to IL-1; 2, receptors to IL-2.

Consider the cell cycle of a lymphocyte presented in Fig. 47 [79, 300]. Recirculating helper T-lymphocytes (H) and effector T-lymphocytes (E) are in the rest phase of the cell cycle. Denote them by H_r and E_r respectively. When H_r and E_r become bound with activated macrophages (M_V), the lymphocytes receive the first activating signal from the antigen (Ag) bound with macrophagal antigens of major hystocompatibility complex [311].

The activation of a macrophage consists of the presentation of antigenic determinants of a virus on the cell surface, production and secretion of lymphocyte activating factor, interleukin 1 (IL-1) by the macrophage [302, 311]. The first activating signal leads to: the expression of receptors by lymphocytes to IL-1 (state H_a), the expression of receptors by lymphocytes to interleukin 2, which is the T-cell growth factor (state E_p) [296].

Interleukin-2 (IL-2) is produced by a certain portion of cells H_a following the interaction of receptors to IL-1 on their surface with IL-1 and their transition into a special phase of the cell cycle G_{1d}, the phase of differentiation (state H_f) [289]. Part of H_a after the interaction with IL-1 express the receptors to IL-2 and are capable of further dividing in response to IL-2 (state H_p) [44, 185]. The lymphocytes E_p can also divide in response to IL-2 and, futhermore, they can differentiate into mature cytotoxic effectors (state E_k) under the influence of differentiating factors [51, 79]. All the states of lymphocytes are unstable. Eventually the lymphocytes pass into the phase G_0 or die due to aging [230, 289, 326].

7.3.2. Mathematical model of antiviral immune response with refined description of T-lymphocyte division. Introduce, in accordance with identified states of cells, the following notations:

$M_V(t)$, concentration of activated macrophages M_V at the moment of time t;

$H_r(t)$, concentration of helper T-lymphocytes H_r in the rest phase G_0;

$H_a(t)$, concentration of activated helper T-lymphocytes in the phase G_{1a};

$H_p(t)$, concentration of helper T-lymphocytes, expressing receptors to IL-2;

$H_f(t)$, concentration of helper T-lymphocytes producing IL-2 in the phase G_{1d};

$E_r(t)$, concentration of effector T-lymphocytes in the rest phase G_0;

$E_p(t)$, concentration of activated effector T-lymphocytes, expressing receptors to IL-2, in the phase G_{1a};

$E_k(t)$, concentration of mature cytotoxic effector T-lymphocytes;

$I_1(t)$, $I_2(t)$, concentrations of soluble factors IL-1, IL-2 respectively.

Let us construct, beginning with the scheme of cellular interactions presented in Fig. 43, the refined model of antiviral immune response on the basis of equations of antiviral immune response model and refined description of T-lymphocytes division process:

$$\begin{aligned}
\frac{dV_f(t)}{dt} &= \nu C_V(t) + nb_{CE}C_V(t)E_k(t) - \gamma_{VM}MV_f(t) - \gamma_{VF}(t)F(t)V_f(t) \\
&\quad - \gamma_{VC}V_f(t)(C^* - C_V(t) - m(t)), \\
\frac{dM_V(t)}{dt} &= \gamma_{VM}MV_f(t) - \alpha_M M_V(t) \\
\frac{dH_r(t)}{dt} &= -b_H M_V(t)H_r(t) + \xi(m(t))\rho_H h_2 I_2(t-\tau_p^H)H_p(t-\tau_p^H) \\
&\quad + q_a H_a(t) + q_p H_p(t) + q_f H_f(t) + \alpha_H\left(H_r^* - H_r(t)\right), \\
\frac{dH_a(t)}{dt} &= -h_1 I_1(t)H_a(t) + b_H M_V(t-\tau_r^H)H_r(t-au_r^H) + q_a\left(H_a^* - H_a(t)\right), \\
\frac{dH_p(t)}{dt} &= -h_2 I_2(t)H_p(t) + A_1 h_1 I_1(t-\tau_{A_1}^H)H_a(t-\tau_{A_1}^H) + q_p\left(H_p^* - H_p(t)\right), \\
\frac{dH_f(t)}{dt} &= A_2 h_1 I_1(t-\tau_{A_2}^H)H_a(t-\tau_{A_2}^H) + q_f\left(H_f^* - H_f(t)\right), \\
\frac{dI_1(t)}{dt} &= -a_H H_a(t)I_1(t) + f_1 M_V(t) - \alpha_1 I_1(t), \qquad (7.3.1) \\
\frac{dI_2(t)}{dt} &= -(p_H H_p(t) + p_E E_p(t))I_2(t) + f_2 H_f(t) - \alpha_2 I_2(t), \\
\frac{dE_r(t)}{dt} &= -b_E M_V(t)E_r(t) + s_p E_p(t) + \alpha_E(E_r^* - E_r(t)), \\
\frac{dE_p(t)}{dt} &= -e_2 I_2(t)E_p(t) + \xi(m(t))\rho_E e_2 I_2(t-\tau_p^E)E_p(t-\tau_p^E) \\
&\quad + b_E M_V(t-\tau_r^E)E_r(t-\tau_r^E) - s_k E_p(t) + s_p(E_p^* - E_p(t)), \\
\frac{dE_k(t)}{dt} &= s_k E_p(t) - b_{EC}C_V(t)E_k(t) + \alpha_k(E_k^* - E_k(t)), \\
\frac{dH_B(t)}{dt} &= b_{H_B}\xi(m(t))\rho_{H_B}M_V(t-\tau_{H_B})H_B(t-\tau_{H_B}) \\
&\quad - M_V(t)H_B(t) - b_p^{H_B}M_V(t)H_B(t)B(t) + \alpha_{H_B}\left(H_B^* - H_B(t)\right), \\
\frac{dB(t)}{dt} &= b_p^B\xi(m(t))\rho_B M_V(t-\tau_B)H_B(t-\tau_B)B(t-\tau_B) \\
&\quad - M_V(t)H_B(t)B(t) + \alpha_B\left(B^* - B(t)\right),
\end{aligned}$$

$$\frac{dP(t)}{dt} = b_p^P \xi(m(t)) \rho_P M_V(t-\tau_P) H_B(t-\tau_P) B(t-\tau_P) + \alpha_P (P^* - P(t)),$$

$$\frac{dF(t)}{dt} = \rho_F P(t) - \gamma_{FV} V_f(t) F(t) - \alpha_F F(t), \qquad (7.3.1)$$

$$\frac{dC_V(t)}{dt} = \sigma V_f(t)(C^* - C_V(t) - m(t)) - b_{CE} C_V(t) E_k(t) - b_m C_V(t),$$

$$\frac{dm(t)}{dt} = b_{CE} C_V(t) E_k(t) + b_m C_V(t) - \alpha_m m(t), \quad \xi(m(t)) = 1 - \frac{m(t)}{C^*}.$$

Estimates of bounds for permissible values of the model's parameters were constructed on the basis of literary data on qualitative and quantitative characteristics of described processes (Table 7).

Literary data on Scatchard-analysis for the values of mean dissociation times of receptor-factor complexes [9, 51, 90, 120, 121, 184–185, 296] were used in order to estimate the rate constants for the interactions of the type of cell-factor: h_1, h_2, p_H, p_E, a_H, e_2.

Estimates for the rest of parameters were obtained as a result of direct measurements.

In order to identify the parameters for system (7.3.1) we use the following approach. It is seen from the structure of equations (7.3.1) that the values of homeostasis for the states H_r, H_a, H_f, H_p, E_r, E_p, E_k, I_2 in the organism ($\hat{H}_r$, ..., $\hat{E}_k$, $\hat{I}_2$, respectively) are some functions of the values $H_r^*, \ldots, E_k^*$. It is easy to obtain concrete form of these functions equating the right-hand sides of (7.3.1) to zero. Moreover, we shall assume that

$$\hat{H}_r = p_r^H \hat{H}_E, \quad \hat{H}_a = p_a^H \hat{H}_E, \quad \hat{H}_f = p_f^H \hat{H}_E, \quad \hat{H}_p = p_p^H \hat{H}_E,$$

$$\hat{E}_r = p_r^E \hat{E}, \quad \hat{E}_p = p_p^E \hat{E}, \quad \hat{E}_k = p_k^E \hat{E},$$

$$p_r^H + p_a^H + p_f^H + p_p^H = p_r^E + p_p^E + p_k^E = 1,$$

where $\hat{H}_E, \hat{E}$ are the homeostasis of helper cells and effector cells in organism.

Table 7

Parameters of the system (7.3.1).*)

Parameter, unit	Biological meaning of the parameter	Permissible range	Initial value, (identified value)
h_1, $\frac{\text{ml}}{\text{mol·day}}$	Rate constant for H_a spent for binding the IL-1	$(2.2-4.2)10^{15}$	$3.2 \cdot 10^{15}$
a_H, $\frac{\text{ml}}{\text{mol·day}}$	Rate constant for IL-1 spent for binding the H_a	$(6.6-12.6)10^{16}$	$9.6 \cdot 10^{16}$
h_2, $\frac{\text{ml}}{\text{mol·day}}$	Rate constant for H_p spent for binding the IL-2	$(0.1-5.0)10^{15}$	$3.0 \cdot 10^{15}$
p_H, $\frac{\text{ml}}{\text{mol·day}}$	Rate constant for IL-2 spent for binding the H_p	$(0.2-50)10^{18}$	$1.8 \cdot 10^{19}$
b_{EC}, $\frac{\text{ml}}{\text{mol·day}}$	Rate constant for T-effectors decaying as a result of destruction of infected cells	$(0.006-400)10^{17}$	$1.6 \cdot 10^{17}$ $(5.4 \cdot 10^{15})$
e_2, $\frac{\text{ml}}{\text{mol·day}}$	Rate constant for E_p spent for binding the IL-2	$(0.1-5.0)10^{15}$	$3.2 \cdot 10^{15}$
p_E, $\frac{\text{ml}}{\text{mol·day}}$	Rate constant for IL-2 spent for binding the E_p	$(0.2-50)10^{18}$	$1.8 \cdot 10^{19}$
τ_r^H, hour	Duration of the phase G_{1a}^I for H_a	2–24	6.0 (2.03)
$\tau_{A_1}^H$, hour	Duration of the phase G_{1a}^{II} for H_p	2–24	6.0 (2.03)
$\tau_{A_2}^H$, hour	Duration of the phase G_{1a}^{II} for H_f	4–24	6.0 (4.60)
τ_p^H, hour	Duration of the phases G_{1b}, S, G_2, M for H_p	10–20	14.4**) (15.8)
τ_r^E, hour	Duration of the phase G_{1a} for E_r	4–5	4.5
τ_p^H, hour	Duration of the phases G_{1b}, S, G_2, M for E_p	17–24	12.0**) (11.5)

*) See Table 4 for missing parameters.
**) Initial value of the parameter is taken from Table 4.

See the continuation

Table 7 (continuation)

Parameters of the system (7.3.1).

Parameter, unit	Biological meaning of the parameter	Permissible range	Initial value, (identified value)
A_1	Part of helper-lymphocytes capable of proliferating after stimulation	0–1	0.25 (0.87)
A_2	Part of helper-lymphocytes capable of producing IL-2 after stimulation	0–1	0.75 (0.13)
ρ_H	Number of descendants of H_p-cell as a result of single division	2	2
ρ_E	Number of descendants of E_p-cell as a result of single division	2	2
q_a, 1/day	Rate constant for the loss of receptors to IL-1 by H_a-cells	0.4–0.5	0.45
q_p, 1/day	Rate constant for the loss of receptors to IL-2 by H_p-cells	0.1–0.5	0.3
q_f, 1/day	Rate constant for the loss of ability to produce IL-2 by H_f-cells	0.3–1.2	0.75
s_p, 1/day	Rate constant for the loss of receptors to IL-2 by E_p-cells	0.07–0.46	0.3
s_k, 1/day	Maturating rate constant for of E_p-cells into effectors	0.5–1.0	0.75
α_M, 1/day	Rate constant for the loss activated state by M_V-cells	1.0–1.5	1.2
α_H, 1/day	Rate constant for natural death of H_r-cells	$(5.5 - 8.4)10^{-3}$	0.007
α_E, 1/day	Rate constant for natural death of E_r-cells	$(5.5 - 8.4)10^{-3}$	0.007
α_1, 1/day	Rate constant for natural death of IL-1 molecules	$(0.0 - 2.5)10^2$	12
α_2, 1/day	Rate constant for natural death of IL-2 molecules	$(0.0 - 2.7)10^2$	12
f_1, 1/day	Rate constant for the secretion of IL-1 by M_V-cells	$(0.75 - 120)10^5$	$6 \cdot 10^6$ $(12 \cdot 10^6)$
f_2, 1/day	Rate constant for the secretion of IL-2 by H_f-cells	$(0.09 - 500)10^4$	$25 \cdot 10^5$ $(170 \cdot 10^5)$

The process of infection of a healthy organism will be defined by the following initial conditions at the instant $t = 0$ (system (7.3.1) is autonomous):

$$V_f(0) = V_f^0 > 0, \quad M_V(0) = M_V^0, \quad H_r(0) = \hat{H}_r, \quad H_a(0) = \hat{H}_a,$$
$$H_p(0) = \hat{H}_p, \quad H_f(0) = \hat{H}_f, \quad E_r(0) = \hat{E}_r, \quad E_p(0) = \hat{E}_p,$$
$$E_k(0) = \hat{E}_k, \quad I_1(0) = 0, \quad I_2(0) = \hat{I}_2, \quad H_B(0) = H_B^*,$$
$$B(0) = B^*, \quad P(0) = P^*, \quad F(0) = \rho_F P^*/\alpha_F, \quad C_V(0) = m_V(0) = 0,$$
$$m(0) = 0, \quad H_p(s) = \hat{H}_p, \quad I_2(s) = \hat{I}_2, \quad E_p(s) = \hat{E}_p,$$
$$s \in \left[-\max[\tau_p^H, \tau_p^E], 0\right].$$

values of the rest of delaying variables on the delay interval are equal to zero.

Let $y(t, \alpha)$ be a solution vector of the initial value problem (7.3.1), (7.3.2) on the values of parameter vector $\alpha = (\alpha_1, \alpha_2, \ldots, \alpha_M)$ obtained as a result of the search for a minimum of the functional $F(Ay(t, \alpha), Y)$ (A is a matrix of linear relation between model's variables): the variables M_V, V_f, H_B, B, P, F, C, m coincide, $H_E(t) = H_r(t) + H_a(t) + H_f(t) + H_p(t)$, $E(t) = E_r(t) + E_p(t) + E_k(t)$.

Initial values of all parameters of the system are shown in Table 7. A part of these values was obtained as a result of the parameter identification for system (7.3.1) using experimental data of generalized picture of acute course of viral hepatitis B of average severity.

While identifying the parameters of model (7.3.1), the set of experimental data on generalized picture of viral hepatitis B (Fig. 40) was supplemented with the boundaries of working ranges (the ranges of concentrations where a factor exhibits its activity IL-1, IL-2): 10^{10} – 10^{12} molecule/ml [9, 51, 222]. We assume that the lower boundary of the working range of activity is attained to the 100th day (the start of the development of immune response), and the upper boundary is attained to the 110th day (the peak of immune response).

As a result of the sequence of identification steps for system (7.3.1), the values of 9 parameters were estimated which describe the processes of production and action of IL-1 and IL-2. Fig. 48 shows the solution of initial value problem (7.3.1), (7.3.2) for the values of parameters after the identification.

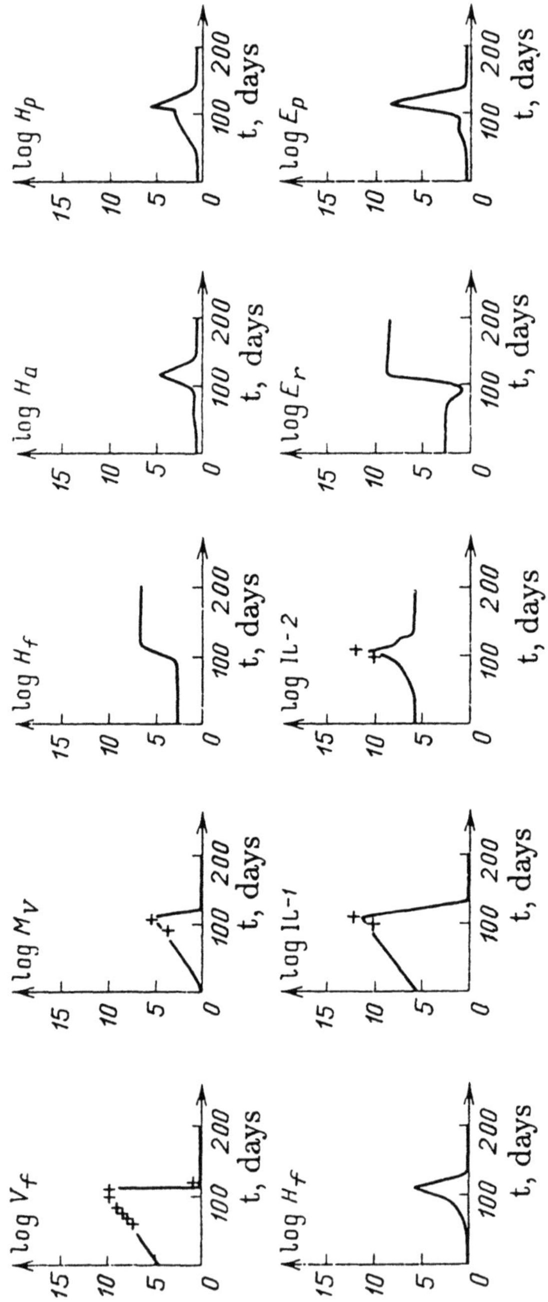
log V_f
log M_V
log H_f
log H_a
log H_p
log H_f
log IL-1
log IL-2
log E_r
log E_p
t, days

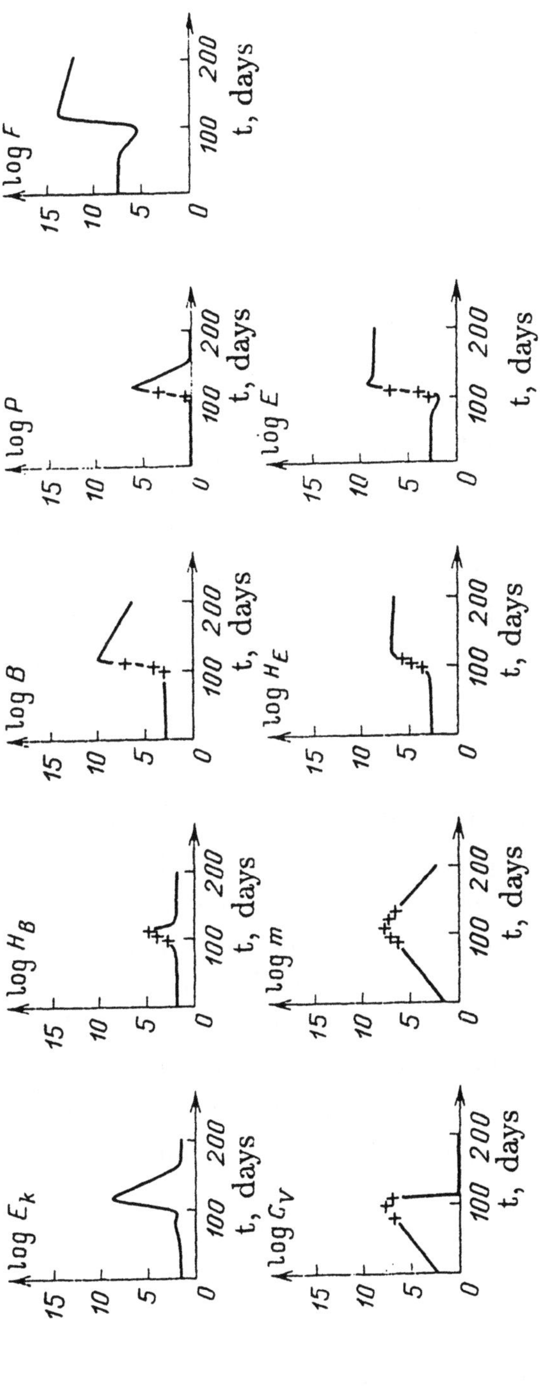

Fig. 48. Solution of the model corresponding to acute course of viral hepatitis B. Measurement units: cell/ml for cells, particle/ml for antibodies, molecule/ml for interleukins.

7.3.3. Modeling the IL-2 action in case of viral hepatitis B. We use the mathematical model of antiviral immune response with refined description of T-lymphocytes proliferation (7.3.1) to study the after-effects of IL-2 treatment in vivo; viral hepatitis B is the infectious disease here. There is a characteristic feature of this disease: the reaction of the immune system to antigen leads to pathological changes in the organism (these changes are reflected in the model as the destruction of infected cells C_V in liver by the effector cells E_k).

There is no reliable immunostimulating antiviral drug at present which could effectively influence the severity and outcome of disease [34]. The application of IL-2 treatment can produce ambiguous results. On the one hand, IL-2 can be considered as a universal stimulator of the immune system (IL-2 enhances the proliferation of cells specific to given virus), and on the other hand, the application of IL-2 treatment during the diseases connected with pathological changes (caused by the immune system reaction) may increase the severity of the disease.

We simulate the IL-2 therapy in the course of the development of the immune response against the background of immunodeficit of helper-cells in organism. In order to simulate the results of IL-2 injections we have used the data from [183]: patients with various forms of cancer (fibrosarcoma, osteosarcoma, melanoma, etc.) were treated with continuous intravenous injections of IL-2 at the rate 100-10,000 $U/(\text{kg}\cdot\text{h})$ during 3–21 days depending on prescribed dose.

The following estimates were obtained in [183]: the average time of IL-2 transition from blood to extracellular liquid ($1/\alpha_+ = 1.54 \cdot 10^{-2}$ day) and back ($1/\alpha_- = 7.01\cdot 10^{-2}$ day); the average time of IL-2 inactivation in blood ($1/\alpha_p = 6.91\cdot 10^{-3}$ day). If V is the rate of IL-2 injection, then the equilibrium concentration of IL-2 in extracellular liquid (IL-2*) to some moment of time (w) from the beginning of injection will be $I_2^* = V\alpha_+/(\alpha_p\alpha_-)$. It takes a number of hours to reach this concentration.

The injection of IL-2 is described by the following equation for IL-2 concentration in lymph nodes:

$$\frac{dI_2(t)}{dt} = -(p_H H_p(t) + p_E E_p(t))I_2(t) + f_2 H_f(t) + \alpha_2(I_2^* - I_2(t)).$$

The rest of the system (7.3.1)'s equations, initial conditions (7.3.2), and values of parameters stay unchanged.

The value of specific activity of IL-2 according to [183] is equal to $(2-6)\cdot 10^6$ U/mg, therefore, $I_2^* = 0.14\cdot 10^{12}$ molecule/ml (this estimate was obtained under following conditions: $V = 1000$ U/(kg·h); weight of a patient is 70 kg; blood volume is 5000 ml). In order to preserve the smoothness of the right-hand side of tyhe system (7.3.1) assume, that

$$V = V(t) = V_0 \exp\left\{-\frac{(t-t^*)^2}{(2w^2)}\right\}, \quad t^* = \begin{cases} t, & t_1 \le t \le t_2, \\ t_1, & t < t_1, \\ t_2, & t > t_2. \end{cases}$$

where t_1, t_2 mark the beginning and the end of injection respectively, w is the average time for the attainment of concentration I_2^* in extracellular liquid (it is assumed that w is sufficiently small in comparison with $t_2 - t_1$: $w = 0.1$ days in calculations).

Let us assume that the level of homeostasis of helper-lymphocytes providing the proliferation of effector lymphocytes is 10% of a norm (such immunodeficiency is frequently encountered in clinical practice [44]). The values H_r^*, H_α^*, H_p^*, H_f^* will change according to the drop in the total level of H_E. A solution of the initial value problem (7.3.1), (7.3.2) in this case is shown in Fig. 49 by a dashed line. One can see that in this case the recovery is not complete (V_f(200 days) $\sim 10^4$ particle/ml), E and H cells are practically not produced, the regeneration rate of a target organ is decreased substantially in comparison with a norm. At the same time, B-cell response increases (the increase in maximal values of H_B, B, P by 8–10 times as compared to a norm).

Since the effector T-lymphocytes are absent, the maximum value of concentration of damaged liver cells m is decreased essentially (by 3.8 times); maximal values of concentrations of cells infected by viruses C_V and of the virus V_f are decreased just a little: by 1.2 times and by 5% respectively.

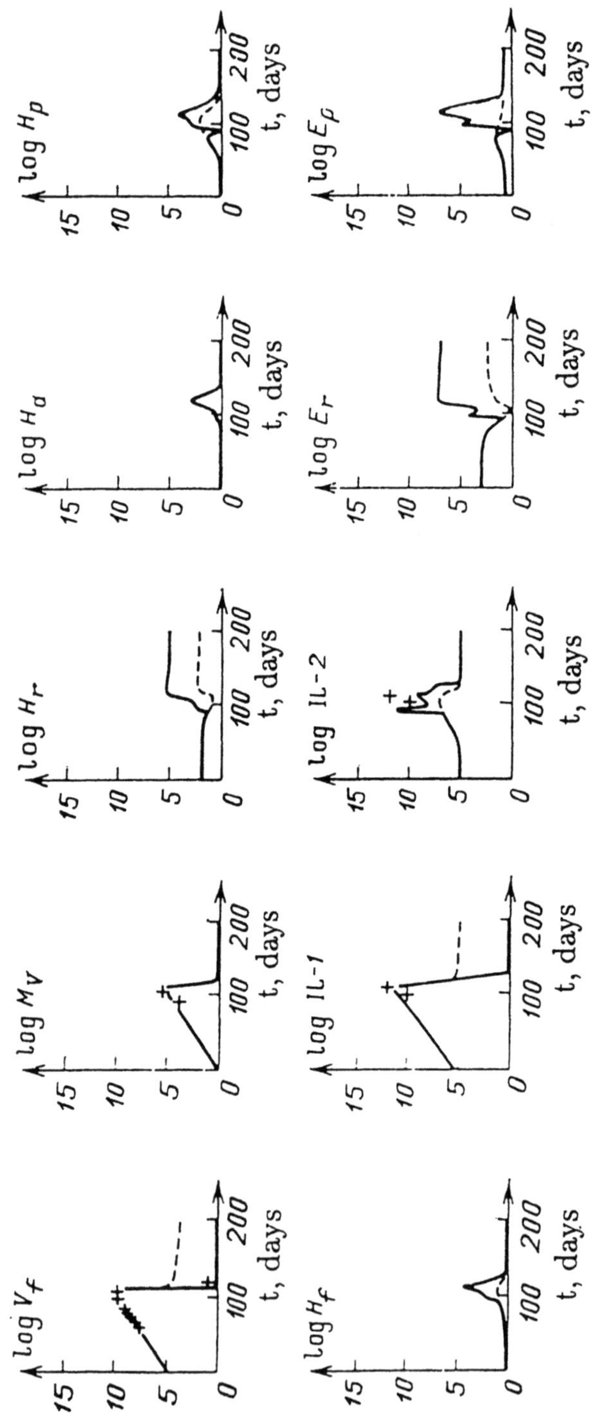
log V_f
log M_V
log H_r
log H_a
log H_p
log H_f
log IL-1
log IL-2
log E_r
log E_p
t, days

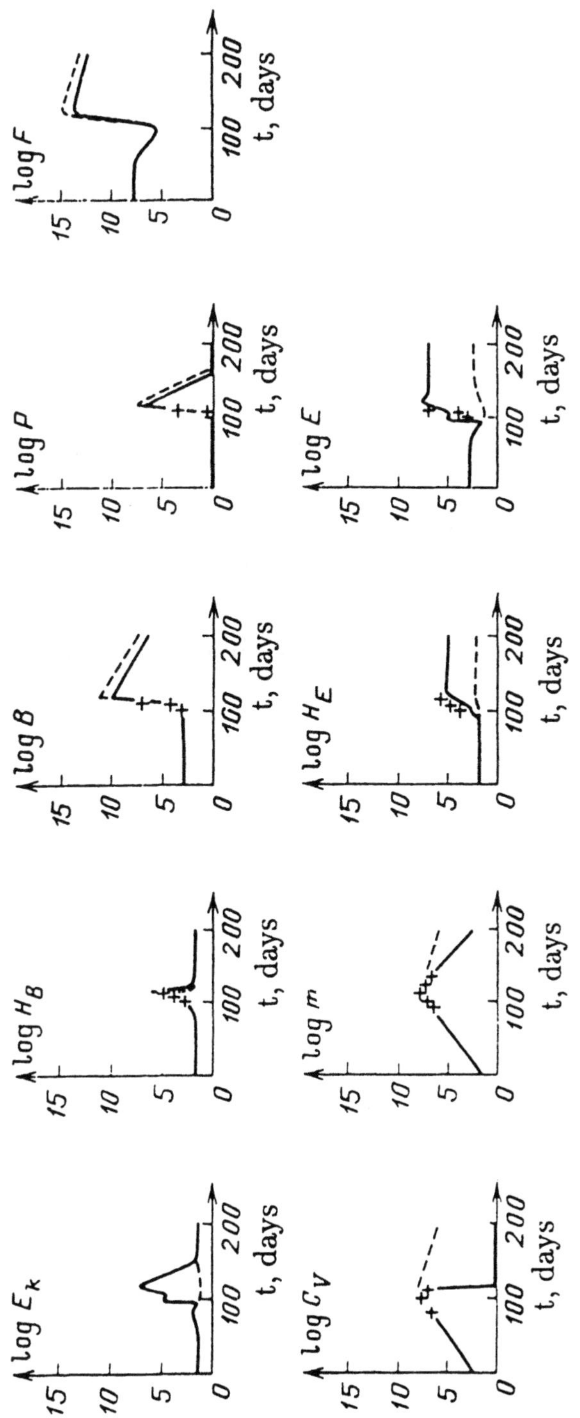

Fig. 49. Modelling of the IL-2 action in viral hepatitis B: case of deficit of T-helpers H_E^*, dashed line; result of IL-2 injection, solid line. Measurement units are the same as in Fig. 48.

Fig. 49 shows that the intravenous injection of IL-2 at the rate of 1000 $U/(\text{kg}\cdot\text{h})$ during 5 days from the 90th to 95th days nearly normalizes the reaction of the immune system, that is, leads to complete elimination of the virus on the 115th day after the infection and to the normalization of functional regeneration rate in liver. Maximal values of V_f, C_V, and m account for 85, 83, 79% of maximum values in a norm respectively, but maximum values of H_E and E remain decreased significantly (by 40–50 times). Thus the IL-2 injections lead to an increase in the degree of damage to a target organ (in comparison with the degree of damage against the background of immunodeficiency) in the case of acute form of viral hepatitis B, but, at the same time, they normalize the immune system reaction and promote complete elimination of the viruses from organism.

We consider the investigations presented in this chapter as the construction of the mathematical model of viral hepatitis B which can be used for the statement and solution of problems related to the simulation of treatment of unfavourable courses of hepatitis B virus infection.

CHAPTER 8

Viral And Bacterial Infections of Respiratory Organs

We discuss in this chapter the questions of modeling the viral and bacterial infections of the upper respiratory tract and lungs, exemplified by influenza and bacterial pneumonia. This study is the necessary stage preparing for the transition from the models of monoinfections to the construction and investigation of mathematical models of mixed (viral-bacterial) infections of respiratory organs.

Mathematical simulation of mixed infections implies the analysis and definition of such concepts as the secondary immunodeficiencies, the competition between different clones during the process of immune reaction for limited resources (see Section 4.5), and the stability of chronic process.

It is necessary, to this end, to clear out quantitative regularities of the development of viral and bacterial infections, and to identify the parameters of mathematical models of antiviral and antibacterial immune response corresponding to these regularities.

Results of the parameter identification of the model of antiviral immune response to the data of generalized picture of uncomplicated influenza A infection are presented in Section 8.1; Section 8.2 contains the results of fitting the model of antibacterial immune response to the data characterizing the acute course of bacterial pneumonia.

8.1. Modeling the Uncomplicated Influenza A

8.1.1. Quantitative characteristics of uncomplicated influenza A. Influenza is one of infections whose inducer (aetiological agent) – the influenza virus – possesses very high variability of surface antigens. This is the cause of periodic epidemics that affect 5–7% of humanity.

Influenza viruses multiply in the cells covering the surface of the upper respiratory tract (epithelial cells). Infection by influenza A viruses leads to various clinical manifestations: from completely an asymptomatic course to serious (hypertoxic) variants of an acute course. We shall consider the case of uncomplicated influenza A in an adult organism. The incubation period of influenza ranges from 24 hours to 4–5 days and depends on the organism's immune status and initial dose of viruses. Influenza A infection is transmitted by aerosol through breathing. A dose of influenza A viruses sufficient for the development of infection in a human is $\geq 10^6 - 10^8$ viruses. Influenza viruses are characterized by a high rate of multiplication in epithelial cells, and in 1–3 days following the infection their concentration increases up to $10^9 - 10^{14}$ particle/ml [210].

The influenza A virus is characterized by a number of surface antigens. Leading among them is hemaglutinin (HA), since it provides the possibility for the infection of epithelial cells. The defence of an organism against the infection implies therefore the production of antibodies specific to hemaglutinin, which neutralize viral particles, and the formation of the clone of cytotoxic T-cells that take part in the destruction of infected cells of epithelium [232].

Mechanisms of nonspecific defence, in particular, the production of interferon by the cells of epithelium of upper respiratory tract, oedema, and intensive infiltration of the site of damage by neutrophils and macrophages, play an important role in limiting the development of infection by influenza viruses [1]. Symptoms of damage to upper respiratory tracts disappear in 10–14 days from the start of disease, and the recovery of epithelial cells is complete in one month [126, 197]. The viruses can be discovered on 10–14th day after the moment of infection, and then disappear completely [188, 210].

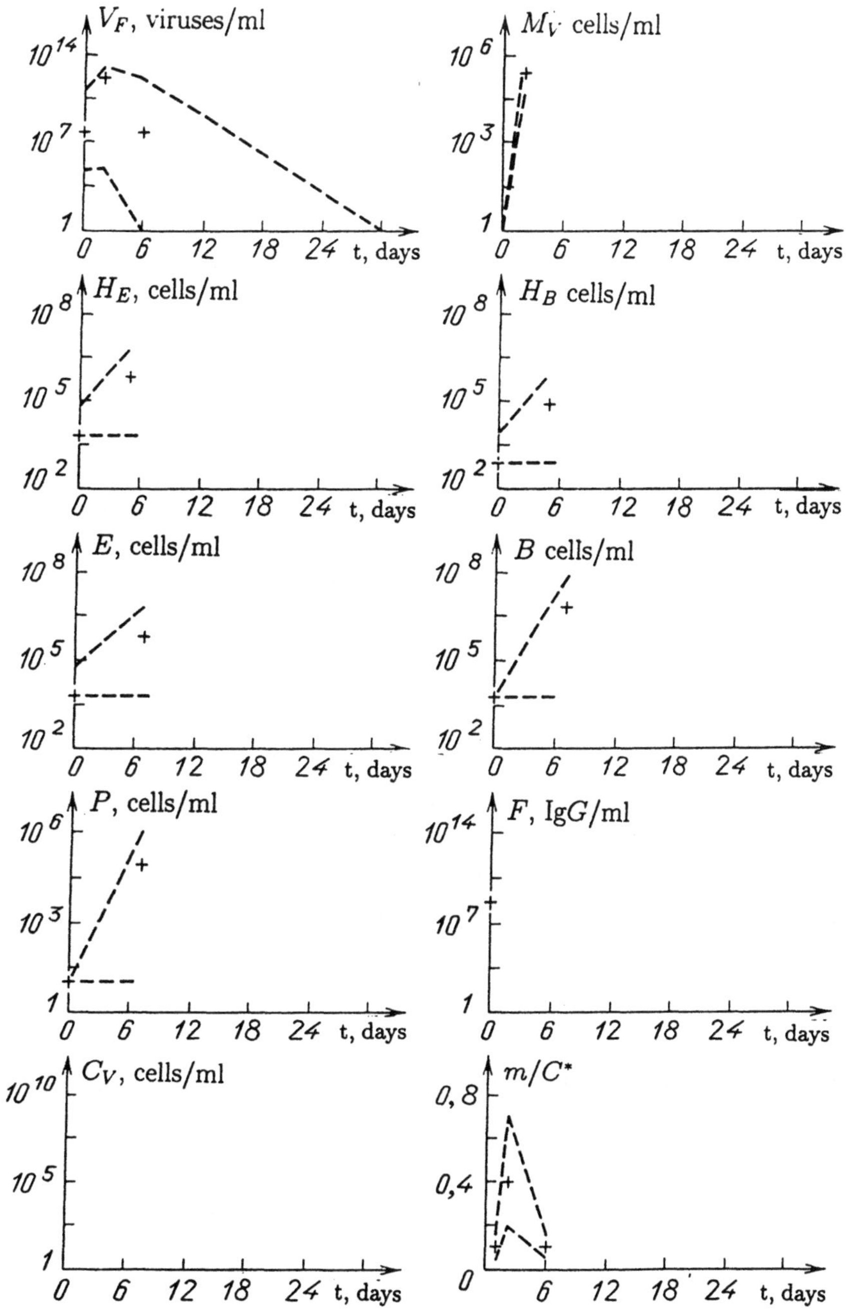

Fig. 50. Generalized picture of uncomplicated influenza A virus infection: +, the estimates of values of model's variables; dashed lines, the boundaries of permissible ranges for the values of model's variables.

While describing the reaction of the immune system to influenza A viruses, we will consider the case of primary infection by a new subtype of influenza A viruses of an organism that was previously infected by other subtypes of this virus.

This assumption implies the increase in the number of specific T-lymphocytes (frequency of occurrence is $10^{-3} - 10^{-4}$), which accounts for the formation of immune memory to slowly changing antigenic determinants of hemaglutinin.

Various clinical and laboratory observations, published immunological and virological data characterizing separate influenza processes were used to construct a generalized picture of uncomplicated influenza A virus infection. This generalized picture describes quantitatively the dynamics of indices corresponding to the variables of model of antiviral immune response. Corresponding data are shown in Fig. 50 (data on the dynamics of F and C_V are absent).

8.1.2. Mathematical model of influenza A. A mathematical model of antiviral immune response was used to construct the mathematical model of influenza which accounts for the peculiarities of immunophysiological reaction during the influenza. Let us consider, first of all, the biological meaning of the model's variables, having "linked" them with concrete processes and their localization in organism.

The processes of infection, multiplication, and destruction of infected epithelial cells are localized in a compartment of the epithelium of the upper and middle parts of respiratory tracts of the volume $\sim 0.6 - 1.0$ ml [291].

The processes of immune reaction occur in a compartment of lymphoid tissue, draining these regions of the lungs of a volume $\simeq 12$ ml. Processes of the neutralization of viruses by antibodies and of the destruction of infected cells by T-effectors occur in mucosa, covering upper and middle parts of respiratory tracts of the volume ~ 0.4 ml [291, 329]. Taking this into account, the model's variables for antiviral immune response to influenza A have the following sense:

$V_f(t)$, concentration of free viruses in 1 ml of mucous compartment (particle/ml);

$M_V(t)$, concentration of stimulated (antigen-presenting and producing IL-1) macrophages in lymphoid tissue, draining the lungs (cell/ml);

$H_E(t)$, concentration of activated (producing IL-2) helper T-cells providing the proliferation of cytotoxic T-cells of lymphoid tissue, draining the lungs (cell/ml);

$H_B(t)$, concentration of activated helper T-cells providing the proliferation of B-lymphocytes in lymphoid tissue, draining the lungs (cell/ml);

$E(t)$, concentration of activated cytotoxic T-cells in lymphoid tissue draining the lungs (cell/ml);

$B(t)$, concentration of B-lymphocytes of the compartment of lymphoid tissue draining the lungs (cell/ml);

$P(t)$, concentration of plasma cells in lymphoid tissue draining the lungs (cell/ml);

$F(t)$, concentration of IgG antibody molecules specific to hemaglutinin of influenza A viruses in mucuous compartment of (molecule/ml);

$C_V(t)$, concentration of ephitelial cells infected by viruses in the compartment of upper respiratory tract ephitelium (cell/ml);

$m(t)$, concentration of damaged cells in 1 ml of the compartment of upper respiratory tract epithelium (cell/ml).

Two features of immune reaction during uncomplicated influenza A must be taken into account, in our opinion, while constructing the mathematical model of influenza.

(1) The reaction time of T-lymphocytes (transition from the phase G_0 of cell cycle into the phase G_1) is the same as the time of the increase in virus population by one order of magnitude. Therefore it is necessary to take into account the initial distribution of T-lymphocytes in lymph nodes over the cell cycle. This distribution is as follows: 90% of T-lymphocytes are in the resting phase G_0 and $\sim$ 10% are in the activated phase G_1 [138].

(2) The comparison of the number of infected epithelial cells in the case of uncomplicated influenza A ($\sim 5 \cdot 10^9$ cells) with the quantity of CTLs generated during the course of immune reaction for the first 5–7 days ($\sim 10^6 - 10^7$ cells) allows us to assume that the effector T-cells together with direct killer action against C_V must initiate the devel-

opment of processes of nonspecific destruction of infected cells [1]. We arrive thus at the necessity of describing the inflammation and oedema phenomenon as one of the manifestations of nonspecific destruction of lungs epithelium, and, correspondingly, of the neutralization of viruses. We use, to this purpose, the description of the oedema phenomenon in the model of antiviral immune response, suggested in Section 4.3.

In order to describe the development of uncomplicated influenza A, consider the initial value problem for the system of delay-differential equations of antiviral immune response model with phenomenological description of oedema phenomenon:

$$\begin{aligned}
\frac{dV_f}{dt} &= {}^{\nu}C_V + nb_{CE}C_V E - \gamma_{VF} f_V(l) F V_f - \gamma_{VM} V_f \\
&\quad - \gamma_{VC}(C^* - C_V - m) V_f,
\end{aligned}$$

$$\frac{dM_V}{dt} = \gamma_{VM} M^* V_f - \alpha_M M_V \tag{8.1.1}$$

$$\begin{aligned}
\frac{dH_E}{dt} &= b_H^{(E)} \left[\xi(m)\rho_H^{(E)} M_V(t-\tau_H^{(E)}) H_E(t-\tau_H^{(E)}) - M_V H_E\right] \\
&\quad - b_p^{(H_E)} M_V H_E E + \alpha_H^{(E)}(H_E^* - H_E),
\end{aligned}$$

$$\begin{aligned}
\frac{dH_B}{dt} &= b_H^{(B)} \left[\xi(m)\rho_H^{(B)} M_V(t-\tau_H^{(B)}) H_B(t-\tau_H^{(B)}) - M_V H_B\right] \\
&\quad - b_p^{(H_B)} M_V H_B B + \alpha_H^{(B)}(H_B^* - H_B),
\end{aligned}$$

$$\begin{aligned}
\frac{dE}{dt} &= b_p^{(E)} \left[\xi(m)\rho_E M_V(t-\tau_E) H_E(t-\tau_E) E(t-\tau_E) - M_V H_E E\right] \\
&\quad - b_{EC} C_V E + \alpha_E(E^* - E),
\end{aligned}$$

$$\begin{aligned}
\frac{dB}{dt} &= b_p^{(B)} \left[\xi(m)\rho_B M_V(t-\tau_B) H_B(t-\tau_B) B(t-\tau_B) - M_V H_B B\right] \\
&\quad + \alpha_B(B^* - B),
\end{aligned}$$

$$\frac{dP}{dt} = b_p^{(P)}\xi(m)\rho_P M_V(t-\tau_P) H_B(t-\tau_P) B(t-\tau_P) + \alpha_P(P^* - P),$$

$$\frac{dF}{dt} = \rho_F P - \gamma_{FV} V_f F - \alpha_F F,$$

$$\frac{dC_V}{dt} = \sigma V_f (C^* - C_V - m) - b_{CE} f_C(l) C_V E - b_m C_V \tag{8.1.1}$$

$$\frac{dm}{dt} = b_{CE} f_C(l) C_V E + b_m C_V - \alpha_m m,$$

$$\xi(m) \equiv 1 - m/C^*, \quad f_V(l) = 1 + \mu_\nu \frac{C_V}{C^*}, \quad f_C = 1 + \mu_C \frac{C_V}{C^*}.$$

Here, the variable $l = C_V/C^*$ serves as a characteristic of oedema, and the functions $f_V(l)$ and $f_C(l)$ describe the intensification of processes of neutralization and destruction of V_f and C_V, respectively.

Notice, that formally the intensification of lymph circulation must lead both to the increase in the rate of destruction of CTLs and to the expenditure of antibodies; however, we neglect these considerations at a given stage of investigation, since the main result of oedema in this concept is the amplification of the action of CTLs and of specific IgG through nonspecific mechanisms of destruction of C_V and V_f.

Let us describe the process of infection of a healthy organism by the following system of initial conditions at the moment $t = t_0 = 0$:

$$V_f(0) = V_f^0, \quad M_V(0) = M_V^0, \quad H_E(0) = a_{H_E}^1 H_E^*, \quad H_B(0) = a_{H_B}^1 H_B^*,$$

$$E(0) = a_E^1 E^*, \quad B(0) = B^*, \quad P(0) = P^*, \quad F(0) = \rho_F P^*/\alpha_F,$$

$$C_V(0) = 0, \quad m(0) = 0, \tag{8.1.2}$$

$$M_V(s) H_E(s) = \varphi_H^{(E)}(s), \quad -\tau_H^{(E)} \le s \le 0,$$

$$M_V(s) H_B(s) = \varphi_H^{(B)}(s), \quad -\tau_H^{(B)} \le s \le 0,$$

$$M_V(s) H_E(s) E(s) = \varphi_E(s), \quad -\tau_E \le s \le 0$$

$$M_V(s) H_B(s) B(s) = 0, \quad -\tau \le s \le 0, \quad \tau = \max\{\tau_B, \tau_P\},$$

where $\varphi_H^{(E)}(t)$, $\varphi_H^{(B)}(t)$, and $\varphi_E(t)$ are functions of a special form used for the description of initial activation of T-lymphocytes. These functions were selected from the class of infinitely differentiable functions $\omega_1(t)$ with a finite carrier:

$$\omega_1(t) = \begin{cases} \dfrac{1}{C_n} \exp\left\{-\dfrac{1}{1-t^2}\right\}, & 0 \le |t| < 1, \\ 0, & |t| \ge 1, \end{cases}$$

so, that $\varphi(t)$ would have the form presented in Fig. 51 and

$$\left[\int_{-\tau_H^{(E)}} \varphi_H^{(E)}(t)dt + a^1_{H_E}\right] = 1, \quad \left[\int_{-\tau_H^{(B)}} \varphi_H^{(B)}(t)dt + a^1_{H_B}\right] = 1,$$
$$\left[\int_{-\tau_E}^{0} \varphi_E(t)dt + a^1_E\right] = 1. \tag{8.1.3}$$

So, assigning $a^1_{H_E} = a^1_{H_B} = a^1_E = 0.1$, we describe the initial distribution of T-lymphocytes over the phases G_0 (90%) and G_1 (10%) of cell cycle.

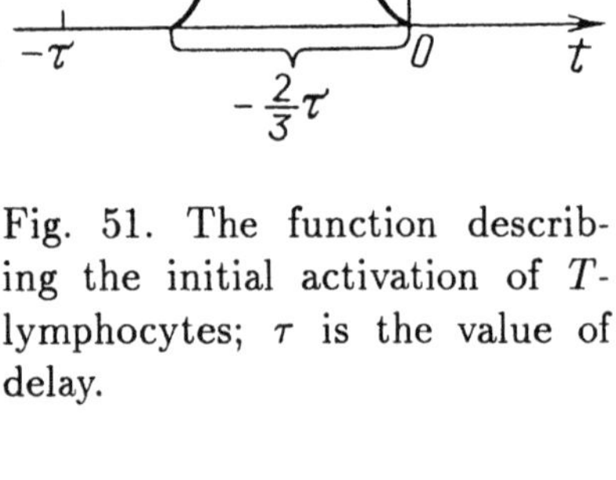

Fig. 51. The function describing the initial activation of T-lymphocytes; τ is the value of delay.

Quantitative fitting of the model (8.1.1)–(8.1.3) to the data of the generalized picture, characterizing the course of uncomplicated influenza A, implies the construction of the regions of permissible values of model's coefficients α. Part of them can be taken from the model of viral hepatitis B since the immune reaction of organism to antigens stereotypic character and practically does not depend on their specificity with respect to initiated pathological processes. Results of estimates of permissible regions are presented in Table 8.

Table 8

Parameter, unit	Biological meaning of the parameter	Permissible range	Initial value
M^*, mol/ml	Concentration of MHC bearing macrophages in a lymph node	$(0.5-3.0)10^{-18}$	10^{-18}
H_E^*, mol/ml	Concentration of specific T-helpers for CTLs in a lymph node	$(1.0-10)10^{-19}$	10^{-19}
H_B^*, mol/ml	Concentration of specific T-helpers for B-lymphocytes in a lymph node	$(1.0-10)10^{-20}$	10^{-20}
E^*, mol/ml	Concentration of specific precursor of CTLs in a lymph node	$(1.0-10)10^{-19}$	10^{-19}
B^*, mol/ml	Concentration of specific B-lymphocytes in lymph node	$(0.5-5.0)10^{-19}$	10^{-20}
P^*, mol/ml	Concentration of specific plasma cells in a lymph node	$(0.3-7.0)10^{-23}$	1.83^{-23}
F^*, mol/ml	Concentration of specific IgG antibodies in mucous compartment	$(0.17-1.7)10^{-15}$	8.5^{-16}
C^*, mol/ml	Number of epithelial cells in upper respiratory tract	$(0.17-1.7)10^{-15}$	1.7^{-14}
α_M, day^{-1}	Rate constant for the loss by a macrophage of stimulated state	$(1.0-1.5)$	1.2
$\alpha_H^{(E)}$, day^{-1}	Rate constant for the loss by T-helpers for E-cells of stimulated state	$(0.8-1.2)$	1
$\alpha_H^{(B)}$, day^{-1}	Rate constant for the loss by T-helpers for B-cells of stimulated state	$(0.8-1.2)$	1
α_E, day^{-1}	Rate constant for natural natural death of CTLs	$(0.33-0.5)$	0.4
α_B, day^{-1}	Rate constant for natural death of B-lymphocytes	$(0.05-0.1)$	0.1

See the continuation

Table 8 (continuation)

Parameter, unit	Biological meaning of the parameter	Permissible range	Initial value
α_P, day^{-1}	Rate constant for natural death of plasma cells	$(0.33 - 0.5)$	0.4
α_F, day^{-1}	Rate constant for natural death of antibodies	0.043	0.043
$\tau_H^{(E)}$, day	Duration of the division cycle for Th1 cells	$(0.4 - 0.8)$	0.6
$\tau_H^{(B)}$, day	Duration of the division cycle for Th2 cells	$(0.4 - 0.8)$	0.6
τ_E, day	Duration of the division cycle for CTLs	$(0.4 - 1.0)$	0.5
τ_B, day	Duration of the division cycle for B-lymphocytes	$(0.4 - 1.0)$	0.5
τ_P, day	Duration of the division and differentiation for B-cells when plasma cells appear	$(0.4 - 1.0)$	0.5
$\rho_H^{(E)}$	Number of Th1 cells created by single division cycle	$(2 - 4)$	4
$\rho_H^{(B)}$	Number of Th2 cells created by single division cycle	$(2 - 4)$	4
ρ_E	Number of CTLs created by division(s series)	$(2 - 4)$	2
ρ_B	Number of B-cells created by division(s series)	$(1.5 - 3)$	3
ρ_P	Number of plasma cells in a clone of B-cells created by division(s series)	$(0.5 - 1)$	1
$b_H^{(E)}$, $\frac{\text{ml}}{\text{mol}\cdot\text{day}}$	Rate constant for the stimulation of Th1 cell	$(0.4 - 3.0)10^{18}$	10^{18}
$b_H^{(B)}$, $\frac{\text{ml}}{\text{mol}\cdot\text{day}}$	Rate constant for the stimulation of Th2 cell	$(0.4 - 3.0)10^{18}$	10^{18}
$b_p^{(E)}$, $\frac{\text{ml}^2}{\text{mol}^2\cdot\text{day}}$	Rate constant for the stimulation of CTLs	$(0.5 - 70)10^{38}$	10^{38}
$b_p^{(B)}$, $\frac{\text{ml}^2}{\text{mol}^2\cdot\text{day}}$	Rate constant for the stimulation of B-cells	$(1 - 100)10^{37}$	10^{38}

See the continuation

Table 8 (continuation)

Parameter, unit	Biological meaning of the parameter	Permissible range	Initial value
$b_p^{(P)}$, $\frac{\text{ml}^2}{\text{mol}^2\cdot\text{day}}$	Rate constant for the stimulation of B-cells resulting in the proliferation and differentiation into plasma cells	$(1-100)10^{37}$	10^{38}
$b_p^{(H_E)}$, $\frac{\text{ml}^2}{\text{mol}^2\cdot\text{day}}$	Expenditure rate of Th1 for the stimulation of CTLs		10^{34}
$b_p^{(H_B)}$, $\frac{\text{ml}^2}{\text{mol}^2\cdot\text{day}}$	Expenditure rate of Th2 for the stimulation of B-cells		10^{34}
γ_{MV}, $\frac{\text{ml}}{\text{mol}\cdot\text{day}}$	Rate constant for antigenic stimulation of macrophages in a lymph node	$10^{11}-4\cdot10^{16}$	10^{12}
γ_{VM}, day^{-1}	Rate constant for the removal of antigen particles by macrophages	$2-4$	2
γ_{FV}, $\frac{\text{ml}}{\text{mol}\cdot\text{day}}$	Rate constant for the binding of one IgG molecule with viral particle	$(0.86-86)10^{14}$	$8.6\cdot10^{14}$
γ_{VF}, $\frac{\text{ml}}{\text{mol}\cdot\text{day}}$	Rate constant for the neutralization of viral particles by IgG molecules	$(0.86-860)10^{13}$	$8.6\cdot10^{14}$
ρ_F^*, $\frac{\text{mol}}{\text{day}}$	Rate constant for the synthesis of IgG molecules by one plasma cell	$(8.5-17)10^{7}$	$1.7\cdot10^{8}$
σ, $\frac{\text{ml}}{\text{mol}\cdot\text{day}}$	Rate constant for the infection of epithelial cells by influenza viruses	$(4.7-12)10^{9}$	$8\cdot10^{9}$
b_{CE}, $\frac{\text{ml}}{\text{mol}\cdot\text{day}}$	Rate constant for the destruction of epithelial cells by CTLs	$(2.4-36)10^{10}$	10^{11}
b_{EC}, $\frac{\text{ml}}{\text{mol}\cdot\text{day}}$	Rate constant for the death of CTLs due to the lytic interaction with infected cells	$(2.4-36)10^{9}$	10^{10}

See the continuation

Table 8 (continuation)

Parameter, unit	Biological meaning of the parameter	Permissible range	Initial value
b_m, day^{-1}	Rate constant for the destruction of infected epithelial cells due to the cytopacycity of viruses	$(0.5-2)$	1
α_m, day^{-1}	Rate constant for the regeneration of epithelial cells	$(1-3)$	3
ν, day^{-1}	Rate constant for the secretion for influenza A viruses by one infected epithelial cell	(10^2-10^4)	$6\cdot10^3$
n	Number of influenza A viruses that appear when epithelial cell is destroyed by CTLs	0	0
γ_{VC}, $\frac{\text{ml}}{\text{mol}\cdot\text{day}}$	Rate constant for the adsorption of influenza viruses by epithelial cells	$(4.7-12)10^{10}$	$8\cdot10^{10}$

The refinement of values of separate model's coefficients was done by solving the identification problem on the observational data of generalized picture. We used the approach described in Sections 5.1 and 7.3., that is, the consecutive fitting of model to data. The solution, so obtained, which describes quantitative characteristics of the immune reaction to uncomplicated influenza A, is shown in Fig. 52 and corresponding values of parameters are presented in Table 9.

Dynamics of the variable $V_f(t)$, i.e., virus concentration, is of special interest. It is characterized by rapid growth and then by the stabilization of the quantity V_f; when the development of immune response and oedema starts, V_f decreases rapidly down to some level, and only after certain quantity of antibodies and T-effectors is produced to the 10–14th day, stable diminution of V_f begins down to zero concentrations. The development of disease leads to significant (by 6 orders) decrease in $F(t)$ because of high infection dose as compared to the level of antibodies F^*.

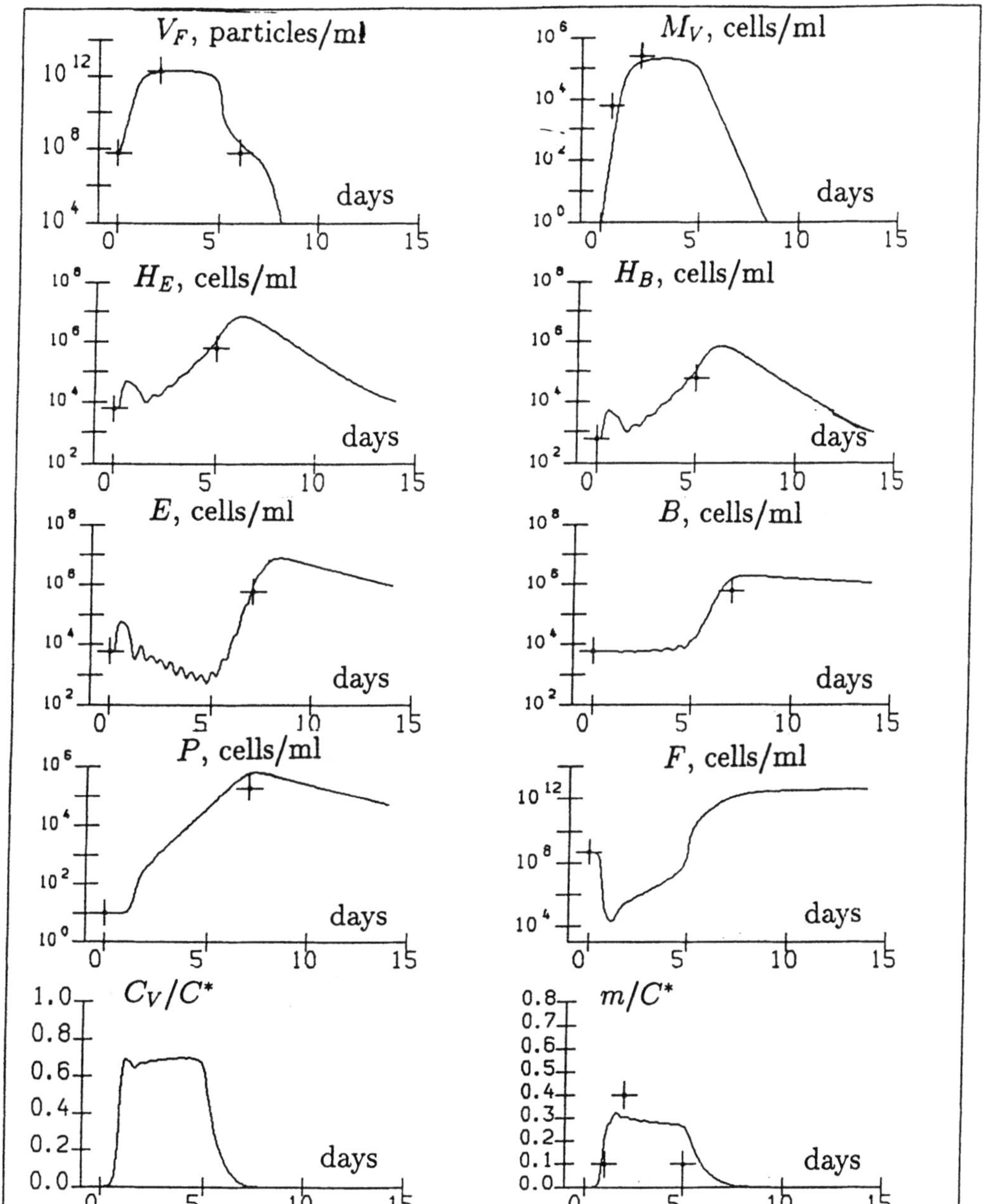

Fig. 52. Model solution obtained as a result of identification of the parameters using the data of generalized picture of uncomplicated influenza A virus infection.

Table 9

Results of the parameter identification

Parameter	Parameter's value	Parameter	Parameter's value	Initial guess
M^*	10^{-18} mol/ml	$\rho_H^{(E)}$	4	
H_E^*	10^{-19} mol/ml	$\rho_H^{(B)}$	4	
H_B^*	10^{-20} mol/ml	ρ_E	2	
E^*	10^{-19} mol/ml	ρ_B	3	
B^*	10^{-20} mol/ml	ρ_P	1	
P^*	$1.83 \cdot 10^{-23}$ mol/ml	ρ_F	$1.7 \cdot 10^8$ molec/(day·)cell	
F^*	$8.5 \cdot 10^{-16}$ mol/ml	$b_H^{(E)}$	$1.8 \cdot 10^{19}$ ml/(mol·day)	10^{18}
C^*	$1.7 \cdot 10^{-14}$ mol/ml	$b_H^{(B)}$	$2.1 \cdot 10^{19}$ ml/(mol·day)	10^{18}
α_M	3.3 day^{-1}	$b_p^{(E)}$	10^{39} $\text{ml}^2/(\text{mol}^2\cdot\text{day})$	10^{38}
$\alpha_H^{(E)}$	1 day^{-1}	$b_p^{(B)}$	$1.2 \cdot 10^{40}$ $\text{ml}^2/(\text{mol}^2\cdot\text{day})$	10^{38}
$\alpha_H^{(B)}$	1 day^{-1}	$b_p^{(P)}$	$1.2 \cdot 10^{38}$ $\text{ml}^2/(\text{mol}^2\cdot\text{day})$	10^{38}
α_E	0.4 day^{-1}	$b_p^{(H_E)}$	10^{31} $\text{ml}^2/(\text{mol}^2\cdot\text{day})$	
		$b_p^{(H_B)}$	10^{31} $\text{ml}^2/(\text{mol}^2\cdot\text{day})$	
α_B	0.1 day^{-1}	σ	$2 \cdot 10^{13}$ ml/(mol·day)	$8 \cdot 10^9$
α_P	0.4 day^{-1}	b_{CE}	$1.9 \cdot 10^8$ ml/(mol·day)	10^{11}
α_F	0.043 day^{-1}	b_m	1.52 day^{-1}	1
V_F^0	$6 \cdot 10^8$ particle/ml	α_m	4.15 day^{-1}	3
	10^{-16} mol/ml	ν	511 day^{-1}	$6 \cdot 10^3$
$\tau_H^{(E)}$	0.6 day	γ_{VC}	$5.5 \cdot 10^{13}$ ml/(mol·day)	$8 \cdot 10^{10}$
$\tau_H^{(B)}$	0.6 day	γ_{VM}	1.7 day^{-1}	2
τ_E	0.5 day	γ_{VF}	$8.6 \cdot 10^{14}$ ml/(mol·day)	
τ_B	0.5 day	γ_{FV}	$7.7 \cdot 10^{17}$ ml/(mol·day)	$8.6 \cdot 10^{14}$
τ_P	0.5 day	γ_{MV}	$1.6 \cdot 10^{11}$ ml/(mol·day)	10^{12}
n	0	b_{EC}	$1.2 \cdot 10^{13}$ ml/(mol·day)	10^{10}
		μ_V	$2.4 \cdot 10^5$	10^4
		μ_C	$2 \cdot 10^5$	10^4

We consider the system of equations (7.1.1) with initial conditions (7.1.2) and the set of coefficients in Table 9 as a basis for future studies in mathematical modelling of influenza infection. In particular, the model could be further refined by considering the reaction of the interferon system that plays a key role in limiting the influenza infection at an early phases [39, 206].

The damage of epithelial cells in the case of influenza creates favorable conditions for the development of secondary bacterial infection induced by staphylococci and other bacteria [197, 198]. The next Section is devoted to the modeling of bacterial infection in lungs exemplified with acute bacterial pneumonia.

8.2. Mathematical modeling of destructive pneumonia

Destructive pneumonia is a pathological process characterized by the disintegration of lung tissue as a result of the action of pathogenic microorganisms. Unlike influenza, when a target is the epithelium of respiratory tracts, in the case of pneumonia the alveolar tissues are destroyed, where the exchange of gases occurs. The pathological process affects 2–3 segments of lungs, which is equal to 10–15% of lungs volume (approximately 500 ml) [197, 291]. The frequency of a lethal outcome is 10–50% of the total number of patients with destructive pneumonia [197].

Various microorganisms may be aetiological agents of destructive pneumonia. It is noted that these microorganisms cannot provoke the inflammatory and destructive processes in lung tissue in case of normal functioning of local and general defence mechanisms. Therefore the malfunction of these mechanisms is necessary for the development of acute destructive pneumonia. This malfunction may be caused by the persistence of viruses in bronchi, influenza, and a number of other factors.

We shall consider destructive pneumonia provoked by S. aureus against the background of a defence mechanism mulfunctioning because of influenza. The disease is caused by the multiplication of microorganisms develops during the period of 10–14 days and in case of the recovery ends as a result of destruction of bacteria by antibodies, macrophages, and neutrophils.

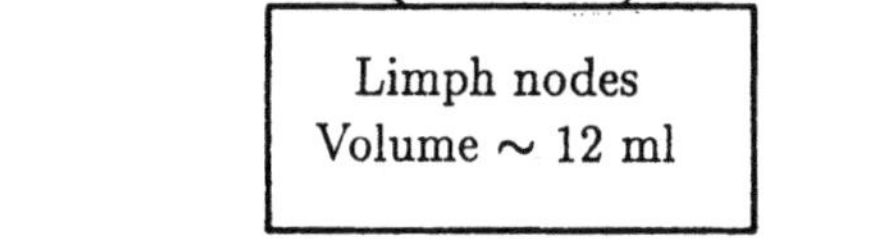

Fig. 53. Structural and functional organization of processes connected with destructive pneumonia.

Consider the structural and functional organization of processes involved in destructive pneumonia (Fig. 53). As we have noticed, the disease develops in two–three lung segments. Macrophages activated as a result of absorption and digestion of bacteria get into lymphoid tissue draining the damaged segments [189]. Expanding infection affects the lymphoid tissue of the spleen. Antibodies (F) are transferred to the site of damage with blood through the alveolar-capillary membrane [354]. Absorption and further destruction of bacteria in the site of damage are mediated by neutrophils and macrophages, which absorb effectively the bacteria opsonized by antibodies [52].

The investigation of opsonization and further destruction of S. aureus by phagocytes [271] revealed that:

- the main target is the peptidoglycan of cellular wall of staphylococci (Pg);
- the opsonization is performed through IgG-class of antibodies;
- the activation of the complement system occurs both in classical and in alternative pathways.

Concentration of IgG antibodies specific to Pg in blood of healthy adult human is 15 mkg/ml, i.e., is about 0.1% of total IgG concentration in blood. It was found that normally the concentration of IgG does not limit the rate of opsonization and absorption of microorganisms. The average amount of bacteria absorbed by one monocyte is about 120 bacteria [181] (for neutrophils it is about 40–50 bacteria [180]).

Multiplying, S. aureus secretes a number of toxins. Among them, α-toxin is the most important for the pathogenesis of pneumonia. This toxin triggers the cascade of pathophysiological reactions which leads to the migration of macrophages and neutrophils into the seat of damage from circulating blood and to the percolating of blood plasma through the alveolar space.

According to conceptions on destructive pneumonia discussed, separate the following major variables of the model:

$K(t)$, concentration of S. aureus bacteria capable of multiplication and secreting toxins in damaged lung segments;

$M_K(t)$, concentration of activated macrophages and dendritic cells in lymph nodes draining the damaged lung segments (called hereafter simply LN). M_K carry on their surface the complex of processed Ag with MHC-II and produce IL-1;

$H_B(t)$, concentration of activated helper T-lymphocytes in LN;

$B(t)$, concentration of B-lymphocytes in LN;

$P(t)$, concentration of plasma cells producing the antibodies of given specificity in lymph nodes of lungs;

$F(t)$, concentration of specific antibodies in blood;

$m(t)$, part of damaged cells of lung parenchyma.

In order to describe the destructive pneumonia we use the system of equations of antibacterial immune response model constructed in Section 4.4. There are some reasons to believe that the process of interaction between the antibodies F and bacteria depends essentially on the rate of their delivery to the seat of damage from blood [339]. To describe this phenomenon, introduce the variable l characterizing the intensity of bacteria caused inflammation. Since the intensity of inflammatory reaction is determined mainly by the portion of damaged cells of lungs parenchyma, i.e., by the value of m,

$$l = k_m m.$$

The rate of delivery of antibodies to the site of inflammation from blood is determined by the function

$$f(l) = 1 + k_l l = 1 + k_l k_m m = 1 + \mu m = \eta(m),$$

where k_m, k_l are certain constants, $\mu = k_l k_m$. The system of equations of the model of antibacterial immune response applied for destructive pneumonia is of the following form:

$$\frac{dK}{dt} = \beta K - \gamma_{KM} KM - \gamma_{KF} KF_{\eta}(m),$$

$$\frac{dM_K}{dt} = \gamma_{MK} MK - \alpha_M M_K,$$

$$\begin{aligned}\frac{dH_B}{dt} &= b_H(\rho_H \xi(m) H_B(t - \tau_H) M_K(t - \tau_H) - H_B M_K) \\ &\quad - b_p M_K H_B B + \alpha_H(H_B^* - H^B),\end{aligned}$$

$$\frac{dB}{dt} = b_p^{(B)}(\rho_B \xi(m) H_B(t - \tau_B) M_K(t - \tau_B) - H_B M_K B) + \alpha_B(B^* - B),$$

$$\frac{dP}{dt} = b_p^{(P)}(\rho_P \xi(m) H_B(t-\tau_P) M_K(t-\tau_P) B(t-\tau_P) + \alpha_P(P^* - P),$$

$$\frac{dF}{dt} = \rho_F P - \gamma_{FK}\eta(m) K F + \alpha_F(F^* - F),$$

$$\frac{dm}{dt} = \sigma K - \alpha_m m \max(0, 1-m),$$

$$\xi(m) = \max(0, 1-m).$$

The process of formation of bacterial inflammatory site can be described by the following initial conditions:

$$M_K(s)H_B(s) = 0, \quad -\tau_H \leq s < 0,$$
$$M_K(s)H_B(s)B(s) = 0, \quad -\max(\tau_B, \tau_P) \leq s < 0,$$
$$K(0) = K_0, \quad M_K(0) = 0, \quad H_B(0) = H_B^*, \quad B(0) = B^*,$$
$$P(0) = P^*, \quad F(0) = \bar{F}, \quad m(0) = 0.$$

In order to fit the model to the data of the generalized picture the regions of permissible values for all parameters of the model of destructive pneumonia were constructed (Table 10), and the problem of parameter identification was solved.

Fig. 54 shows the model solution corresponding to the refined set of parameters (Table 10). The solution so obtained reproduces quantitative characteristics of destructive pneumonia. These results are preliminary and provide a certain basis for further modelling.

We regard the construction of the models of influenza and pneumonia as an initial stage in the solution of complex problems of modeling and controlling the mixed infections on the basis of mathematical model (see Section 4.5), which accounts for the limitations of total intensity of immune reactions to several antigens.

This approach opens the way to the analysis of such phenomena as secondary immunodeficiencies and immune homeostasis.

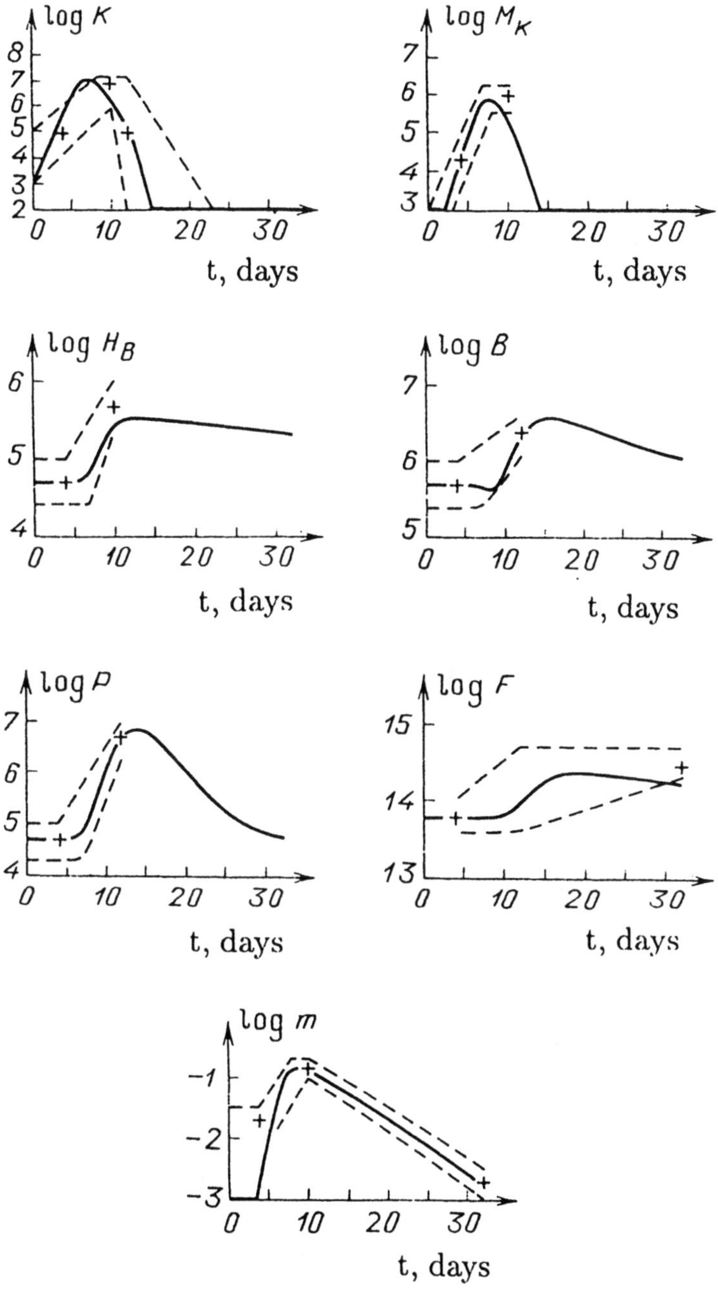

Fig. 54. Solution of the model corresponding to acute course of destructive pneumonia. +, estimations of the model's variables; dashed lines, limits of admissible values of the model's variables. Units of measurements: cell/ml for K, M_K, H_B, B, P; molecule/ml for F; part of damaged cells for m.

Table 10

Parameter, unit	Biological meaning of the parameter	Permissible range	Initial (refined) value
β, day^{-1}	Constant for the rate of multiplication of bacteria	$(1-30)$	4.0 (2.7)
γ_{KM}, $\frac{\text{ml}}{\text{mol}\cdot\text{day}}$	Rate constant for nonspecific destruction of bacteria	$(0.6-30)10^{16}$	$7.2\cdot 10^{16}$ $(3.6\cdot 10^{16})$
M, mol/ml	Alveolar macrophages concentration	$(1.7-5.0)10^{-18}$	$3.4\cdot 10^{-18}$
γ_{KF}, $\frac{\text{ml}}{\text{mol}\cdot\text{day}}$	Rate constant for the destruction of bacteria enhanced by specific antibodies	$(0.6-3)10^{10}$	$2\cdot 10^{10}$ $(1\cdot 10^{10})$
μ	Acceleration rate constant for the delivery of plasma blood proteins and neutrophiles	$(1-100)$	10 (15)
K_0, mol/ml	Initial bacteria concentration	$(0.17-170)10^{-21}$	$1.7\cdot 10^{-21}$
γ_{MK}, $\frac{\text{ml}}{\text{mol}\cdot\text{day}}$	Rate constant for the activation of macrophages	$(0.6-600)10^{16}$	$2.3\cdot 10^{16}$ $(5\cdot 10^{16})$
α_M, day^{-1}	Rate constant for the loss by a macrophage of stimulated state	$(1-3)$	1.2 (2.5)
b_H, $\frac{\text{ml}^2}{\text{mol}^2\cdot\text{day}}$	Rate constant for the stimulation of Th2 cells	$(0.4-3)10^{18}$	10^{18}
ρ_H	Number of descendants of Th2 cells created by single division	2	2
b_p, $\frac{\text{ml}^2}{\text{mol}^2\cdot\text{day}}$	Coefficient for the expenditure of Th2 cells to stimulate B-cells	$(0.1-100)10^{34}$	$5\cdot 10^{34}$ $(6\cdot 10^{33})$
α_M, day^{-1}	Rate constant for natural death of Th2 cells	$(0.01-0.1)$	0.03
H_B^*, mol/ml	Concentration of specific Th2 cells in lymphoid tissue	$(1.7-25)10^{-20}$	$8.4\cdot 10^{-20}$
τ_H, day	Duration of the division cycle for Th2 cells	$(0.4-0.8)$	0.6 (0.4)
$b_p^{(B)}$, $\frac{\text{ml}^2}{\text{mol}^2\cdot\text{day}}$	Rate constant for the stimulation of B-cell	$(0.01-10)10^{37}$	10^{37} $(7.2\cdot 10^{35})$

See the continuation

Table 10 (continuation)

Parameter, unit	Biological meaning of the parameter	Permissible range	Initial (refined) value
ρ_B	Number of B-cells created by single division	$(10-18)$	16
α_B, day^{-1}	Rate constant for natural death of B-cells	$(0.05-0.1)$	0.1
B^*, mol/ml	Concentration of specific B-cells in lymphoid tissue	$(1.7-25)10^{-19}$	$8.4 \cdot 10^{-19}$
τ_B, day	Duration of the division cycle for B-cells	$(2-3)$	2 (3)
$b_p^{(P)}$, $\frac{\text{ml}^2}{\text{mol}^2 \cdot \text{day}}$	Rate constant for the stimulation of B-cell while describing the number of plasma cells	$(0.01-10)10^{37}$	10^{37} $(1.4 \cdot 10^{37})$
ρ_P	Number of plasma cells created by B-cell in division(s series)	$(2-5)$	3
α_P, day^{-1}	Rate constant for natural death of plasma cells	$(0.33-0.5)$	0.4
P^*, mol/ml	Concentration of specific plasma cells in a lymph node	$(1.7-25)10^{-20}$	8.4^{-20}
τ_P, day	Duration of division and differentiation for B-cells before plasma cells appear	$(3-4)$	3
ρ_F, day^{-1}	Rate constant for the secretion of antibodies by plasma cells	$(1-50)10^6$	$2 \cdot 10^6$ $(5.1 \cdot 10^6)$
γ_{FK}, $\frac{\text{ml}}{\text{mol} \cdot \text{day}}$	Rate constant for bacteria clearance	$(1.2-6) \cdot 10^{11}$	$4.3 \cdot 10^{11}$
α_F, day^{-1}	Rate constant for natural death of IgG antibodies	0.043	0.043
F^*, mol/ml	Concentration of specific antibodies synthesized in lymphoid tissue intact by infection	$(0.83-25)10^{-11}$	9.5^{-11}
$\overline{F}$, mol/ml	Concentration of specific antibodies in blood	$(0.17-2.5)10^{-10}$	10^{-10}
σ, $\frac{\text{ml}}{\text{mol} \cdot \text{day}}$	Rate constant for the infection in lung tissue	$(0.6-30)10^{15}$	$6 \cdot 10^{15}$ $(4.1 \cdot 10^{15})$
α_M, day^{-1}	Rate constant for the regeneration of lung tissue	$(0.1-0.4)$	0.15 (0.2)

CHAPTER 9

Model of Experimental Influenza Infection

This chapter presents investigation results related to the action of antiviral drugs on the processes in organisms of animals during the influenza infection. The methodology of statistical analysis of the data of immunological experiments was discussed in Section 5.2. We begin with the construction of the mathematical model of processes under study, and then we use this model to analyse the experimental data.

The results discussed in this chapter were obtained as a result of a joint research effort of the Institute for Numerical Mathematics, the Russian Academy of Sciences, and of the Institute of Experimental Medicine, Russian Academy of Medical Sciences (Laboratory of Professor R.Y. Polyak, Leningrad), published in [203, 356].

9.1. Mathematical Model of Experimental Influenza Infection

Two sections of this chapter deal with the mechanisms of influence of antiviral drugs on the defence processes in organisms of animals in the case of acute influenza infection. The antiviral effect of these drugs was established experimentally and is based on the fact that the application of these drugs decreases the mortality in the group of lethally infected animals. In order to elaborate a strategy and tactics of treatment we need to investigate the mechanisms of the action of a drug on the level of organism's cellular population. To this end we construct a mathematical model of the process of acute influenza infection accounting for the basic factors that determine the outcome of disease.

The construction of the mathematical model implies, naturally, a certain idealization of a real scheme of interaction between a virus and a host's organism. Therefore, in accordance with contemporary concepts of the process under study, we consider the following basic factors: quantity of viruses in a damaged organ; influence of the system of interferon; cellular and humoral systems of immunity.

While constructing the model we will not separate α, β, and γ interferons, M, G, and A classes of immunoglobulins, etc. We describe the main defence factors taking into account their effect. So, the variable "interferon" will be considered as the factor which makes a cell insensitive to viral infection; "antibodies" as the factor that provides for specific neutralization of extracellular viruses; "specific effector-lymphocytes" as the population of effector-lymphocytes (with no division into T and B subpopulations) connected with the processes of production of antibodies and with the elimination of cells infected by viruses.

A number of processes of the last type are connected with specific lymphocytes just indirectly: the latter can activate, for example, the series of nonspecific cellular mechanisms (NK-cells, macrophages), which directly kill the cells infected by viruses. Processes of the lysis of infected cells by specific lymphocytes belong here too.

While constructing the model we shall consider all the interactions as homogeneous reactions between cellular–humoral components. We take into account the fact that the measurements of interacting components are made in those organs of the animal where the basic reservoirs are located for the corresponding components (virus – lungs; immunoglobulins – blood, lung secretions; cellular system – spleen, lungs). Having made these comments, proceed with the construction of the mathematical model.

Let us introduce the following variables:

t, time after the infection (in days);

$C_V(t)$, population of cells infected by viruses;

$V(t)$, population of infectious viral particles;

$F(t)$, immunoglobulins;

$L_E(t)$, specific effector-lymphocytes, which are final product of the differentiation of specific lymphocyte precursors $L_P(t)$, i.e., lymphocytes that have corresponding specificity to viral antigen and that proliferate and differentiate (in L_E) under the action of stimulus.

Assume that the total quantity of sensitive cells in a lung is approximately constant (we study acute but sublethal infection when ~10% of lung cells are damaged), and that $C_I \approx qC_V$ where C_I is the quantity of population of the cells that become insensitive to infection under the interferon action.

Select the equations for C_V and V in the following form:

$$\begin{aligned} \frac{dC_V}{dt} &= \underset{\text{I}}{k_{C_V} V(C_0 - C_I - C_V)} \; \underset{\text{II}}{-\gamma_{C_V L_E} C_V L_E} \; \underset{\text{III}}{-\mu_{C_V} C_V}, \\ \frac{dV}{dt} &= \underset{\text{IV}}{k_V C_V} \; \underset{\text{V}}{-\gamma_{VF} VF} \; \underset{\text{VI}}{-\mu_V V}. \end{aligned} \tag{9.1.1}$$

Here the term I corresponds to the appearance of new infected cells as a result of interaction between infectious particles V with the intact sensitive cells $(C_0 - C_I - C_V)$; C_0 is a value characterizing total quantity of the cells potentially sensitive to virus (mainly the epithelium of lungs); the term II corresponds to the reduction in the number of cells infected by viruses due to specific effector-lymphocytes (integral effect); the term III is related to the death of C_V cells due to their destruction by viruses which are synthesized in them, i.e., it is not connected with specific effector-lymphocytes; the term IV: the synthesis of new virus particles; the term V: the neutralization of extracellular viral particles by immunoglobulins; the term VI: the process of elimination of viral particles from extracellular space which is not caused by specific humoral factors; this process involves the adsorption and penetration of viruses into cells, the loss of infectious properties by viruses under the action of various "background" humoral factors: inhibitors, temperature, acidity, etc. Notice, that it is the infectious virus itself, that is the effective stimulator for the immune system [1].

the above processes manifest clearly both spatial and temporal hierarchy. The size of viral particles is ~ 80 – 100nm [188], cells of animals have the size of ~ 10^{-5}m [328], immunoglobulins (diameter of

F_{ab}-fragment) are $\sim 30 - 40$Å in diameter [148]. One viral particle is capable of infecting a cell [188], and then $\sim 10^3$ particles are produced during the generation cycle which takes ~ 8 hours (the cell dies after that time). The duration of sorption and penetration of virus into a cell is about 10–20 minutes. Free viruses disappear quickly.

Molecular reactions under study – binding the viruses by immunoglobulins, absorption and penetration of a virus into new cells – have characteristic duration of some minutes to dozens of minutes, whereas the complex of cellular reactions, that provide the destruction of the cells infected by viruses, is characteristic of a longer duration. Therefore, neglecting the detailed consideration of the processes that last for more than dozens of minutes or hour (for example, the "eclipse" phase of a virus after the infection) and using the procedure of "adiabatic" elimination [124, 333], one can assume approximately that

$$V = \frac{k_V C_V}{\mu_V + \gamma_{VF} F}. \tag{9.1.2}$$

Now, rewrite equation (9.1.1) in the form:

$$\frac{dC_V}{dt} = \frac{k_{C_V} k_V C_V (C_0 - (1+q)C_V)}{\mu_V + \gamma_{VF} F} - \gamma_{C_V L_V} C_V L_V - \mu_{C_V} C_V. \tag{9.1.3}$$

We assume, as an initial condition, that

$$C_V(0) = C_V^0,$$

where C_V^0 is a certain effective quantity of infected cells, which triggers the infection (transition processes in the system stop in 1–2 hour, and dynamic equilibrium between V and C_V is attained according to (9.1.2)).

In order to describe the dynamics of the cells of the immune system, we approximate real processes [106] by the following model scheme. Resting precursor-cells (phase G_0 of the cell cycle) with the specificity corresponding to a given type of virus, become activated under the action of antigenic stimulus and pass into the stage of proliferating precursor-cells L_P which differentiate into final effector-cells L_E.

The separation of these very variables is connected with the assumption, that specific lymphocyte mechanisms play a key role in the recovery process of organism [1]. The large number of various mechanisms neutralizing the viruses in extracellular space and mechanisms of destruction of the cells infected by viruses are either caused directly by the effector-lymphocytes (specific cytotoxic lymphocytes, immunoglobulins), or controlled and triggered by these cells (NK-cells, macrophages, K-cells [1, 106]). The variables L_P and L_E are the leading ones, since they reflect integrally the mechanisms of defence.

Write the following relationship for the variable L_P:

$$\frac{dL_P}{dt} = L_{P_0}\delta(t - t^*) + U(C_V)(\alpha_{L_P}L_P - \beta_{L_P}L_P), \tag{9.1.4}$$

and for L_E:

$$\frac{dL_E}{dt} = U(C_V)\beta_{L_P}L_P - \mu_{L_E}L_E, \tag{9.1.5}$$

$$U(C_V) = 1 - \exp\left\{\frac{C_V}{q_u}\right\}. \tag{9.1.6}$$

Let us explain these equations: $L_{P_0}\delta(t - t^*)$ corresponds to the activation of resting precursor-cells whose number in time t^* after antigen stimulus is equal to L_{P_0}; $\delta(t)$ is the delta-function. It is supposed that the processes of proliferation and differentiation are of threshold character: they proceed at a maximum rate while $C_V \geq q_u$ and stop when $C_V < q_u$.

The expression $-\mu_{L_E}L_E$ in (9.1.5) reflects the processes of the loss of activity and/or natural death of effector-cells. The corresponding term in (9.1.4) is absent since the variable L_P should correspond to the forming clone of memory-cells, whose lifetime is 1–2 years [1, 10]. It exceeds greatly time interval of the development of acute process.

The following equation is written for the immunoglobulins F:

$$\frac{dF}{dt} = \rho L_E - \gamma_{FV}FV - \mu_F F. \tag{9.1.7}$$

With regard to (9.1.2) we have

$$\frac{dF}{dt} = \rho L_E - \frac{\gamma_{FV}k_V F C_V}{\mu_V + \gamma_{VF}F} - \mu_F F. \tag{9.1.8}$$

Preliminary calculations have shown that the second term is small as compared to the first one in the expression

$$\ln V = \ln\left(\frac{k_V}{\mu_V}C_V\right) - \ln\left(1 + \frac{\gamma_{VF}}{\mu_V}F\right),$$

therefore we assume hereinafter that

$$V = \frac{k_V}{\mu_V} = C_V. \tag{9.1.9}$$

The dependence on F remains in the right-hand sides of equations (9.1.3), (9.1.8).

Here are some characteristic values for equations (9.1.4)–(9.1.8): $t^* \approx 1$–3 days, i.e., is of the same order of magnitude with the time that is required for the reactions leading to the activation of lymphocytes [106]. The maximal rate of cellular division for mammals is ~ 8–12 hours [353], hence one can estimate α_{L_P} and β_{L_P}. The half-life time for the effector-cells is ~ 1–2 days [273], and for molecules of immunoglobulin (IgG of mice) is ~ 5–6 days; therefore, it is possible to estimate μ_{L_E} and μ_F. Total quantity of lymphocytes in the organism of a mouse is $\approx 10^9$ cells; The frequency – probability to encounter a cell specific to certain strain of influenza virus – is $\sim 1/10^5$–$1/10^6$ [1, 10, 353], hence $L_{P_0} \approx 10^3$–10^4 cells. After the infection this frequency increases by 50–100 times.

Let us write down the system of model equations so obtained as it is used for the investigation of dynamics of influenza infection:

$$\frac{dV}{dt} = \frac{\alpha_1 + \alpha_{10}}{1 + \alpha_9 F}V - \alpha_2 V^2 - \alpha_3 V L_E - \alpha_{10} V, \tag{9.1.10}$$

$$\frac{dL_P}{dt} = L_{P_0}\delta(t - t^*) + \alpha_{11} L_P[1 - \exp\{-\alpha_{12} V\}], \tag{9.1.11}$$

$$\frac{dL_E}{dt} = \alpha_4 L_P[1 - \exp\{-\alpha_{12} V\}] - \alpha_5 L_E, \tag{9.1.12}$$

$$\frac{dF}{dt} = \alpha_6 L_E - \frac{\alpha_7 V F}{1 + \alpha_9 F} - \alpha_8 F, \tag{9.1.13}$$

$$V(0) = V^0, \quad L_P(0) = L_E(0) = 0, \quad F(0) = 0. \tag{9.1.14}$$

Equations (9.1.10)–(9.1.14) were written with regard to relation (9.1.9). The system was used for the data analysis, since in experimental system there is a possibility to trace $V(t)$ but not $C_V(t)$.

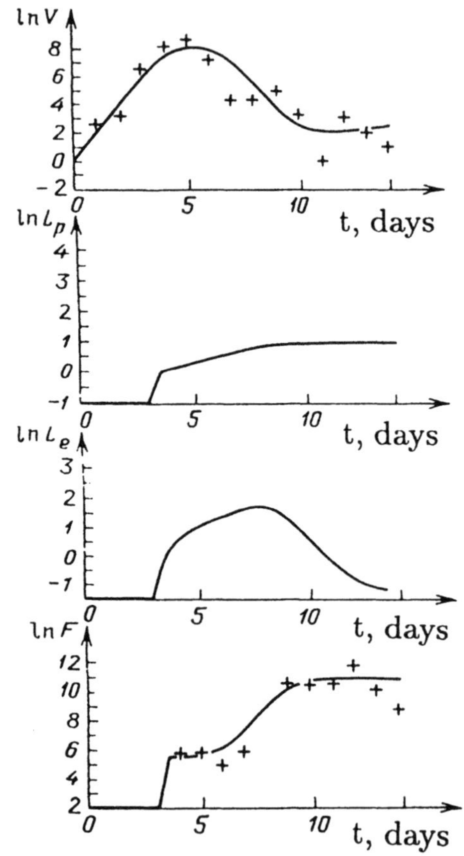

Fig. 55. Model solution for the set of coefficients obtained by the identification using experimental data on the control group of animals; (+), experimental data. Units for the concentrations: V, the number of 50% embryonic infectious doses; F, the level of immunoglobulins (IgM+IgG) in blood plasma of animals.

In case of the process under study, that is, acute uncomplicated viral infection, one can presume a certain scenario of the realization of defence mechanisms. So, the structure of model solutions corresponding to the process under study is such that the system of interferons ($-\alpha_2 V^2$) and generalized cytotoxic action of effector-lymphocytes ($-\alpha_3 V L_E$) are the mechanisms which provide for the saturation in the dynamics of $V(t)$ (maximum at 4–5th day) and the subsequent sharp drop in the quantity of infectious virus in lungs (5–8 days).

There is practically no influence of the variable F on the solution in this case. This effect starts to work later (8–14 days), when F reaches larger values, when the quantity of infectious virus V is already much smaller than the maximal one, which makes the term $-\alpha_2 V^2$ negligible on this intervals of time. It is quite possible that such a structure of solution reflects the real scheme of action of defence mechanisms in time. By the virtue of the above considerations, equation (9.1.10) slightly differs in structure from (9.1.3): there is no dependence on F in the term $-\alpha_2 V^2$, which is left just in the term $(\alpha_1 + \alpha_{10} V)/(1 + \alpha_9 F)$.

The mathematical model we have constructed fits experimental data reasonably well. This can be seen in Fig. 55 where solid lines depict the solutions of model equations obtained with the vector of coefficients α. This vector of coefficients was found as a result of the identification of the model parameters by experimental data.

We emphasize that the coefficients so determined and another characteristics of solutions agree in general with known quantitative parameters of the process.

9.2. Application of the Model for the Investigation of Drugs Influence Mechanisms on the Dynamics of Influenza Infection

The mathematical model constructed in the previous Section is the system of differential equations in the form of (5.2.3). Its state variables describe the dynamics of viruses and "generalized" cellular populations in an organism of animal, and the coefficients are interpreted as the parameters of corresponding interactions. Therefore, estimating the model's coefficients on the basis of data obtained as a result of tracing model's variables in experiments we have the possibility of judging, what are the influences of drugs on the parameters of the process under study. Not all parameters must be observed for the solution of this problem; just the observations over the dynamics of antibodies F and of the number of a virus V are necessary.

Measurements of these quantities were performed in the Institute of Experimental Medicine, Russian Academy of Medical Sciences, using three groups of animals (mice F1 (CBA×C57 black)): no drugs (control group); treated with ionol; treated with ε-aminocaproic acid. The influenza virus A/PR8/34 was used in all groups. Sublethal infection was induced by intranasal doses (3.0–3.5 $\lg EID_{50}$) of the virus. Observations were carried out daily using not less than 3–5 animals in each experimental session. With these data and a numerical algorithm for the determination of model's parameters on observational data (see Section 5.2) we can estimate the parameters of modeled process in each of three groups of animals. Comparing these parameters and establishing the differences between them one can formulate hypothesis on the mechanisms of drugs action.

Notice that the comparison is performed on the basis of statistical criteria obtained in Section 5.2 for the verification of correlation between the model and data, and on the basis of the hypothesis on the equality of parameters of processes in different independent groups of animals.

Solid lines in Fig. 55 depict the model (9.1.10)–(9.1.14) solution corresponding to the vector of coefficients calculated with observational data over the control group that was given no preparatus. The observational data are shown in this picture by crosses. Fig. 56 and Fig. 57 show similar results obtained with experimental data for animals treated with ionol and ε-aminocaproic acid, respectively. Calculated values for the coefficients of the model (9.1.10)–(9.1.14) are presented in Table 11.

Application of criteria for the verification of statistical hypothesis (see Section 5.2) has shown that the changes in the coefficient vector under the action of drug are established with a high statistical significance level. However, the statistical data in our possession were not sufficient to establish the significance of differences between each component of the coefficient vector, which is not difficult in case of large amount of experimental data. So, our conclusions are not more than the hypotheses and serve to illustrate the methodology for the solution of problems of this kind.

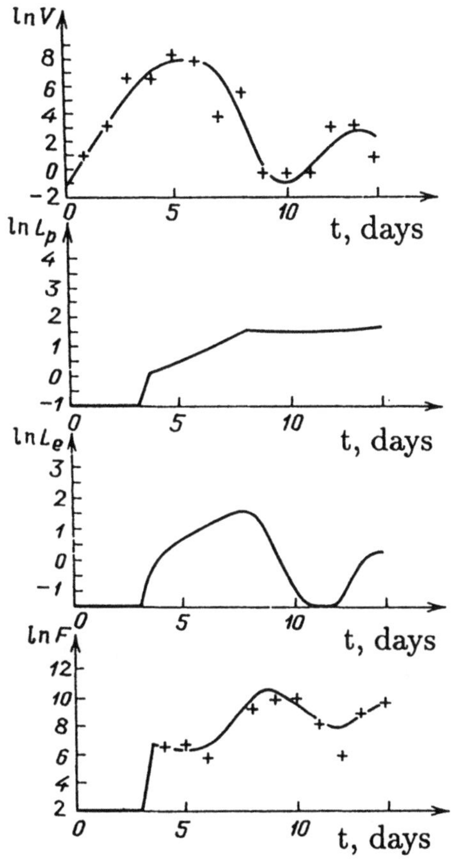

Fig. 56. Model solution for the set of parameters obtained as a result of identification by the data of experiments on animals treated with ionol during the infection. Units of measurements are the same as in Fig. 58.

The analysis of the possibility of identifying the coefficients of the model (9.1.10)–(9.1.14) using the dynamics of $V(t)$, $F(t)$, $t \in [0,T]$ with the methods presented in Section 5.2, shows that α_4 and L_{P_0}, as well as α_3 and α_6 cannot be determined uniquely just through $V(t)$ and

$F(t)$. However, the dynamics of $\alpha_6 L_E(t)$ and $\alpha_3 L_E(t)$ can be determined uniquely by these data. We fixed the parameters α_4 and L_{P_0} in computations (see Table 11), the parameter α_{10} was also fixed, since this parameter is easily determined by the data of experiments in vitro on the infection of a cell culture by a virus. In order to simplify further analysis we assume that fixed coefficients do not change from one group to another.

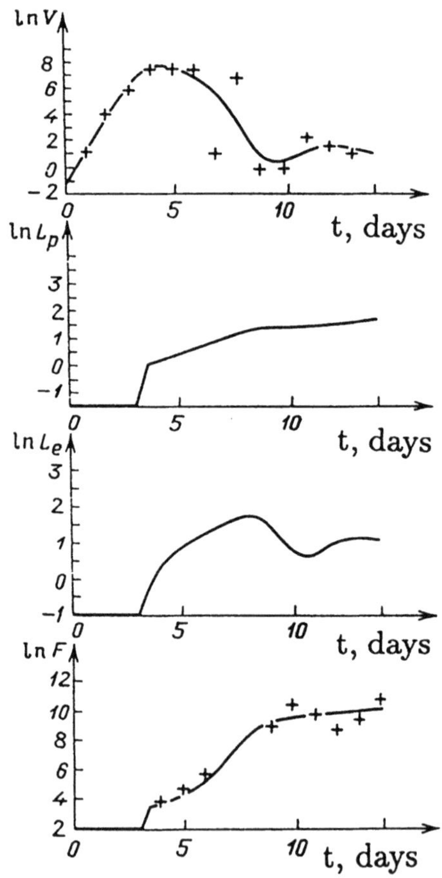

Fig. 57. Model solution for the set of parameters obtained as a result of identification by the data of experiments on animals treated with ε-aminocaproic acid during the infection. Units of measurements are the same as in Fig. 58.

It is seen from Table 11 that the coefficients α_1, α_3, α_5 remain practically unchanged in three groups of animals. One can suggest, therefore, that such parameters of the process as integral cytotoxic effects of effector-lymphocytes on viral population, intensity of multiplication of viruses, and the rate of elimination from organism and/or of inactivation of effector-cells, remain practically unchanged.

Table 11

Identification of the model's (8.1.10)–(8.1.14) coefficients on the experimental data

Coefficients	Values corresponding to the group of animals that were not given the preparation	Values corresponding to the group of animals treated with the preparation*)	
		ionol	ε-amincapron acid
α_1	2.0	2.328	2.581
α_2	0	0	$4.3 \cdot 10^{-4}$
α_3	0.958	0.972	0.950
$\alpha_4^{**)}$	1.8	1.8	1.8
α_5	0.891	1.51	4.0
α_6	$1.53 \cdot 10^4$	$2.178 \cdot 10^4$	$1.411 \cdot 10^3$
$\alpha_7^{**)}$	$2.19 \cdot 10^{-2}$	$2.19 \cdot 10^{-2}$	$2.19 \cdot 10^{-2}$
α_8	0	1.834	10
α_9	$4.73 \cdot 10^{-6}$	$2.226 \cdot 10^{-5}$	0
$\alpha_{10}^{**)}$	7.0	7.0	7.0
α_{11}	0.201	0.331	0.294
α_{12}	$3.870 \cdot 10^{-3}$	$1.115 \cdot 10^{-2}$	$7.664 \cdot 10^{-2}$
$(t^*)^{**)}$	3 days	3 days	3 days
$\ln V^0$	$5.037 \cdot 10^{-2}$	-1.139	-1.394
L_{P_0}	1	1	1

*) Coefficient dimensions are the same as in the model (8.1.10)–(8.1.14) with regard to $V(t)$, $F(t)$ units and the choice of $L_{P_0} = 1$.
**) These coefficients are fixed similar for all three groups

In the third group, treated with ε-aminocaproic acid, the parameter α_6 is smaller by an order of magnitude than for the control group. This parameter characterizes the intensity in the production of immunoglobulins. It was obtained for the same group that $\alpha_9 = 0$, which may be interpreted as a very weak influence of immunoglobulins on the process under study against the background of other mechanisms of defence. In the case of the third group $\alpha_2 \neq 0$, though for the control group and for the first (ionol) group $\alpha_2 = 0$. This may be interpreted in the following way: the maximum of viral infection and its subsequent drop (inhibition of multiplication of viruses) for the first and the control groups is caused by the cytotoxic mechanisms of effector-lymphocytes L_E. It seems that in the case of the third group some saturation mechanism is added, apparently owing to the action of interferon.

Both groups treated with drugs have the coefficient α_{12}^{-1} several times smaller than for the control group. It may be the reflection of the decrease in the threshold value for the quantity of viruses that triggers the processes of proliferation and differentiation with maximal rate. The investigations show that these changes are very significant. Besides, the coefficient α_{11} have increases by 1.5 times in these groups; it may be interpreted as more intensive processes of proliferation and formation of immune memory, which is of prime importance too. It is possible that these very changes are responsible for the antiviral effect of studied drugs observed in the experiment. Some biological information on these drugs is presented in [282].

These results demonstrate the possibility of using the suggested approach in the study of the influence of chemical drugs (and another factors) on the process of acute infection in organism. Furthermore, the application of the mathematical model allows one to obtain quantitative characteristics of inner processes using *in vivo* data. A large volume of experimental data should give the possibility of defining with statistical significance the changes in the mechanisms of the studied processes which are provoked by chemical drugs, and to calculate the most important components of the defence, i.e., to construct and testify the "scenarios" of the development of infectious disease in organism.

CHAPTER 10

Adjoint Equations and Sensitivity Study for Mathematical Models of Infectious Diseases

We present in this chapter an approach developed on the basis of adjoint equations for the sensitivity study of the functionals of problems arising in the modelling of infectious diseases. The models are formulated as the systems of differential equations with a delaying argument.

There exist theoretical and practical problems both in the estimation of deviations of immunological parameters in patients with unfavourable course of disease, and in the prognosis of variations in the course of disease for known deviations in the values of parameters of the immune system.

The influence of parameters of immune response on variations in the characteristics of the course of disease and of infection's severity may be investigated through the sensitivity analysis of mathematical models of infectious diseases. On the other hand, considering the deviations from normal course of viral and bacterial infections as a sign of the secondary immunodeficiency, it is possible, solving an inverse problem, to identify the values of corresponding deviations in coefficients of models as related to a norm.

Mechanisms of immunodeficiencies are analysed in Section 10.1 and their qualitative description is presented in the framework of the parametric structure of mathematical models. An approach to the estimation of sensitivity of the functionals to variations in parameters is formulated in Section 10.2 for multi-parametric mathematical models on the basis of the perturbations theory and adjoint equations.

10.1. Analysis and Description of the Mechanisms of Secondary Immunodeficiencies in Mathematical Models of Infectious Diseases

It was discovered, in the process of development of modern concepts on mechanisms of immunity, that the immunodeficiencies of various degree are observed in many people which makes them susceptible to such diseases as complications of infections, tumor, autoimmune disorders. Clinical solution of the immunodeficiency problem implies the diagnosis, identification of etiological factors which cause the immunodeficiency, and the elaboration of adequate therapy [197, 198, 270, 310].

The concept of an immunodeficient state of an organism is related to deviations in the character of the immune reaction to a certain infection from some typical picture. There exists, in this connection, a problem of the estimation of deviations of immunological parameters in patients with unfavourable course of infectious disease, and the problem of the prognosis of severity of the course of disease for known deviations in immunoreactivity parameters.

In clinical practice, a patient with deviations in the immunoreactivity is referred to as an immunocompromized or immunodeficient host. Various physical and biological processes can be the factors that reduce nonspecifically the immunoreactivity of organism: environment pollution, stresses, viral and bacterial infections. The immunodeficiency in human organism is accompanied by various clinical manifestations, in particular, by the rise of frequency and increase of severity and duration of infectious diseases [197, 198, 270].

10.1.1. Phenomenology of immunodeficiencies. At present a system of classification is accepted in immunology for the immunodeficiency in terms of primary and secondary immunodeficiencies [270, 273, 297]. Modern concepts on primary and secondary immunodeficiencies are the bases of the diagnosis, their classification is published in the WHO reports. Primary immunodeficiencies that appear because of the loss or malfunction of one gene are rear diseases for which the character of immunodeficiencies is well established. They are divided into the following major groups: defects in humoral and cellular immunity, disfunction of the phagocyte system, and deficiency of the complement system.

Most of the immunodeficiencies observed in adults are of secondary origin [218]. The secondary immunodeficiencies have the character of systemic, metabolic, general membrane or nutritional defects and affect various complex parts of immune response or separate functions of the cells of immune system. They are characteristic of the malfunction of various parts of the immune system of a host, unlike the primary immunodeficiencies, when just one component of the immune system is affected or a part of cellular function is lost completely.

Very often the secondary immunodeficiencies appear in the course of viral, bacterial, or parasitic infections. The brightest example is the damage to the immune system in the course of the infection by the viruses of immunodeficiency which manifests itself in the initial phase of infection by functional defect of T-helpers and then by a quantitative decrease of their level. The first display of the appearing defect of immune system at the level of $CD4^+$ T-cells is a selective decrease of proliferative reaction to various antigens against the background of normal reactions to mitogens and alloantigens.

The list of aetiological factors that are capable of provoking the acquired defects of immunoreactivity is large and is increasing continuously. Table 12 contains the main factors and the secondary immunodeficiencies they induce.

10.1.2. Mechanisms of immunodeficiencies and their description in a model. The acquired immunodeficiencies of the immune system are characterized by their influence rather on a number of components of the host's defence system than on just one component. The secondary immunodeficiencies vary in the degree of manifestation and not often are as complete as primary immunodeficiencies. It is believed that the defects of a specific immune system explain 80–90 % of all cases of clinically important immunodeficiencies.

Table 12

Aetiological factor, causes and character of secondary immunodeficiencies

No.	Aetiological factor	Character of immunodeficiency
1.	Viral infections 1.1. Influenza viruses	· suppression of the response of T- and B-lymphocytes to mitogens; · changes in the number of circulating T- and B-lymphocytes; · perturbations in the circulation of T- and B-lymphocytes; · decrease in the chemotaxis and phagocytosis of the monocyte–macrophage system; · defects in the cooperation processes between macrophages, T- and B-lymphocytes;
	1.2. Hepatitis B virus	· changes in the number of circulating T- and B-lymphocytes;
	1.3. Human immuno-deficiency virus	· functional weakening of T-helper activity; · decrease in the number of circulating $CD4^+T$-lymphocytes;
2.	Lympholcukosis (increase in infection frequency, sepsis are observed)	· suppression of the responses of T- and B-lymphocytes to specific or nonspecific mitogens; · delayed type of hypersensitivity reactions; · hypogammaglobulinemia; · Disfunction of phagocytic and bactericidal activity of neutrophiles; · decrease in the number of circulating neutrophiles and monocytes;
3.	Pharmacological treatment -immunosuppressants 3.1. Adrenocortical steroids (development of infectious complications)	· delayed proliferative reactions of T- and B-cells; · decrease in the production of IL-2; · changes in the parameters of interaction between IL-2 and lymphocyte receptors; · decrease in the synthesis of antibodies; · suppression of the chemotaxis of neutrophiles and monocytes; · suppression of inflammatory reactions;

See the continuation

Table 12, (continuation)

No.	Aetiological factor	Character of immunodeficiency
	3.2. Purine antagonists (frequent viral infections, for high doses – bacterial and fungus infections	· suppression of humoral immunity; · depression of delayed type of hypersensitivity reactions · suppression of allograft reactions;
	3.3. Anticancer chemotherapy	· suppression of the most of of immune reactions; · suppression of the antibody synthesis; · suppression of transplant versus host reactions;
	3.4. Antibiotics	· immunosuppression;
4.	The burns	· decrease in the level of immunoglobulins of all classes; · suppression of the chemotaxis and bactericidal functions of macrophages and neutrophiles; · suppression of T-cellular immunity;
5.	Surgical interventions	· decrease in the number of T- and B- cells; · suppression of cellular reactions;
6.	Trauma and shock	· suppression of the chemotaxis, phagocytosis, and bacterial function of neutrophiles;
7.	Malnutrition	· suppression of phagocytosis and bactericidal functions; · suppression of lymphocyte response to mitogens; · suppression of humoral immunity; · drop of the complement level.

Immune reactions during infectious diseases represent a complex set of regulated processes of activation, division, and differentiation of T- and B-lymphocytes. Defects of immune defence can be localized on the level of an antigen presenting system, T-cellular part, B-cellular part, phagocyte system, and complement system. Defects of different components of immune system manifest themselves in various ways. So, immunodeficiency in the system of B-lymphocytes increases the rate of respiratory and intestinal diseases, development of tumors. Defects of T-cellular system are characteristic of opportunist infections, activation of latent viral infections, increase in the severity of acute viral infections. Defects of the phagocyte system lead to the development of skin staphylococcal infections. Deficits of components in the complement system enhance the development of relapsing bacterial infections.

The secondary immunodeficiencies can, first of all, manifest themselves in the changes of cellular immune status. The number of circulating peripheral T- or B-cells, and NK-cells decreases. The decrease in the number of lymphocytes can be observed also in tissues; for example, plasma cells localized under mucous surfaces of organs (intestines, upper respiratory tract) and secreting IgA and IgG2. A deficit on the level of cellular immune status can be described in a model through the changes in parameters of teh homeostasis level of lymphocytes or macrophages (H_E^*, H_B^*, E^*, B^*, P^*, M^*) as compared to normal values.

Specific immune reactions develop as a result of a series of sequential and parallel interactions between antigen presenting cells and regulating and T- and B-lymphocytes. The interactions are either of contact character or mediated by the secretion of monokines and cytokines which control the state of T- and B-lymphocytes through their surface receptors. Therefore the immunodeficiencies can appear as a result of deviations in the characteristics of interactions on the level of any of these processes taking part in the formation of immune reactions. The molecular background of these deviations is related to the changes in the degree of expression of receptors on the surface of cells or certain protein structures, such as MHC-molecules, glycoproteins, the secretion rate of cytokines, processes that control the activity of genes [218].

Let us separate the main types of immunodeficiencies manifesting themselves in the decrease of immunoreactivity. The latter characterizes the ability to react to antigens with activation and division and is estimated as a rule by maximal level of inclusion of [^{3}H]-thymidine or by the number of synthesized immunoglobulins.

First of all we separate a large group of secondary immunodeficiencies existing on the level of processes of T-cellular activation. This group includes the disfunction of the monocyte population as related to their ability to maintain the process of activation of T-cells. In particular, the deviations in the level of expression of molecules of MHC-complex on the surfaces of macrophages belong here, which are the most important molecules in intercellular interactions of T-lymphocytes and antigen presenting macrophages (their number on the surfaces of cells can drop by 10–1000 times).

Such defects can be described in a model in a number of ways, for example, through the decrease in the value of the parameter γ_{MV} characterizing the activation rate of macrophages of a lymph tissue and determining the number of antigen presenting macrophages. Another possible parameterization of this defect is the variation of coefficients b_H^E, b_H^B, b_p^E, b_p^B. Initial uncertainty in the localization of a defect of interaction characteristics for immunocompetent cells manifests itself in teh ambiguous choice of parameters of mathematical model capable of description of given immunodeficiency.

The second group of defects leading to the deviations in the character of proliferative reactions *in vitro* to the mitogens, antigens, or allogenic cells is connected with the functional activity of T-helpers producing various lymphokines which control the division and differentiation of T- and B-lymphocytes. The expression of T-helper deficiency is estimated by the inclusion of [^{3}H]-thymidine mark on the basis of these reactions. The kinetics of Th-lymphocytes's reaction is determined by the rate parameters of stimulation and division in above models of immune response. Respectively, the defect of T-helpers reactivity may be described through the decrease in the values of b_H^E, b_H^B, ρ_H^E, ρ_H^B or increase in the values of τ_H^E, τ_H^B. Similar reasoning is applicable to the characteristics of proliferation activity of T- and B-lymphocytes.

Defects in the synthesis of various classes of immunoglobulins (IgM, IgG) are investigated in the tests on polyclonal stimulation of samples of peripheral blood lymphocytes by mitogens. A deficit may be localized both on the level of the synthesis of IL-2 and on the level of mechanisms controlling the differentiation of B-lymphocytes and the switching of the synthesis of antibodies from IgM to IgG. It can be described in the models by the variation of parameters b_H^P, b_p^B, b_p^P, ρ_B/ρ_P, τ_P. Widespread immunodeficiencies include the changes in functional activity of macrophages, granulocytes: chemotaxis, phagocytosis, bactericide activity etc. This defect may be taken into account through the decrease in the values of coefficients γ_{VM} or γ_{WM} in the models of antiviral and antibacterial immune response.

Results of the analysis are presented in the Table 13 which contains the description of mechanisms of the secondary immunodeficiencies in the terms of parameters of mathematical models of viral and bacterial infections.

Table 13

Major types of the secondary immunodeficiencies

No	Model parameters representing possible mechanisms of immunodeficiency	Character of changes	
1.	Change of cellular immune status at the level of macrophages, T- or B-lymphocytes	M^*, H_E^*, H_B^*, E^*, B^*, P^* (characteristics of cellular homeostasis)	reduction
2.	Reduction of the immunoglobulins level	P^*, F^*	
3.	Dysfunction of antigen presenting cells as related to the activation of T-cells	γ_{MV}, γ_{MK}	reduction
4.	Functional reduction of the activity of helper T-cells	b_H^E, b_H^B, b_p^E, b_p^B, b_p^P	reduction
5.	Weakening of proliferative reactions of T-cells, CTLs and B-lymphocytes	b_H, b_p^E, b_p^B ρ_H, ρ_E, ρ_B τ_H, τ_E, τ_B	reduction reduction rise
6.	Reduction of the synthesis of antibodies	b_p^P, ρ_F	reduction
7.	Change of the indicators of the phagocytosis and chemotaxis of macrophages	γ_{MV}, γ_{MK}	reduction
8.	Suppression of the inflammation	μ_V, μ_C	reduction
9.	Deficiency of the complement system	γ_{MK}	reduction

These models describe the processes where the deviations are localized and anticipated qualitative character of changes in the parameters' values. It should be noted that quantitative and functional deviations of the components of immune defence from a norm are established nowadays rather efficiently. The ambiguity appears in the localization of a mechanism of immunodeficiency on the level of processes enhancing the development of immune reactions. It determines both clinical problems connected with the correction of immunodeficiencies which implies the intervention into immune processes in immuno-compromised subjects, and ambiguity in the choice of parameters of mathematical models that are most suitable for the description of given immunodeficiency.

The infection provoked by the influenza A viruses is one of widespread infections with high epidemic potential. Severe and lethal form of influenza are often observed in immunodeficient people. The investigation of the influence of immunodeficiencies on the variations of characteristics of the course and severity of viral infections may be carried out through the sensitivity analysis for mathematical models on the basis of adjoint equations.

10.2. Adjoint Equations and Models of Immune Response

The sensitivity analysis for mathematical models related to the functionals of problems is the most important tool for the investigation of the qualitative influence of processes described by the models on the characteristics of modelled phenomena.

It is possible to construct, on the basis of the methods of perturbation theory and adjoint equations, efficient and economical methods of estimating sensitivity of functionals of solutions of multiparameter models of large dimension [195, 196, 283].

10.2.1 The method of small perturbations. Let the mathematical model of a process be formulated as an initial value problem for the system of delay-differential equations:

$$\begin{cases} \dfrac{dy(t)}{dt} = f(\alpha, y(t), y(t-\tau)), & 0 \le t \le T, \\ y(t) = \varphi_0(t), & -\tau \le t \le 0, \end{cases} \tag{10.1}$$

where the vector function $f(\alpha, y, z)$ in the right-hand side is sufficiently smooth with respect to each of arguments; $\alpha \in \mathrm{R}_+^L$ is the vector of parameters; $y(t) \in \mathrm{R}^N$. Let a functional be assigned which depends on the values of the vector function of solution $y(t)$ on the interval $[0, T]$, in the following form:

$$J(y) = \int_0^T \langle y, y \rangle dt \equiv \int_0^T \langle y(t,\alpha), y(t,\alpha) \rangle dt. \tag{10.2}$$

Let us denote through α_* some basic values of parameters of the unperturbed problem and the corresponding solution through $y(t, \alpha_*)$. It is required to estimate the sensitivity of the functional $J(y)$ to variations ($\delta\alpha$) of model's parameters. Regarding $\delta\alpha$ as a small perturbation of the parameter, we may use the perturbation theory to estimate the sensitivity of a functional of nonlinear problem. By virtue of the assumption on the correctness of statement of the initial value problem, small perturbations of the system's parameters $\alpha_* + \delta\alpha$ correspond to the solution of perturbed problem which is representable as a series in powers of $\delta\alpha$:

$$y(t,\alpha) = y(t,\alpha_*) + y^{(\alpha)}(t,\alpha_*)\delta\alpha + \dots\ .$$

We may examine, with no loss of generality, an effect of the perturbation on the value of ith parameter α_i. The vector function $y^{(\alpha_i)}(t,\alpha_*) \equiv \partial y(t,\alpha_*)/\partial\alpha_i$ is a solution to the system of linear approximation:

$$Ay^{(\alpha_i)} = \frac{\partial f}{\partial \alpha_i}, \quad 0 \le t \le T,$$

$$y^{(\alpha_i)}(t,\alpha_*) = 0, \quad -\tau \le t \le 0.$$

The operator A acts in some Hilbert space H with domain of definition $\mathcal{D}(A)$ and has the form

$$A \equiv \frac{d}{dt} - \left[\frac{\partial f}{\partial y}\right]_t - \left[\frac{\partial f}{\partial y_\tau}\right]_t D_\tau,$$

where $[\cdot]_t$ denotes a matrix of partial derivatives for the moment t, D_τ is a translation operator backward in time by τ. The following variation of the functional correspond to the variation of solution $\delta y(t,\alpha_*)$:

$$\delta J = J(y_* + \delta y) - J(y_*) = 2\int_0^T \langle y(t,\alpha_*), \delta y(t,\alpha_*) \rangle dt.$$

It is representable in the form

$$\delta J = 2\int_0^T \langle y(t,\alpha_*), y^{(\alpha_i)}(t,\alpha_*)\delta\alpha_i\rangle dt$$

by virtue of the assumption on small perturbations in the values of parameters $\delta\alpha$. Just the first correction was taken into account while estimating the variation δJ for each of the model's parameters α_i, $i = \overline{1,L}$.

Let us introduce an operator A^* adjoint to the operator A according to classical definition based on the Lagrange identity $(Aq,p) = (q, A^*p)$ where $(\cdot,\cdot)$ is an inner product in H, $q \in \mathcal{D}(A)$, $p \in \mathcal{D}(A^*)$,

$$A^* \equiv -\frac{d}{dt} - \left[\frac{\partial f}{\partial y}\right]_t^T - \left[\frac{\partial f}{\partial y_\tau}\right]_{t+\tau}^T D_{-\tau}. \tag{10.3}$$

The adjoint operator A^* acts in H with the definition domain $\mathcal{D}(A^*)$. Let $p(t,\alpha_*)$ be a solution to the adjoint system of equations

$$A^*p = y(t,\alpha_*), \quad 0 \le t \le T,$$

$$p(t) = 0, \quad -T \le t \le T+\tau,$$

with zero initial conditions

$$-\frac{dp(t)}{dt} - \left[\frac{\partial f}{\partial y}\right]_t^T p(t) - \left[\frac{\partial f}{\partial y}\right]_{t-\tau}^T p(t+\tau) = y_*(t,\alpha_*), \tag{10.4}$$
$$0 \le t \le T, \quad p(t) = 0, \quad T \le t < T+\tau.$$

It is possible, using $p(t)$ and performing some transformations based on the conjugation relationship, to obtain the following representation for the value of variation of the functional:

$$\begin{aligned}\delta J &= 2\int_0^T \langle y(t,\alpha_*), \delta y\rangle dt = 2\int_0^T \langle A^*p, \delta y\rangle dt \\ &= 2\int_0^T \langle p, A\delta y\rangle dt = 2\int_0^T \langle p, \frac{\partial f}{\partial \alpha_i}\delta\alpha_i\rangle dt.\end{aligned}$$

This relationship permits us to obtain in the limit for $\delta\alpha_* \to 0$ an estimate of the components of the first derivative, that is, of a gradient of the functional under consideration with respect to parameters:

$$\frac{\partial J}{\partial \alpha_i} = 2\int_0^T \langle p, \frac{\partial f}{\partial \alpha_i}\rangle dt.$$

So, in order to find a variation of the functional under study, conditioned by the changes in the values of all L parameters of the model, it is sufficient to solve only once initial value problem for the main system (10.1) and then to solve only once adjoint problem (10.4). Having calculated the gradient of the functional in the space of parameters with the relationships $\frac{\partial J}{\partial \alpha_i} = 2\int_0^T \langle p, \frac{\partial f}{\partial \alpha_i}\rangle dt$, it is possible to estimate the influence of small perturbations with the formula

$$\delta J = \sum_{i=1}^{L} \frac{\partial J}{\partial \alpha_i}\delta\alpha_i$$

10.2.2. Adjoint system for simple model of infectious disease. A simple mathematical model of infectious disease was formulated as a four-dimensional system of nonlinear differential equations with delaying argument for the vector of variables $V(t)$, $C(t)$, $F(t)$, $m(t)$ (see Chapter 3):

$$\begin{aligned}
\frac{dV}{dt} &= (\beta - \gamma F)V, \\
\frac{dC}{dt} &= \xi(m)\alpha V(t-\tau)F(t-\tau) - \mu_c(C - C^*), \\
\frac{dF}{dt} &= \rho C - (\mu_f + \eta\gamma V)F, \\
\frac{dm}{dt} &= \sigma V - \mu_m m.
\end{aligned} \tag{10.5}$$

The system of equations adjoint to this main system of equations is constructed on the basis of the Lagrange identity. We write down its final form:

$$\frac{dV^*}{dt} = -(\beta - \gamma F)V^* + \eta\gamma F F^* - \alpha F \xi(m(t+\tau))C^*(t+\tau) - \sigma m^*,$$

$$\frac{dC^*}{dt} = \mu_C C^* - \rho F^*,$$

$$\frac{dF^*}{dt} = \gamma V V^* + \eta\gamma V F^* - \alpha V \xi(m(t+\tau))C^*(t+\tau) + \mu_F F^*, \qquad (10.6)$$

$$\frac{dm^*}{dt} = \mu_m m^* - \alpha \frac{\partial \xi}{\partial m} V(t-\tau)F(t-\tau)C^*,$$

where $V^*(t)$, $C^*(t)$, $F^*(t)$, $m^*(t)$ are variables of the vector function for adjoint problem.

10.2.3. Adjoint system for the model of antiviral immune response. The mathematical model of antiviral immune response was formulated as a system of ten nonlinear delay differential equations. The adjoint system is to be constructed on the basis of the Lagrange identity in the framework of above approach. Let us present the adjoint system skipping the layings out:

$$\begin{aligned} \frac{dV_F^*}{dt} &= \left[\gamma_{VF} F \left(1 + \frac{\mu_V C_V}{C^0}\right) + \gamma_{VM} + \gamma_{VC}(C^0 - C_V - m)\right] V_F^* \\ &\quad - \gamma_{MV} M_V^* + \gamma_{FV} F F^* - \sigma(C^0 - C_V - m) C_V^*, \end{aligned}$$

$$\begin{aligned} \frac{dM_V^*}{dt} &= \alpha_M M_V^* - \rho_H^E b_H^E H_E \xi\left(m(t+\tau_H^E)\right) H_E^*(t+\tau_H^E) \\ &\quad - \rho_H^B b_H^B H_B \xi\left(m(t+\tau_H^B)\right) H_B^*(t+\tau_H^B) \\ &\quad - \rho_E b_P^E H_E E \xi\left(m(t+\tau_E)\right) E^*(t+\tau_E) \\ &\quad - \rho_B b_P^B H_B B \xi\left(m(t+\tau_B)\right) B^*(t+\tau_B) \\ &\quad - b_P^P H_B B \xi\left(m(t+\tau_P)\right) P^*(t+\tau_P) + \left(b_H^E H_E + b_P^{H_E} H_E E\right) H_E^* \\ &\quad + \left(b_H^B H_B + b_P^{H_B} H_B B\right) H_B^* + b_P^E H_E E E^* + b_P^B H_B B B^*, \end{aligned}$$

$$\begin{aligned}\frac{dH_E^*}{dt} &= -\rho_H^E b_H^E M_V \xi\left(m(t+\tau_H^E)\right) H_E^*(t+\tau_H^E) \\ &- \rho_E b_P^E M_V E \xi\left(m(t+\tau_E)\right) E^*(t+\tau_E) \\ &+ \left(b_H^E M_V + b_P^{H_E} M_V E + \alpha_H^E\right) H_E^* + b_P^E M_V E E^*,\end{aligned}$$

$$\begin{aligned}\frac{dH_B^*}{dt} &= -\rho_H^B b_H^B M_V \xi\left(m(t+\tau_H^B)\right) H_B^*(t+\tau_H^B) \\ &- \rho_B b_P^B M_V B \xi\left(m(t+\tau_B)\right) B^*(t+\tau_B) \\ &- b_P^P M_V B \xi\left(m(t+\tau_P)\right) P^*(t+\tau_P) \\ &+ \left(b_H^B M_V + b_P^{H_B} M_V B + \alpha_H^B\right) H_B^* + b_P^B M_V B B^*,\end{aligned}$$

$$\begin{aligned}\frac{dE^*}{dt} &= -nb_{CE} C_V V_F^* + b_P^{H_E} M_V H^E H_E^* + (b_P^E M_V H_E E^* + b_{EC} C_V + \alpha_E) E^* \\ &- \rho_E b_P^E M_V H_E \xi\left(m(t+\tau_E)\right) E^*(t+\tau_E) \\ &+ b_{CE} C_V \left(1 + \mu_C \frac{C_V}{C^0}\right) C_V^* - b_{CE} C_V \left(1 + \mu_C \frac{C_V}{C^0}\right) m^*,\end{aligned}$$

$$\begin{aligned}\frac{dB^*}{dt} &= +b_P^{H_B} M_V H_B H_B^* + (b_P^B M_V H_B + \alpha_B) B^* \\ &- \rho_B b_P^B M_V H_B \xi\left(m(t+\tau_B)\right) \times B^*(t+\tau_B) \\ &- b_P^P M_V H_B \xi\left(m(t+\tau_P)\right) P^*(t+\tau_P),\end{aligned}$$

$$\frac{dP^*}{dt} = \alpha_P P^* - \rho_F F^*,$$

$$\frac{dF^*}{dt} = \gamma_{VF}\left(1 + \mu_V \frac{C_V}{C^0}\right) V_F V_F^* + \left(\gamma_{FV} V_F + \alpha_F\right) F^*,$$

$$\begin{aligned}\frac{dC_V^*}{dt} &= -\left(\nu + nb_{CE} E + \gamma_{VF} \mu_V \frac{C_V}{C^0} V_F F\right) V_F * + b_{EC} E E^* \\ &+ \left(\sigma V_F + b_{CE} E \left(1 + 2\mu_C \frac{C_V}{C^0}\right) + b_m\right) C_V^* \\ &- \left(b_{CE} E \left(1 + 2\mu_C \frac{C_V}{C^0}\right) + b_m\right) m^*,\end{aligned}$$

$$\begin{aligned}\frac{dm^*}{dt} &= -\gamma_{VC}V_F V_F^* + \sigma V_F C_V^* + \alpha_m m^* \\ &\quad - \frac{\partial\xi}{\partial m}\left[\rho_H^E b_H^E M_V(t-\tau_H^E)H_E^*(t-\tau_H^E)H_E^*\right. \\ &\quad + \rho_H^B b_H^B M_V(t-\tau_H^B))\, H_B^*(t-\tau_H^B)H_B^* \\ &\quad + \rho_E b_P^E M_V(t-\tau_E)H_E(t-\tau_E)E(t-\tau_E)E^* \\ &\quad + \rho_B b_P^B M_V(t-\tau_B)\times H_B(t-\tau_B)B(t-\tau_B)B^* \\ &\quad \left. + b_P^P M_V(t-\tau_P)H_B(t-\tau_P)B(t-\tau_P)P^*\right].\end{aligned}$$

Initial conditions are assigned with the following relationships:

$$\begin{array}{ll} V_F^*(T)\delta V_F(T) - V_F^*(0)\delta V_F(0) = 0, & \\ M_V^*(T)\delta M_V(T) - M_V^*(0)\delta M_V(0) = 0, & \\ \delta H_E(t) = 0, & -\max\left(\tau_H^E, \tau_E\right) \le t \le 0, \\ H_E^*(t) = 0, & T \le t \le T + \tau_H^E, \\ \delta H_B(t) = 0, & -\max\left(\tau_H^B, \tau_B, \tau_P\right) \le t < 0, \\ H_B^*(t) = 0, & T \le t < T + \tau_H^B, \\ \delta E(t) = 0, & -\tau_E \le \tau \le 0, \\ E^*(t) = 0, & T < t \le T + \tau_E, \\ \delta B(t) = 0, & -\max(\tau_P, \tau_B) \le t \le 0, \\ B^*(t) = 0, & T < t \le T + \max(\tau_P, \tau_B), \\ \delta P(t) = 0, & \\ P^*(t) = 0, & T < t \le T + \tau_P, \\ F^*(T)\delta F(T) - F^*(0)\delta F(0) = 0, & \\ C_V^*(T)\delta C_V(T) - C_V^*(0)\delta C_V(0) = 0, & \\ m^*(T)\delta m(T) - m^*(0)\delta m(0) = 0. & \end{array}$$

10.3. Numerical Methods of Sensitivity Analysis for the Functionals of Problems as Related to Parameters of Models

10.3.1. Numerical method for the solution of adjoint problem. The adjoint problem is formulated as the initial value problem for the system of adjoint equations:

$$\begin{cases} \dfrac{dy^*(t)}{dt} = \left[\dfrac{\partial f}{\partial y}\right]^T y^*(t) - \left[\dfrac{\partial f}{\partial y_\tau}\right]^T y(t+\tau), & t_0 + T \geq t \geq t_0, \\ y^*(t) = 0, & T \leq t < T + \tau, \end{cases} \tag{10.7}$$

where the right-hand side of system (10.7) is determined by the solution $y(t)$ of the main problem (10.1) on the interval $[t_0, t_0+T]$. The systems of the main and adjoint equations are the systems of differential equations with a delay. We used the adaptive algorithm DIFSUB-DEL (see Chapter 6) to solve them, which is based on linear multi-step methods (ψ_p^s, π_q) using the representation of the solution of the difference problem at a grid point t_n as a vector $\bar{y}_n$ whose components are the values of the first p derivatives of a polynomial which approximates continuously a solution of differential problem on the interval $[t_{n-p}, t_n]$. The initial value problem for the system of equations of the mathematical model of antiviral immune response is solved numerically forward in time with the positive integration step from t_0 to $t_0 + T$, and for the adjoint system backward in time with negative integration step from $t_0 + T$ to t_0. The solution of initial value problem for the main system of equations of model (10.1) on the interval $[t_0, t_0 + T]$ results in the set of values $\{t_n, \bar{y}_n\}$ in the grid points of an irregular grid of integration which are used while solving the adjoint problem to form the right-hand side of system (10.7). Notice that grid points of the integration algorithms are not the same for the main and adjoint problems. Therefore it is necessary to approximate the model solution between the grid points. We used the Nordsieck form for $\bar{y}_n$ which provided the interpolation whose accuracy was naturally correlated with the accuracy of the solution of the main problem.

While solving the adjoint problem we have taken into account the existence of discontinuity points for the derivatives of the solutions both primary and induced by the jumps of derivatives of a solution of the main problem. So, the set of discontinuity points for the initial value problem (10.1) of the first p derivatives of a solution contains a join $\{t_0 + j\tau\}_{j=0}^{p+1} \cup \{t_0 + T - j\tau\}_{j=0}^{p+1}$. Gradually, with passing through the first j points of discontinuity $\{t_0 + T - j\tau\}_{j=0}^{p+1}$ the solution of the adjoint problems smooths out, though, beginning with $t_0 + (p+1)\tau$ and up to t_0 the inverse process works, that is, the smoothness order of the solution $y^*(t)$ of adjoint problem decreases from p in the points $t_0 + p\tau$ to 1 in the point t_0.

10.3.2. Method for the computation of the functionals' sensitivity. The problem of estimation of a gradient's value of the functionals of model solutions consists of the following computational stages:

1. Solve the initial value problem on $[t_0, t_0+T]$ for the main system of equations of the model (10.1) using linear multistep methods (ψ_p^s, π_q):

$$\bar{y}_{n+1} = SC_{n+1}\bar{y}_n + h_{n+1}\bar{\psi}(t_n, C_{n+1}, \bar{y}_n, \pi_{q,n+1}(V_{\sigma_1,\sigma_\nu}), h_{n+1}).$$

2. Solve the initial value problem on $[t_0 + T, t_0]$ for the adjoint system (10.7) with initial conditions at the right end of the interval using the method (ψ_p^s, π_q) with negative integration step $-|h_n| < 0$, i.e., the adjoint method:

$$\bar{y}_{n+1} = SC_{n+1}\bar{y}_n - |h_{n+1}|\bar{\psi}(t_n, C_{n+1}, \bar{y}_n, \pi_{q,n+1}(V_{\sigma_1,\sigma_\nu}), -|h_{n+1}|).$$

Approximate elements of matrices in the right-hand side of (10.7) at each integration step of adjoint system using the Nordsieck vector of the solution of the main system of equations in the grid points $\{t_n, \bar{y}_n\}$. Preserve the approximations $\bar{y}^*$ of the solution $y^*(t)$ of adjoint problem and its first derivatives so obtained in the integration grid points t_n^* of the adjoint problem.

3. Determine the gradient $\partial J/\partial\alpha_i$ of the functional of the model solution on $[t_0, t_0+T]$ for each model's parameter α_i, $i = \overline{1, L}$ integrating the following equations:

$$\frac{d}{dt}\left(\frac{\partial J}{\partial \alpha_i}\right) = 2\langle y^*(t), \frac{\partial f}{\partial \alpha_i}(t)\rangle dt, \quad t \in [t_0, t_0 + T], \quad \frac{\partial J}{\partial \alpha_i}|_{t=0} = 0. \quad (10.8)$$

These differential equations are the sequence of the above relationship for the first variation of the quadratic functional

$$\delta J = 2\sum_{i=1}^{L}\langle y^*(t), \frac{\partial f}{\partial \alpha_i}\rangle dt\delta\alpha_i.$$

It is possible to use simple one-step integrators for the numerical solution of (10.8), for example the Runge-Kutta method with estimation of an error at each step of integration. The computation process must be organized with regard to the existence of discontinuities of solutions and their first derivatives of the main and adjoint systems of equations. The right-hand side of system (10.8) is approximated at each integration step with numerical solution in the Nordsieck form of the main $\{t_n, \bar{y}_n\}$ and adjoint $\{t_n^*, \bar{y}_n^*\}$ problems.

The sensitivity study for the functionals to the parameters of models helps to separate the most important processes which determine the variations of observed characteristics of infections. We present below some results of computing the sensitivity for various functionals of solutions to the mathematical model of the antiviral immune response.

10.3.3. Example of sensitivity study for the model of influenza. Let us consider the quadratic functional $J(y) = \int\limits_0^T V_f(t) \cdot V_f(t)dt$ characterizing total number of viruses in mucosa of the upper respiratory tract during the infection.

Fig. 58 shows the solution of the adjoint problem corresponding to the system of equations of mathematical model of antiviral immune response and to influenza parameters. Values of the derivatives $\partial J/\partial \alpha_i$ are presented in Table 14. It is possible to separate the expressed influence, though of opposite sign, of the replication rate of viruses and of the parameter of nonspecific clearance of viruses on total number of free viruses. the sensitivity of the functional is rather higher for characteristics of humoral immunity than to those of cellular immunity.

The results we obtained suggest that the variations in the severity of the course of influenza depend most of all on the cytopathicity of viruses and the rate of regeneration processes in epithelium of lungs. Intensification of cytotoxic reactions increases the severity of infection, while the intensification of humoral reactions leads to decrease in the severity of influenza.

Observed variations in the severity of influenza may be caused by deviations of various scales in the values of parameters of immune response and viral characteristics with the ratio from 1 to 10^5.

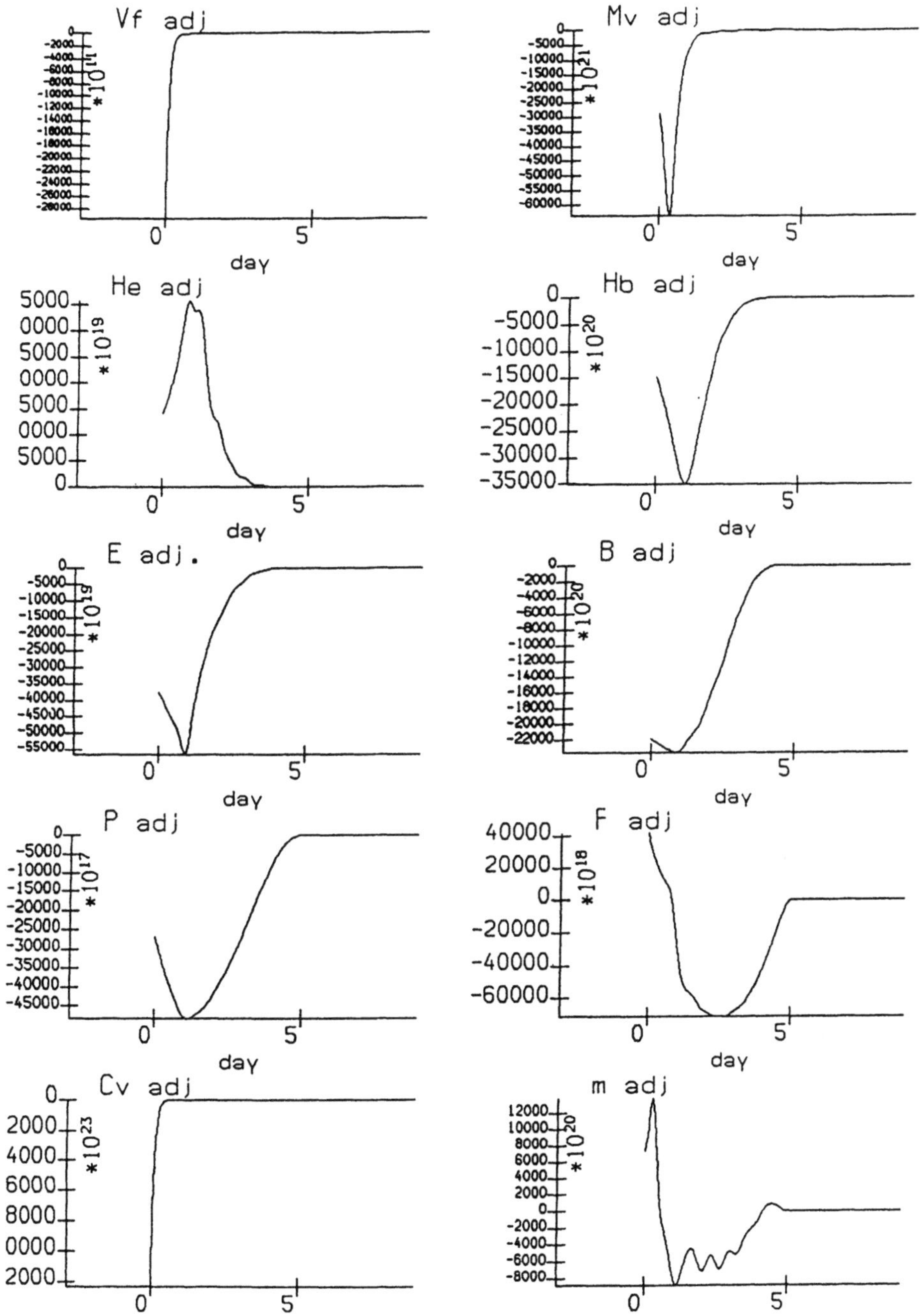

Fig. 58. Solution of the adjoint problem.

These regularities correspond to modern conceptions of the mechanisms of pathogenesis, defence and recovery during viral infections, in particular, during influenza [107, 148, 171, 352]. At the same time, obtained estimates of sensitivity permit direct quantitative interpretation concerning the contribution of various processes described by the model of antiviral immune response into a variation of disease severity indices.

Table 14

Sensitivity of the functional $J(y) = \int_0^T V_f(t) \cdot V_f(t)dt$ to variations of the model's parameters

Parameter α	Value of $\frac{dJ}{d\alpha} \cdot \alpha$ (+) $(\cdot 10^{-25})$	Parameter α	Value of $\frac{dJ}{d\alpha} \cdot \alpha$ (-) $(\cdot 10^{-25})$
ν	2.0	γ_{VM}	-2.0
γ_{FV}	0.3	ρ_H^B	-0.6
α_H^B	0.2	γ_{VF}	-0.3
α_m	0.2	μ_C, μ_V	-0.3
α_M	0.1	ρ_B	-0.3
α_P	0.06	b_p^P	-0.3
ρ_H^E	0.05	ρ_F	-0.3
b_{EC}	0.02	γ_{MV}	-0.2
b_H^E	0.01	b_H^B	-0.2
α_E	0.008	ρ_E	-0.1
b_p^B	0.006	H_B^*	-0.006
H_E^*	0.005	B^*	-0.006
$b_p^{H_B}$	0.00002	b_{CE}	-0.005
		γ_{VC}	-0.002
		α_H^E	-0.002
		b_p^E	-0.001
		σ	-0.0005
		E^*	-0.0004
		α_B	-0.0002
		P^*	-0.00006
		b_m	-0.00002
		α_F	-0.000005
		$b_p^{H_E}$	-0.0000001

The approaches to the determination of a norm for the course of a disease and to the sensitivity analysis of models make it possible to pass to the solution of the localization problem for the mechanisms of an unfavourable course of disease in terms of deviations from a norm in the parameters of immune, viral, and bacterial processes. In fact, if we have a definition for the norm of the course of disease in terms of model's variables $y_N(t, \alpha_*)$ and parameters α_*, then it is possible, in case of unfavourable variant in the course of infection $y(t, \alpha)$, to describe total deviation in the dynamics from the norm with the functionals of $J = \langle y - y_N, y - y_N \rangle$ type.

In this connection, use of adjoint equations for the estimation of a gradient of the functional J opens the possibility of constructing effective numerical procedures for the solution of the inverse problem for large-scale models, that is, the identification of deviations in the values of model's parameters corresponding to an unfavourable course of the infectious disease.

Bibliography

1. Ada G.L., Jones P.D. The immune response to influenza infection, *Current topics in microbiology and immunology.* 1986. V. 128. P. 1-54.

2. Adomian G. and Adomian G.E. Solution of the Marchuk Model of Infectious Disease and Immune Response. *Mathematical Modelling* N. 7. (1986) P. 803-807.

3. Agur Z. Fixed Points of Majority Rule Cellular Automata with Application to Plasticity and Precision of the Immune System. *Complex Systems.* 1991. N. 2. P. 351-357.

4. Agur Z.A. Theoretical Analysis of the Immune response to Antigenically Varying Pathogens, *Mathematics Applied to Biology and Medicine*, Eds. J. Demongeot, V. Capasso. Mathematica Biology Ser. Winnipeg.-Canada: Wuerz Publishing Ltd., 1993. P. 237-242.

5. Albrecht P. Numerical treatment of O. D. Es.: The theory of A-Methods,*Numerical Maths.* 1983. V. 47, Fasc. 1. P. 59-87.

6. Anderson R.W. and Perelson A.S. A Cayley Tree Immune Network Model with Antibody Dynamics, *Preprint* of the Los Alamos National Laboratory. LA-UR-92-3416.

7. Asachenkov A.L., Belykh L.N. Investigation of a mathematical model of viral disease, *Mathematical methods in clinical practice* Novosibirsk: Nauka, 1978, P. 19-26 (in Russian).

8. Asachenkov A.L., Marchuk G.I., Mohler R.R., Zuev S.M. *Disease dynamics.* Birkhauser – IIASA. Boston–Berlin–Tokyo. 1994. 360 p.

9. Ashwell J.D., Robb R.J., Malek T.B. Proliferation of T-lymphocytes in response to interleukin 2 varies with their state of activation, *J. Immunol.* 1986. V. 137. N. 8. P. 2572-2578.

10. Askonas B.A., Mullbacher A., Ashman R.B. Cytotoxic T memory cells in virus infection and the specificity of helper T-cells, *Immunology.* 1982. V. 45. P. 79-84.

11. Auben J.P. European Science Foundation. Final Report of Activities of the ESF Scientific Network "Dynamics of Complex Systems in Bio-Sciences", 1994. 92 p.

12. Baker C.T.H., Bocharov G.A., Paul C.A.H. and Romanyukha A.A. Delay Effects in Cell Proliferation. N.A. Report 266. Mathematics Department, University of Manchester (May, 1995).

13. Baker C.T.H. and Paul C.A.H. Pitfalls in Parameter Estimation for Delay Differential Equations. (1996). *SIAM J. Sci. Comp.* (In press).

14. Baker C.T.H., Paul C.A.H., Wille D.R. Issues in the Numerical Solution of Evolutionary Delay Differential Equations. *Adv. Comput. Math.* 3, 1996. P. 171-176.

15. Bard Y. *Nonlinear Parameter Estimation.* New York. Academic Press. 1974.

16. Bard J.B.L. A quantitative theory of liver regeneration in rat. II. Matching and improved mitotic inhibitor model to the data, *J. Theor. Biol.* 1979. V. 79. P. 121-136.

17. Behn U., van Hemmen J.L., Sulzer B. Memory B-cells stabilize cycles in a repressive network. In: (Perelson, Weisbuch, eds.) Theoretical and Experimental Insights into Immunology. NATO ASI Series H. Heidelberg: Springer-Verlag.1992. Vol. 66. P. 250-260.

18. Bell G. Mathematical model of clonal selection and antibody production. I *J. Theor. Biol.* 1970. V. 29. P. 191-232.

19. Bell G. Mathematical model of clonal selection and antibody production. II *J. Theor. Biol.* 1971. V. 33. P. 339-378.

20. Bell G. Mathematical model of clonal selection. III. The cellular basis of immunological paralysis, *J. Theor. Biol.* 1971. V. 33. P. 379-398.

21. Bell G. Prey-predator equations simulating an immune response, *Math. Biosci.* 1973. V. 16. P. 291-314.

22. Bell G. Lymphocyte traffic patterns and cell–cell interactions, *Theoretical Immunology*, Ed. G. Bell, A. Perelson, G. Pimbley. N.Y., 1978. P. 341-375.

23. Bellen A.J., Zennaro M. Numerical solution of delay differential equations by an implicit Runge-Kutta method, *Numer. Math.* 1985. V. 47. P. 301-316.

24. Belykh L.N. Mathematical model of joined disease, *Mathematical models of diseases and methods of medical information processing* Novosibirsk: Nauka, 1979. P. 32-39 (in Russian).

25. Belykh L.N. Mathematical model of biinfection and treatment of chronic form with exacerbation, *Mathematical modeling in immunology and medicine.* Novosibirsk: Nauka, 1979. P. 32-38. (in Russian).

26. Belykh L.N. Analysis of some mathematical problems in immunology Moscow: Nauka, 1988 (in Russian).

27. Belykh L.N., Kaljaev D.V. Investigation of hypertoxic forms of a disease in the framework of mathematical model, *Numerical processes and systems. Issue 3.* Moscow: Nauka, 1985. P. 180-187 (in Russian).

28. Benacerraf B., Unanue E.R. *Textbook of Immunology.* Baltimore: Williams and Whilkinson, 1979. 298 p.

29. Berningen M., Hammer M., Hoyer B., Gerin J.L. An assay for the detection of the DNA genome of hepatitis B virus in serum, *J.Med.Virol.* 1982.V.9. P.57-68.

30. Bertuzzi A., Bruni C., Gandolfi A., Koch G. Maximum likelihood identification of an immune response model. Modeling and optimization of complex systems, *Proceedings of the IFIP TC7 Working Conference*, Novosibirsk, Ed. G. I. Marchuk. Berlin etc.: Springer, 1979. P. 1-14.

31. Bersini H., Varela F.J. Hints for adaptive problem solving gleaned from immune networks. In: Parallel Problems solving from Nature, Eds. Manner, Schwefel. *Lecture Notes in Computer Sciences.* Springer-Verlag: Berlin. 1991. P. 343-354.

32. Bickart T.A. P-stable and P$[\alpha, \beta]$-stable integration/interpolation methods in the solution of retarded differential equations, BIT. 1982. V. 22. P. 464-476.

33. Bluger A.Ph. *Viral Hepatitis.* Riga: Zvaigzne, 1978. 397 p. (in Russian).

34. Bluger A.Ph., Novitskij I.N. *Practical Hepatology.* Riga, 1984. 405 p. (in Russian).

35. Bocharov G.A., Romanyukha A.A. Numerical solution of delay-differential equations by Runge-Kutta-Fehlberg methods. *Preprint* OVM AN USSR N. 99. Moscow, 1985. (in Russian).

36. Bocharov G.A., Romanyukha A.A. Numerical solution of delay-differential equations by linear multistep methods. *Preprints* OVM AN USSR NN. 116, 117. Moscow, 1986. (in Russian).

37. Bocharov G.A., Marchuk G.I., Romanyukha A.A. Numerical solution by LMMs of delay-differential equations modeling an immune response. *Numer. Mathem.* 73: (1996) P. 131-148.

38. Bocharov G.A., Romanyukha A.A. Mathematical modelling of immune response during acute viral infections, Theoretical and Experimental Insights into Immunology, Eds. Perelson, Weisbuch. NATO ASI Series H.-Heidelberg: Springer-Verlag. 1992. V. 66. P. 309-321.

39. Bocharov G.A., Romanyukha A.A. Mathematical model of antiviral immune response. III. Influenza A Virus Infection, *J. Theor. Biol.* 1994. V. 167. P. 323-360.

40. Bocharov G.A., Romanyukha A.A. Numerical treatment of parameter identification problem for delay-differential systems arising in immune response modelling, *Appl. Num. Math.* 1994. V. 15. P. 307-326.

41. Bock H.G., Schloder J. Numerical solution of retarded differential equations with state dependent time lags, *Zeitschrift fur Angew. Math. und Mech.* 1981. B 61, Heft. 5. P. 260-271.

42. Brass A., Grencis R., Else K.J. A Cellular Automata Model for helper T-cell subset polarization in Chronic and Acute Infection. *J. Theor. Biol.* 1994.

43. Brendel V., Perelson A.S. Kinetic analysis of absorption processes. *SIAM J. Appl. Math.* 1987. V. 47 P. 1306-1319.

44. Brondz B.D. T-lymphocytes and their receptors in immunological recognition. Moscow: Nauka, 1987. 471 p. (in Russian).

45. Bruni C., Gandolfi A., Germani A. Theoretical investigation and numerical procedure for the estimation of antibody affinity distribution from binding data *Mathematical models in immunology and medicine*, Ed. G.I. Marchuk, L.N. Belykh. Amsterdam: North Holland, 1993. P. 97-108.

46. Bruni C., Giovenco M., Koch G., Strom R. A dynamical model of humoral immune response, *Math. Biosci.* 1975. V. 27. P. 191-212.

47. Bruni C., Giovenco M., Koch G., Strom R. Modeling of immune response: a system approach, *Theoretical Immunology.* N. Y.: Marcel Dekker, 1978. P. 379-414.

48. Bruni C., Giovenco M., Koch G., Strom R. Optimal derivation of antibody distribution in the immune response from noisy data, *Proc. of the 8th IFIP Conference on Optimization Techniques.* Heidelberg: Springer-Verlag, 1978. P. 411-420.

49. Burnet F.M. *The Clonal Selection Theory of Acquired Immunity.* Cambridge: The University Press, 1959.

50. Byrne G.D., De Gregoria A.J., Salane D.E. A program for fitting rate constants in gas phase chemical kinetics models, *SIAM J. Sci. Stat. Comput.* 1984. V. 5. P. 642-657.

51. Cantrell D.A., Smith K.A. The interleukin 2 T cell system: a new cell growth model, *Science.* 1984. V. 224. P. 1312-1326.

52. Celada F., Seiden P. A computer model of cellular interactions in the immune system. *Immunol. Today,* 1992. V. 13 P. 56-62.

53. *Cell Surface Dynamics: Concepts And Models,* Eds. A.S. Perelson, C. De Lisi, F.W. Wiegel. New York: Marcel Dekker, Inc., 1984.

54. Chazov E.I., Smirnov V.N. et al. Classification of pathological isoenzyme spectrum of serum lactatdehydrogenase using computer. *Cardiologija,* 1969. N. 12. (in Russian).

55. Chernukh A.M. *Inflammation. An essay on pathology and experimental therapy.* Moscow: Medicina, 1979. 448 p. (in Russian).

56. Chowdhury D. A unified model of immune response II: Continuum approach, *J. theor. Biol.* 1993, V. 165. P. 135-159.

57. Cohen I.R. The cognitive principle challenges clonal selection. *Immunology Today.* 1992, V. 13. P. 441-444.

58. Crouzeix M., Lisbona F.J. The convergence of variable-stepsize, variable-formula, multistep methods, *SIAM J. on Numer. Anal.* 1984. V. 21. N. 3. P. 512-534.

59. De Boer R.J. Information processing in immune systems: Clonal selection versus idiotypic network models, *Cell to Cell Signalling. From Experiments to Theoretical Models.* Academic Press Limited. 1989. P. 285-302.

60. De Boer R.J. Poor repertoire selection in symmetric idiotypic models, *Immunology Letters.* 1989. V. 22. P. 101-112.

61. De Boer R.J., Boerlijst M.C. Diversity and virulence thresholds in AIDS. *Proc. Natl Acad. Sci.* USA. 1994. V. 94. P. 544-548.

62. De Boer R.J., Hogeweg P. Interactions between macrophages and T-lymphocytes: Tumor sneaking through intrinsic to helper T-cell dynamics, *J. Theor. Biol.* 1986. V. 120. P. 331-351.

63. De Boer R.J., Hogeweg P. Immunological discrimination between self and non-self by precursor depletion and memory accumulation, *J. Theor. Biol.* 1987. V. 124. P. 343-369.

64. De Boer R.J., Hogeweg P., Dullens H.F.J., De Weger R.A., Den Otter W. Macrophage T-lymphocyte interactions in the anti-tumor immune response: a mathematical model., *J. Immunol.* 1985. V. 134. N. 4. P. 2748-2758.

65. De Boer R.J., Hogeweg P. Tumor escape from immune elimination: simplified bound cytotoxicity models. *J. Theor. Biol.* 1985, V. 113. P. 719-736.

66. De Boer R.J., Kevrekidis I.G., Perelson A.S. Immune network behavior. I. From stationary states to limit cycle oscillations. (Submitted to: *Bull. Math. Biol.* 1992a).

67. De Boer R.J., Kevrekidis I.G., Perelson A.S. Immune network behavior. II. From oscillations to chaos and stationary states. (Submitted to: *Bull. Math. Biol.* 1992b).

68. De Boer R.J., Neumann A.U., Perelson A.S., Segel L.A., Weisbuch G. Recent Approaches to Immune Networks, *Mathematics Applied to Biology and Medicine,*

Eds. J. Demongeot, V. Capasso. Mathematical Biology Ser. Winnipeg-Canada: Wuerz Publishing Ltd., 1993. P. 243-261.

69. De Boer R.J., Perelson A.S. Size and connectivity as emergent properties of a developing immune network, *J. theor. Biol.* 1991. V. 149. P. 381-424.

70. De Boer R.J., Perelson A.S. How diverse should the immune system be? *Proc. R. Soc.* Lond. B. 1993, V. 252. P. 171-175.

71. De Boer R.J., Segel L.A., Perelson A.S. Pattern Formation in One- and Two-dimensional Shape-space Models of the Immune System, *J. Theor. Biol.* 1992. V. 155. P. 295-233.

72. De Boer R.J., van der Laan J.D., Hogeweg P. Randomness and pattern scale in the immune network. A cellular automata approach, *Thinking about biology.*, Eds. F.J. Varela, W.D. Stein. SFI lectures in complex system. Addison-Wesley, 1992.

73. Deinhardt E., Zuckerman A.J. Immunization against hepatitis B: Report on WHO meeting on viral hepatitis in Europe, *J. Med. Virology.* 1985. V. 17. N. 3. P. 209-217.

74. De Lisi C. Detection and analysis of recognition and selection in the immune response, *Bull. Math. Biol.* 1977. V. 39. P. 705-719.

75. De Lisi C. Some mathematical problems in the initiation and regulation of the immune response, *Math. Biosci.* 1977. V. 35. P. 1-26.

76. De Lisi C., Perelson A. The kinetics of aggregation phenomena, *J. Theor. Biol.* 1976. V. 62. P. 159-210.

77. De Lisi C., Rescigno A. Immune surveillance and neoplasia. I. Minimal mathematical model, *Bull. Math. Biol.* 1977. V. 39. P. 204-221.

78. De Lisi C., Rescigno A. Immune surveillance and neoplasia. II. Two-stage mathematical model, *Bull. Math. Biol.* 1977. V. 39. P. 487-497.

79. De Weck A., Kristensen F., Jocourt F., Bettens F., Walker C., Wang Y. Lymphocyte receptors in aging and various clinical condition, *Springer Semin. Immunopathol.* 1984. V. 7. P. 273-279.

80. Dibrov B.F., Livshits M.A., Volkenstein M.V. The effect of time lag in the immune reaction, *Proc. of IFIP Working Conference on Modeling and Optimization of Complex Systems.* Berlin a. o.: Springer-Verlag, 1979. P. 87-94.

81. Dibrov B.F., Livshitz M.A., Volkenstein M.V. Mathematical Models of Immune Processes, *J. Theor. Biol.* 1977. V. 65. P. 609-631.

82. Dibrov B.Ph., Livschitz M.A., Vol'kenstein M.V. Mathematical model of immune reaction. *Biophisica*, 1976. N. 21. P. 905-909. (in Russian).

83. Djery I., Pertsev N.V. On the stability of equilibrium states of functional delay-differential equations possessing the property of mixed monotonicity, *DAN*, 1987. V. 297. N. 1. P. 23-25. (in Russian).

84. Doherty P. C. T cells and viral infections , *Br. Med. Bull.* 1985. V. 41. N. 1. P. 7-14.

85. Doherty P.C. Some problem areas in the interaction between viruses and the immune system , *Immunol. Cell. Biol.* 1986. V. 65. P. 279-286.

86. Dongarra J.J., Bunch J.R., Moler C.B., Stewart G.W. LINPACK *Users Guide*, Soc. Ind. Appl. Math. Philadelphia, 1979.

87. Eddleston A.L W.F., Williams R. The role of immunological mechanisms in chronic hepatitis, *Annals of Clinical research.* 1976. V. 8. P. 162-173.

88. El'sgolz L.E., Norkin S.B. *Introduction to the theory of delay-differential equations.* Moscow: Nauka, 1971. (in Russian).

89. Emanuel N.M., Knorre D.G. *Course on chemical kinetics.* Moscow: Vysshaya schkola, 1984. (in Russian).

90. Endo J., Matsushima K., Oppenheim J.J. Mechanisms of *in vitro* antitumor effect of interleukin 1 (IL-1), *Immunobiol.* 1986. V. 172. P. 316-322.

91. Ermentrout G.B., Edelstein-Keshet L. Cellular Automata Approaches to Biological Modelling. *J. Theor. Biol.* 1993. V. 160. P. 97-133.

92. Ermol'eva Z.V. *Antibiotics, interferon, bacterial polysaccharides.* 2-nd ed. Moscow: Medicine, 1968. 384 p.

93. Essunger P., Perelson A.S. Modeling HIV Infection of CD4+ T-cell Population, *J. Theor. Biol.* 1994, V. 170. P. 367-391.

94. Farmer J.D., Kauffman S.A., Packard N.H., Perelson A.S. Adaptive dynamic networks as models for the immune system and autocatalytic sets. *Ann. N.Y. Acad. Sci.*. 1987. V. 504. P. 118-131.

95. Faro J., Velasco S. Studies on a Recent Class of Network Models of the Immune System, *J. Theor. Biol.* 1993, V. 164. P. 271-290.

96. Faro J., Velasco S. Numerical Analysis of a Model of Ligand-Induced B-Cell Antigen-Receptor Clustering. Implications for Simple Models of B-Cell Activation in an Immune Network, *J. Theor. Biol.* 1994 V. 167. P. 45-53.

97. Faro J., Velasco S. Crosslinking of membrane-immunoglobulins and B-cell activation: A simple model based on percolation theory, *Proc. R. Soc.* Lond. B 1993, V. 254. P. 139-145.

98. Farooqi Z.H., Mohler R.R. Distribution Models of Recirculating Lymphocytes. *IEEE Trans. on Biomed. Engineer.* 1989. V. 36. P. 355-362.

99. Feldman M. Cell interactions in the immune response *in vitro*, *J. Exp. Med.* 1972. V. 136. P. 3-14.

100. Fishman M.A., Perelson A.S. Modeling T Cell-Antigen Presenting Cell Interactions *J. Theor. Biol.* 1993. V. 160 P. 311-342.

101. Fishman M.A., Perelson, A.S. Th1/Th2 Cross-Regulation. *Preprint* of the Santa Fe Institute 93-08-051 (Submitted to J. theor. Biol.)

102. Fletcher R. A new approach to variable metric algorithms, *The Computer J.* 1970. V. 13. N. 3. P. 317-322.

103. Forrest S., Perelson A.S. Genetic Algorithms and the Immune System. In: *Parallel Problems Solving from Nature. Lecture Notes in Computer Science*, Eds. Schwefel, Manner. Springer Verlag: Berlin. 1991. P. 320-325.

104. Forsythe G.E., Malcolm M.A., Moler C.B. *Computer methods for mathematical calculations.* Prentice-Hall, Inc., 1977.

105. Freedman H., Gatica J. A threshold model simulating humoral immune response to replicating antigens, *Math. Biosci.* 1977. V. 37. P. 113-134.

106. *Fundamental Immunology*, Ed. W.E. Paul. N. Y.: Raven Press, 1984.

107. *Fundamental Virology*, Eds. B.N. Fields, D.M. Knipe. N. Y.: Raven Press, 1986.

108. Gandolfi A., Strom R. Cell-mediated antigen presentation to T cells: mathematical model. R. No 315, IASI (CNR) Roma, 1990.

109. Gatica J. Qualitative properties of a threshold model for antibody production, *Tech. Report University of Iowa.* 1977. 10 p.

110. Gatica J., Waltman P. A singular functional differential equation arising in an immunological model, *Proc. of the Conf. on Theory of Ordinary and Partial Differential Equations*, 1976. Springer-Verlag, 1977.

111. Gavrilov V.I., Semenov B.Ph., Zhdanov V.M. *Chronic viral infections and their modelling.* Moscow: Medicine, 1974. 233 p. (in Russian).

112. Gear G.W. *Numerical initial value problems in ordinary differential equations.* N. Y.: Prentice-Hall, Englewood Cliffs, 1971.

113. Gear G.W. The automatic integration of ordinary differential equations, *Commun. of the ACM.* 1971. V. 14. N. 3. P. 176-179.

114. Gear G.W. DIFSUB for solution of differential equations D2. Algorithm 407, *Commun. of the ACM.* 1971. V. 14. N. 3. P. 185-190.

115. Gear G.W., Tu K.W. The effects of variable mesh size on the stability of multistep methods, *SIAM J. Num. Anal.* 1974. V. 11. P. 1025-1043.

116. Gear G.W., Vu T. Smooth numerical solutions of ordinary differential equation systems, *Numerical treatment of Inverse Problems in Differential and Integral equations*, Eds. P. Deuflhard, E. Hairer. Birkhäuser, Basel, 1983. P. 2-12.

117. Gear G.W., Watanabe D.S. Stability and convergence of variable order multistep methods, *SIAM J. Num. Anal.* 1974. V. 11. P. 1044-1058.

118. Gill Ph.E, Murray W., Wright M.H. *Practical optimization.* Academic Press. London, 1981.

119. Good R.A. Disorders of the immune response, *Immunobiology.* Stanford (USA), 1971. P. 3-16.

120. Grabstein K., Dower S. et al. Expression of interleukin 2, interferon-γ, and the IL-2 receptor by human peripheral blood lymphocytes, *J. Immunol.* 1986. V. 136. P. 4503-4508.

121.Greene W.C. Characterization of the human IL-2 receptor at the DNA, RNA and protein level, *Immunobiol.* 1986. V. 172. P. 357-364.

122. Gunter N., Hoffman G.W. Qualitative dynamics of network model for regulation of immune system: a rationale for the IgM to IgG switch, *J. Theor. Biol.* 1982. V. 94. P. 815-855.

123. Hairer E., Norsett S., Wanner.G. *Solving ordinary differential equations. I. Nonstiff problems.* Springer-Verlag: Berlin etc. 1987.

124. Haken H. *Synergetics. An Introduction. Second Enlarged Edition.* Springer-Verlag, Berlin, 1978.

125. Hale J. *Theory of functional differential equations.* Spr.-Ver. N.Y. 1977.

126. Hall W.J. et al. Pulmonary mechanics after uncomplicated influenza A infection, *Am. Rev. Respir. Dis.* 1976. V. 113. N. 1. P. 141-147.

127. Hantmacher F.R. *Theory of matrices.* 4th ed. Moscow: Nauka, 1988. 543 p. (in Russian).

128 Hege J.S., Cole G. A mathematical model relating circulating antibody and antibody forming cells, *J. Immunol.* 1966. V. 97. P. 34-40.

129. Hellriegel B. Modelling the immune response to malaria with ecological concepts: short-term behavior against long-term equilibrium, *Proc. R. Soc.* Lond. B 1992 V. 250. P. 249-256.

130. Hightower R., Forrest S., Perelson A.S. The evolution of Secondary Organization in Immune System gene Libraries. *Preprint* of the Santa Fe Institute 93-01-002. 1993. (Submitted to the European Conference on Artificial Life).

131. Himmelblau D.M., Jones C.R., Bischoff K.B. Determination of rate constants for complex kinetics models, *Ind. Eng. Chem. Fund.* 1967. V. 6. N. 4. P. 539-543.

132. Hoffman G.A. A symmetrical model for the regulation of the immune system, *IFIP Working Conference on Modeling and Optimization of Complex Systems.* Novosibirsk, 1978. P. 38-41.

133. Hoffmann G.W. A theory of regulation and self-nonself discrimination in an immune network. *Eur. J. Immunol.* 1975 V. 5 P. 638-647.

134. Hoffmann G.W. On network theory and H-2 restriction, In: Warner N. (ed) Contemporary topics in immunobiology. vol. II. New York: Plenum Press, 1980. P. 185-226.

135. Hoffman G.W., Benson M.W., Bree G.M., Kinahan P.E. A teachable neural network based on an unorthodox neuron, *Physica D.* 1986. V. 22. N. 1-3. P. 233-246.

136. Hoffman G.W., Cooper-Willis A. Symmetry, complexity and stability in immune system network theory, *Mathematical Modeling in Immunology and Medicine*, Eds. G.I.Marchuk, L.N.Belykh. North-Holland Publ. Co., Amsterdam, 1983. P. 31-42.

137. Hoffmann G.W., Hraba T. (Eds.) Immunology and epidemiology. Berlin: Springer-Verlag, 1986.

138. Hofman F.M. et al. Distribution of cells bearing the Tac antigen during the ontogeny of human lymphoid tissue, *J. Immunol.* 1985. V. 134. N. 6. P. 3751-3755.

139. Hollinger F.B., Melnick J.L. Features of Viral Hepatitis. Part IV: Prevention and control, *Virology*, Ed. by B.N.Fields et al. N. Y.: Raven Press, 1985. P. 1472-1494.

140. Hollinger F.B., Troisi C., Heiberg D., Sanches Y., Dreesman G.R., Melnick J.L. Response to a hepatitis B polypeptide vaccine in micelle form in a young adult population, *J. Med. Virol.* 1986. V. 19. N. 3. P. 229-240.

141. Hoofnagle J.H., Seef L.B. Natural history of chronic type B hepatitis, *Progress in Liver Diseases*, Ed. H.Popper. N. Y.: Grune and Stratton, 1982. P. 469-479.

142. Hosten L.H. A comparative study of short cut procedures for parameter estimation in differential equations, *Comp. Chem. Eng.* 1979. V. 3. P. 117-126.

143. Hsu Ih-Ding, Kasarinoff N. Existence and stability of periodic solutions of a third order nonlinear autonomous system simulating immune response in animals, *J. Roy. Soc.* Edinburg. 1977. V. 77A. P. 163-175.

144. Van der Houwen P.J., Sommeijer B.P. Stability in linear multistep methods for pure delay equations, *J. of Comput. and Appl. Math.* 1984. V. 10. P. 55-63.

145. Van der Houwen P.J., Sommeijer B.P., Baker C.T.H. On the stability of predictor-corrector methods for parabolic equations with delay, *Report* NM-R8410, 1984, Centre for Mathematics & Computer Science, The Netherlands.

146. Hoyle L. *The Influenza Viruses.* Wien–N. Y.: Springer, 1986. 375 p.

147. *Immunoglobulins*, Ed. G. Litman, R. Good. N.Y.–London.: Plenum Medical Book Co., 1978.

148. *Immunological Aspects of Infectious Diseases.* Ed. J. Dick. 1980.

149. *Immunology and Epidemiology*, Ed. by G.W.Hoffman and T.Hraba. Springer-Verlag, 1986, *Lecture Notes in Biomathematics.* V. 65.

150. *Immunology: In 3 vol*, Ed. W. Paul. New York.: Raven Press, 1984.

151. In't Hout K.J. A new interpolation procedure for adapting Runge–Kutta methods to delay differential equations. *BIT* 32, 1992. P. 634-649.

152. Intrator N., Deocampo G.P., Cooper L.N. Analysis of Immune System Retrovirus Equations, Theoretical Immunology Part Two, SFI Studies in the Science of Complexity, Ed. A.S. Perelson. Addison-Wesley, Redwood, CA. 1988. V. III. P. 85-100.

153. Ivanov K.P. *Bioenergetics and temperature homeostasis.* Leningrad: Nauka, 1972. (in Russian).

154. Iwarson S., Tabor E., Thomas H.C., Goodall A., Waters J., Snoy P., Shin J.W.-K., Gerety R.J. Neutralization of hepatitis B virus infectivity by a murine monoclonal antibody: an experimental study in the chimpanzee, *J. Med. Virol.* 1985. V. 16. N. 1. P. 89-96.

155. James F., Roos M. MINUIT-A system for function minimization and analysis of the parameters errors and correlations, *Computer Physics Communications.* 1975. V. 10. P. 343-367.

156. Jerne N. The immune system, *Scientific American.* 1973. N. 229. P. 52-60.

157. Jilek M. The number of immunologically activated cells after repeated immunization (a mathematical model), *Folia Microbiol.* 1971. V. 16. P. 12-23.

158. Jilek M. On contact of immunocompetent cells with antigen (Note on probability model), *Folia Microbiol.* 1971. V. 16. P. 83-87.

159. Jilek M., Sterzl J. A model of differentiation of immunological competent cells, *Developm. Aspects Antib. Form. Struct.* Prague–N. Y., 1970. P. 963-981.

160. Jilek M., Ursinova Z. The probability of a contact between immunocompetent cell and antigen, *Folia Microbiol.* 1970. V. 15. P. 294-302.

161. Jilek M., Ursinova Z. On the distribution of the first contact of immunocompetent cell with antigen, *Folia Microbiol.* 1970. V. 15. P. 492-499.

162. Kalman R.E. Identification of noisy systems, UMN. 1985. V. 40. Iss. 4 (244). P. 27-41. (in Russian).

163. Karpov A.V., Romanyukha A.A. Mathematical Modelling of Destructive Pneumonia, Immunological and Metabolic Systems Mathematical Models and Methods of Investigations, Eds. G.I. Marchuk, A.Werynski. Lecture Notes of the ICB Seminars. Biosystems. Warsaw: ICB, 1992. P. 50-67.

164. Kaufman M., Thomas R. Model analysis of the bases of multistationarity in the humoral immune response, *J. Theor. Biol.* 1987. V. 129. P. 141-162.

165. Kaufman M., Urbain J., Thomas R. Towards a logical analysis of the immune response, *J. Theor. Biol.* 1985. V. 114 P. 527-561.

166. Kauffman S.A., Weinberger E.D., Perelson, A.S. Maturation of the immune response via adaptive walks on affinity landscapes. Theoretical Immunology Part I (Perelson, A.S., ed.) Reading, MA: Addison Wesley. 1988. 349-382.

167. Kepler T.B., Perelson A.S. Somatic Hypermutation in B Cells: An Optimal Control Treatment, *J. Theor. Biol.* 1993. V. 164 P. 37-64.

168. Kepler T.B., Perelson A.S. Cyclic re-entry of germinal center B-cells and the efficiency of affinity maturation. *Immunology Today.* 1993, 14. P. 412-415.

169. Khaitov R.M. Results and prospects of investigations on the construction of artificial vaccine, *Immunologija.* 1982. N 6. P. 35-40. (in Russian).

170. Kilburn E. D. *Influenza viruses and influenza.* Moscow: Medicine, 1978, 584 p. (in Russian).

171. Kingsbury D.W. Orto- and paramyxoviruses replication. In: *Fundamental Virusology,* Eds. B.N. Fields, D.M. Knipe. N. Y.: Raven Press, 1986.

172. Klein P., Hraba T., Dolezal J. Mathematical Model of B lymphocyte replacement kinetics: its application to the recovery from tolerance in adult mice, *Math. Biosci.* 1985, 73. 228 p.

173. Kondo S., Shimizu A., Saito Y. Molecular basis for two different affinity states of the IL-2 receptor: affinity conversional model, *Cell Biol.* 1986. V. 83. P. 9026-9029.

174. Kuznetsov V.A. Dynamics of immune processes during tumor growth, Moscow: Nauka. 1992, 340 p. (in Russian).

175. Kuznetsov V.A., Alekseyeva Ye.I. (1991) Computer analysis of the dynamics of two tumor foci with reference to circulation of immune cells. *Biofizika,* (Pergamon Press, transl. form russ.) 1991. V. 36. P. 109-113.

176. Kuznetsov V.A., Zhivoglyadov V.P., Stepanova L.A. Kinetic Approach and Estimation of the Parameters of Cellular Interaction Between the Immunity System and a Tumor, *Archivum Immunologiae et Therapiae Experimentalis* 1993. V. 41 T. 21-31.

177. Lambert J.D. *Computational methods in ordinary differential equations.* London: Wiley, 1973.

178. Langman R.E., Cohn M. The E-T (elephant-tadpole) paradox necessitates the concept of a unit of B-cell function: the Protection, *Molecular Immunology.* 1987. V. 24. P. 675-697.

179. Lefevre O. and Parisi G. Learning and Memory in Spinn-glass like Immune Networks., *Network.* 1994. V. 4. P. 39-65.

180. Leijh P.C.J. et al. Kinetics of phagocytosis of Staphylococcus aureus by human granulocytes, *Immunology.* 1979. V. 37. P. 343.

181. Leijh P.C.J., van der Barselaar M.Th., van Furth R. Kinetics of phagocytosis and intracellular killing of Staphylococcus aureus and Escherichia coli by human monocytes, *Scand. J. Immunol.* 1981. V. 13. P. 159-174.

182. Levi M.I., Smirnova O.A. Cyclic kinetics and mathematical expression of the primary immune response to soluble antigen: VII. The conveyer hypothesis and its mathematical expression, *Folia Microbuologica.* 1977, 22. P. 117-127.

183. Lotze M.T., Matory Y.L. et al. In vivo administration of purified human IL-2. II. Half-life, immunologic effects and expansion of peripheral lymphoid cells *in vivo* with recombinant IL-2, *J. Immunol.* 1985. V. 135. N. 4. P. 2865-2874.

184. Lowenthal J. W., MacDonald H.R. Expression of interleukin 1 receptors is restricted to the L3T4+ subset of mature T lymphocytes, *J. Immunol.* 1987. V. 138. N. 1. P. 1-3.

185. Lowenthal J.W., Corthesy P., Touque C., Lees R. et al. High and low affinity receptors: analysis by IL-2 dissociation rate and reactivity with monoclonal anti-receptor antibody PC61, *J. Immunol.* 1985. V. 135. N. 6. P. 3988-3994.

186. Lowenthal J.W., Tougue C., MacDonald H.R., Smith K.A., Nabholz M. Antigen stimulation regulates the expression of IL-2 receptors in a cytolytic T lymphocyte clone, *J. Immunol.* 1985. V. 134. N. 2. P. 931-939.

187. Lumb J.R. Lymphocyte differentiation, repertoire development and migration: The need for mathematical models, *Computers and Mathematics with Applications.* 1987. V. 14. P. 657-697.

188. Luria S.E., Darnell J., Baltimor D., Campbell A. *General virology* 3rd. Edition. John Willey and Sons. N.Y. 1978.

189. Lyons C.R., Lipscomb M.F. Alveolar macrophages in pulmonary immune I. Role in initialization of primary immune response and in selective recruitment of T lymphocytes to the lung, *J. Immunol.* 1983. V. 130. P. 1113-1119.

190. Macken C.A., Perelson A.S. A multistage model for the action of cytotoxic T lymphocytes in multicellular conjugates, *J. Immunol.* 1984. V. 132. P. 1614-1624.

191. Macken C.A., Perelson A.S. Branching Processes Applied to Cell Surface Aggregation Phenomena, *Lecture Notes in Biomathematics.* V. 58. Berlin: Springer-Verlag, 1985.

192. Maizels R.M., Bundy D.A.P., Selkirk M.E., Smith D.F., Anderson R.M. Immunological Modulation and Evasion by helminth parasites in human populations, *Nature.* 1993. 365. P. 797-805.

193. Marchuk G.I. Some mathematical models in immunology, *Proc. of the IFIP Conference on Optimization Techniques.* Heidelberg: Springer-Verlag, 1978. P. 41-62.

194. Marchuk G.I. Mathematical immune response models and their interpretation, *Proc. of the IFIP Working Conference on Modelling and Optimization of Complex Systems.* Berlin a. o.: Springer-Verlag, 1979. P. 114-129.

195. Marchuk G.I. it Adjoint Equations and Analysis of Complex Systems. Dordrecht: Kluwer Academic Publishers, 1995. 466 p.

196. Marchuk G.I., Agoshkov V.I., Shutyaev V.P. *Adjoint equations and perturbation algorithms in nonlinear problems of mathematical physics*, CRC Press, 1996 (to be published). 250 p.

197. Marchuk G.I., Berbentsova E.P. Acute pneumonia: immunology, estimation of gravity, clinics, treatment. Moscow: Nauka, 1989. 306 p. (in Russian).

198. Marchuk G.I., Berbentsova E.P. *Chronic Bronchitis: Immunology, Estimation of Severity, Clinics, Treatment.* Moscow: UFN, 1955. 480 p. (in Russian).

199. Marchuk G.I., Nisevich N.I., Zubikova I.I. Biochemical index and estimation of functional state of liver during viral hepatitis in children using mathematical methods, *Mathematical Methods in Clinical Practice*, Ed. by G.I. Marchuk, N.I. Nisevich. Oxford: Pergamon Press Ltd., 1980.

200. Marchuk G.I., Petrov R.V. The mathematical model of the antiviral immune response, *Mathematical Modelling in Immunology and Medicine*, Eds. G.I. Marchuk, L.N. Belykh. North-Holland, Amsterdam 1983. P. 161-173.

201. Marchuk G.I., Petrov R.V. Viral damage of organ and immunophysiological defence reactions (mathematical model), *Problems and perspectives of modern immunology*, Ed. Petrov R. V. Novosibirsk: Nauka, 1988, P. 100-108. (in Russian).

202. Marchuk G.I., Petrov R.V., Romanyukha A.A., Bocharov G.A. Mathematical model of antiviral immune response. I. Data analysis, generalized picture construction and parameters evaluation for hepatitis B, *J. Theor. Biol.* 1991. 151. P. 1-40.

203. Marchuk G.I., Poljak R.Y., Zuev S.M., Kalyaev D.V. Mathematical modeling of interaction in the system virus–host (experimental influenza infection), *J. All-Union D. I. Mendeleev's Chem. Society.* 1988. P. 537-544. (in Russian).

204. Marchuk G.I., Romanyukha A.A., Bocharov G.A. Mathematical model of antiviral immune response. II. Parameters identification for acute viral hepatitis B, *J. Theor. Biol.* 1991. 151. P. 41-70.

205. Marchuk G.I., Romanyukha A.A., Bocharov G.A. Mathematical Model of acute hepatitis B infection, *Immunological and Metabolic Systems Mathematical Models and Methods of Investigations*, Eds. G.I. Marchuk, A. Werynski. Lecture Notes of the ICB Seminars. Biosystems. Warsaw: ICB, 1992. P. 9-49.

206. Marchuk G.I., Romanyukha A.A., Bocharov G.A. Mathematical models of immune response during infectious diseases, *Mathematics Applied to Biology and Medicine*, Eds. J. Demongeot, V. Capasso. Mathematica Biology Ser. Winnipeg.-Canada: Wuerz Publishing Ltd., 1993. P. 263-273.

207. Martz E., Heagy W., Grokowski S.H. The mechanism of CTL-mediated killing: monoclonal antibody analysis of the roles of killer and target-cell membrane proteins, *Immunol. Rev.* 1983. 72. P. 73-96.

208. *Mathematical Modelling in Immunology and Medicine*, Eds. G.I. Marchuk, L.N. Belykh. North-Holland, Amsterdam, 1983.

209. McCartny J.E., Kozak J.J. A dynamic model for *in vivo* virus replication, *J. Theor. Biol.* 1981. V. 90. N. 2. P. 265-281.

210. McLaren C., Grubbs G.E., Ennis E.A. Detection of cells producing surface antigen-specific antibody to influenza viruses, *J. Clin. Microbiol.* 1978. V. 8. N. 4. P. 438-444.

211. McLean A.R. Modelling T Cell memory. *J. Theor. Biol.* 1994. 170, 63-74.

212. McLean A.R., Kirkwood T.B.L. A model of HIV infection in T-helper cell clones, *J. Theor. Biol.* 1990. 147. P. 177-203.

213. McLean A.R., May R.M. Understanding the immune system: the role of mathematical models. *J. NIH Res.* 1992, 4 P. 101-103.

214. McLean A.R., Michie K. Viral burden in AIDS, *Nature* 1993. 365. 301 p.

215. Merrill S. Mathematical model of humoral immune response, *Techn. Rep. of the Univ. of Iowa*, 1976. 40 p.

216. Merrill S. A mathematical model of B-cell stimulation and humoral immune response, *Techn. Rep. of the Univ. of Iowa*, 1976. 70 p.

217. Merrill S.J. AIDS: Background and the Dynamics of the Decline of Immunocompetence , *Theor. Immun. Part Two*, SFI Studies in the Science of Complexity, Ed. A.S. Perelson. Addison. Wesley, Redwood, CA. 1988. V. III. P. 59-75.

218. Meuer S.C., Qiao L., Schirren A., Volpel H., Köhler H. Vaccination in Immunodeficiency, Symposium in Immunology, Eds. M. Eibl, C. Huber, H. Peter, U. Wahn. Berlin, Heidelberg: Springer-Verlag, 1993.

219. Michie C.A., McLean A.R., Alcock C., Beverly P.C.L. Lifespan of human lymphocyte subsets defined by CD45 isoforms. *Nature.* 1992, 360. P. 264-265.

220. Miller J.F., Mitchell C.E. Cell-to-cell interaction in the immune response, *J. Exp. Med.* 1968. N. 128. P. 801-820.

221. Minkoff M. Approaches to optimization/simulation problems, *Appl. Numer. Math.* 1987. V. 3. P. 453-466.

222. Mizel S.B., Mizel D. Purification to apparent homogeneity of murine interleukin 1., *J. Immunol.* 1981. V. 126. P. 834-837.

223. *Modern numerical methods for ordinary differential equations*, Ed. G. Hall, J.M. Watt. Oxford (U.K.) Oxford University Press, 1976.

224. Mohler R. Bilinear control structures in immunology, *Proc. of the IFIP Working Conference on Modeling and Optimization of Complex Systems.* Berlin a. o.: Springer-Verlag, 1979. P. 58-67.

225. Mohler R., Barton C. Compartmental control model of the immune process, *Proc. of the 8th IFIP Conference on Optimization Techniques.* Heidelberg: Springer-Verlag, 1978. P. 421-430.

226. Mohler R., Barton C., Hsu C. System theoretic control in immunology, Oregon State Univ., 1975. 12 p.

227. Mohler R.R., Bruni C., Gandolfi A. A systems approach to immunology, *Proc. IEEE.* 1980. 68. P. 964-990.

228. Mohler R. et al. T- and B-cell models in the immune system, *Theoretical Immunology.* Marcel Dekker, 1978. P. 421-430.

229. Molecular Evolution on Rugged Landscapes: Proteins, RNA and The Immune System., The Proceeding of the Workshop on Applied Molecular Evolution and The Maturation of The Immune Response, Eds. A.S. Perelson, S.A. Kauffman. Santa Fe Institute. Studies in the Science of Complexity, vol. IX. Addison-Wesley Publishing Company. 1991.

230. Moretta A. Frequency and surface phenotype of human T lymphocytes producing IL-2. Analysis by limiting dilution and cell cloning, *Eur. J. Immunol.* 1985. V. 15. P. 148-155.

231. Muraille E., Leo O., Kaufman M. The Role of Antigen Presentation in the Regulation of Class-Specific (TH1/TH2) Immune Responses *J. Biol. Syst.* 1995, V. 3. P. 397-408.

232. Murthy B.R., Webster R.G. Influenza Viruses, *Virology*, Ed. by B. N. Fields et al. N. Y.: Raven Press, 1985. P. 1179-1239.

233. Nee S. How populations persist, *Nature*, 1994, 367. P.123-124.

234. Nelder J.A., Mead R. A simplex method for function minimization, *The Computer Journal.* 1965. V. 7. P. 308-313.

235. Nelson G.W., Perelson A.S. A mechanism of immune escape by slow-replicating HIV strains. *J. of AIDS*, 5, 1992. P. 82-93.

236. Nelson G.W., Perelson A.S. Modeling Defective Interfering Virus Therapy for AIDS: Conditions for DIV Survival. *Preprint of the Santa Fe Institute*, 1993. P. 93-08-050.

237. Neves K.W., Tompson S. Software for the numerical solution of systems of functional differential equations with state-dependent delays. *Appl. Numer. Math.* 9, 1992. P. 385-401.

238. Neumann A.U., Weisbuch G. Window automata analysis of population dynamics in the immune system, *Bull. Math. Biol.* 54, 21-44. (1992), 54, 699-726.

239. Neumann A.U. and Weisbuch G. Dynamics and topology of immune networks, *Bull. Math. Biol.* (1992), 54, 699-726.

240. Nisevich N.I., Marchuk G.I., Zubikova I.I., Pogozhev I.B. *Mathematical modeling of viral hepatitis B.* Moscow: Nauka, 1981. 350 p. (in Russian).

241. Nossel G. *Antibodies and immunity.* Moscow: Medicine, 1973. 176p. (in Russian).

242. Nowak M.A. Variability of HIV infections. *J. Theor. Biol.* 1992, 155. P. 1-20.

243. Nowak M.A., Anderson R.M., McLean A.R., Wolfs T.F.W., Goudsmit J., May R.M. Antigenic diversity thresholds and the development of AIDS, *Science.* 1991. V. 254. N. 5034. P. 963-969.

244. Nowak U., Deuflhard P. Numerical identification of selected rate constants in large chemical reaction systems, *Appl. Numer. Math.* 1985. V. 1. P. 59-75.

245. Nowak M.A., May R.M. Coexistence and Competition in HIV Infections. *J. Theor. Biol.* 1992, 159. P. 329-342.

246. Nowak M.A., May R.M. Mathematical Biology of HIV infections: antigenic variation and diversity threshold. *Math. Biosci.* 1991, 106. P. 1-21.

247. Nowak M.A., May R.M., Anderson R.M. The evolutionary dynamics of HIV-1 quasispecies and the development of immunodeficiency diseases. AIDS 1990, 4. P. 1095-1103.

248. Nowak M.A., McLean A.R. Mathematical Models for the Pathogenesis of AIDS, *Mathematics Applied to Biology and Medicine*, Eds. J. Demongeot, V. Capasso. Mathematica Biology Ser. Winnipeg. Canada: Wuerz Publishing Ltd., 1993. P. 275-284.

249. Oberle H.J., Pesch H.J. Numerical treatment of delay differential equations by Hermite interpolation., *Numer. Math.* 1981. V. 37. P. 235-255.

250. Oppelstrup J. One-step methods with interpolation for Volterra functional differential equations, TRITA-NA-7623, Roy. Inst. Tech. Stockholm, 1976.

251. Oppelstrup J. The RKFHB 4 method for delay-differential equations, *Numerical treatment of differential equations*, Eds. R. Bulirsh, H.D. Grigorieff, J. Schroder. *Proc. of Conf. held at Oberwolfach*, 1976. Lecture Notes in Math. Berlin-Heidelberg N. Y.: Springer. 1978. V. 631. P. 133-146.

252. Pandey R.B. On a discrete approach to cell dynamics in context to immune response against HIV, *J. Phys. A. France*. 1991, 1. P. 1709-1713.

253. Parisi G. A simple model for the immune network. *Proc. Natl. Acad. Sci. USA*. 1990, 87. P. 429-433.

254. Paul C.A.H. A user guide to ARCHI – an explicit Runge–Kutta code for solving delay and neutral differential equations. NA Report 283, Mathematics Department, University of Manchester ISSN, 1995 P. 1360-1725.

255. Paul W.E. Between two centuries: Specificity and regulation in immunology, *J. Immunol.* 1987. V. 139. N. 1. P. 1-6.

256. Payne R.J.H., Nowak M.A., Blumberg B.S. Analysis of a Cellular Model to Account for the Natural History of Infection by Hepatitis B Virus and its Role in the Development of Primary Hepatocellular Carcinoma, *J. Theor. Biol.* 1992, 159. P. 215-240.

257. Percus J.K., Percus O.E., Perelson A.S. Predicting the size of the T-cell receptor and antibody combining region from consideration of efficient self-nonself discrimination. *Proc. Natl. Acad. Sci. USA*. 1993, 90, P. 1691-1695.

258. Perelson A. The IgM–IgG switch looked at from a control theoretic viewpoint, *Proc. of the 8th IFIP Conference on Optimization Techniques*. Heidelberg: Springer-Verlag, 1978. P. 431-440.

259. Perelson A.S. Towards a realistic model of the immune network, *Theoretical Immunology Part Two, SFI Studies in the Science of Complexity*, Ed. Perelson A.S. Addison-Wesley, Redwood, CA. 1988. V. III. P. 337-401.

260. Perelson A.S. Immune networks: A topological view, *Cell to Cell Signalling. From Experiments to Theoretical Models*. Academic Press Limited. 1989. P 261-272.

261. Perelson A.S. Immune Network Theory. *Immunol. Rev.* 1989, 110 P. 5-36.

262. Perelson A.S. Two theoretical problems in immunology: AIDS and Epitopes, *Complexity: Methaphors, Models and Reality*, Eds. Cowan, Pine, Metzer. Addison-Wesley, Reading, MA. 1994 P. 185-197.

263. Perelson A.S., Bell G.I. Delivery of lethal hits by cytotoxic T lymphocytes in multicellular conjugates occurs sequentially but at random times, *J. Immunol.* 1982. V. 129. P. 2796-2801.

264. Perelson A.S., Farmer J.D., Packard N.H. The immune system, adaptation and machine learning, *Physica D*. 1986. V. 22. N. 1-3. P. 187-204.

265. Perelson A.S., Macken C.A., Grimm E.A., Roos L.S., Bonavida B. Mechanism of cell-mediated cytotoxicity at the single cell level. VIII. Kinetics of target cells bound by more than one cytotoxic T lymphocyte, *J. Immunol.* 1984. V. 132. P. 2190-2198.

266. Perelson A., Mirmirani M., Oster G. Optimal strategies in immunology. I. B-cell differentiation and proliferation, *J. Math. Biol.* 1978. V. 5. P. 213-256.

267. Perelson A.S., Weisbuch G. Modeling immune reactivity in secondary lymphoid organs, *Bull. Math. Biol.* 1992. 54. P. 649-672.

268. Perelson A.S., Weisbuch G. Immunology for Physicists, *Review in Modern Physics.* 1995 (Submitted).

269. Pertsev N.V., Marchuk G.I. Mathematical models of Hemopoiesis, *Mathematical Modelling in Immunology and Medicine*, Eds. G.I. Marchuk, L.N. Belykh. North-Holland, Amsterdam 1983. P. 197-209.

270. Peter H.H., Armbraster M., Banmert E., Buscher H.P., Dräger R., Herbst E.W., Jahreis A., Kliche A., Rump J.A., Vaith P., Wolff-Vorbeck G., Schleiser M. Clinical and Immunological Approach to the Diagnosis of Immunodeficiencies., Symposium in Immunology, Eds. M. Eibl, C. Huber, H. Peter, U. Wahn. Berlin, Heidelberg: Springer-Verlag, 1993. P. 1211-1250.

271. Peterson P.K. et al. The key role of peptidoglycan in the opsonization of *Staphylococcus aureus*, *J. Clin. Invest.* 1978. V. 61. N. 3. P. 597-609.

272. Petrov R.V. *Immunology and Immunogenetics.* Moscow: *Medicine*, 1976. 336 p. (in Russian).

273. Petrov R.V. *Immunology.* Moscow: Medicine, 1983. 368 p. (in Russian).

274. Picker L.J., Siegelman M.H. Lymphoid tissues and organs, *Fundamental immunology, Third edition*, Ed. W. Paul. N.Y.: Raven Press. 1993. P. 145-197.

275. Pigarevskij V.E. Toxic hemorrhagic oedema of lung during influenza, *Clin. Medicine.* 1959. T. 37. N. 12. P. 58-63. (in Russian).

276. Pimbley G.H. Periodic solutions of predator–prey equations simulating an immune response. I, *Math. Biosci.* 1974. V. 20. P. 27-51.

277. Pimbley G.H. Periodic solutions of predator–prey equations simulating an immune response. II, *Math. Biosci.* 1974. V. 20. P. 251-277.

278. Pimbley G.H. Periodic solutions of third order predator–prey equations simulating an immune response. I, *Arch. Rat Mech. Anal.* 1974. V. 55. P. 93-123.

279. Plaa G.L., Hewitt W.R. Quantitative evaluation of indices of hepatotoxicity, *Toxicology of the Liver*, Eds. G.L. Plaa, W.R. Hewitt. N.Y.: Raven Press. 1982. P. 103-120.

280. Pogozhev I.B., Usmanov R.N., Zuev S.M. Models of the Processes in an Organism and the Population Characteristics. *Russ. J. Numer. Anal. Math. Modelling.* 1993, V. 8. P. 441-452.

281. Poljak B.T. *Introduction to optimization.* Moscow: Nauka, 1983. 384 p. (in Russian).

282. Poljak R.Y. et al. System analysis of viral damage to an organism and antiviral action of chemical drugs using a model of influenza, *Infestant strategy in host*, Eds. A.A. Totoljan, R.Y. Poljak. Leningrad, 1987, P. 10-25. (in Russian).

283. Pontryagin L.S., Boltyanski V.G., Gamkrelidze R.V., Mischenko E.V. *Mathematical Theory of Optimal Processes.* Moscow: Fizmatgiz, 1976. (In Russian).

284. Popper H., Dienstag J.L., Feinstone S.M., Alter H.J., Purcell R.H. Lessons from the pathology of viral hepatitis in chimpanzees, *Virus and the Liver*, Eds. Bianchi L. et al. England: MTP Press, 1980. P. 137-150.

285. Prikrylova D., Jilek M., Dolezal J. A model of proliferation control in immune response, *Kybernetika.* 1984. V. 20. P. 37-46.

286. Prikrylova D., Waniewskii J. Some Interesting Results of Mathematical Modelling of the Immune Response, *Mathematics Applied to Biology and Medicine*, Eds. J. Demongeot, V. Capasso. Mathematica Biology Ser. Winnipeg. Canada: Wuerz Publishing Ltd., 1993. P. 285-290.

287. Prince A.M., Brotman B., Purcell R.H., Gerin J.L. A final report on safety and immunogenicity of a bivalent aqueous subunit HBV vaccine, *J. Med. Virol.* 1985. V. 15. N. 4. P. 399-419.

288. Prince A.M., Stephan W., Brotman B. β-propiolactone/ultraviolet irradiation: A review of its effectiveness for inactivation of viruses in blood derivatives, *Rev. Infect. Dis.* 1983. V. 5. N. 1. P. 92-107.

289. Prystowsky M.B., Otten G., Pierce S.K. et al. Lymphokine production by cloned T lymphocytes, *Lymphokines.* 1985. V. 12. P. 13-38.

290. Purcell R.H., Gerin J.L. Hepatitis B vaccines: a status report, *Viral Hepatitis*, Eds. G.N. Vyas, S.N. Cohen & R. Schmidt. Philadelphia: Franclin Institute Press, 1978. P. 491-505.

291. Putov N.V. et al. *Handbook on pulmonology.* Leningrad: Medicine, 1978. (in Russian).

292. Ralston A. *A first course in numerical analysis.* N. Y.: McGraw-Hill, 1965. 573 p.

293. *Regulation of immune response dynamics*, Eds. C. De Lisi, J.R.J. Hiernaux. CRC Press Inc., 1982. V. 182.

294. Reklaitis G., Ravindran A., Ragsdell K. *Engineering Optimization.* New York: Woley, 1983.

295. Richter P. A network theory of the immune system, *Eur. J. Immunol.* 1975. V. 5. P. 350-354.

296. Robb R.J., Rusk C.M. High and low affinity for IL-2: implications of pronase phorbol ester and cell membrane studies upon the basis for differential ligand activities., *J. Immunol.* 1986. V. 137. N. 1. P. 142-149.

297. Roitt I. *Essential Immunology.* Sixth Edit. Oxford: Blackwell Scientific Publications, 1988.

298. Romanovski Yu.M., Stepanova N.V., Chernavski D.S. *Mathematical Modelling in Biophysics.* Moscow: Nauka, 1975. 343 p.

299. Romanyukha A.A., Bocharov G.A., Marchuk G.I. Mathematical modelling of infectious diseases, *Mathematical Aspects of Human Diseases*, Ed. G. Da Prato. Applied Mathematics Monographs. Pisa: Giardini Editori E Stampatori, 1992. V. 3. P. 126-148.

300. Ronot X., Adolphe M. Update of the concept of the cell cycle: the contribution of flow cytometry, *Biology of the Cell.* 1986. V. 58. N. 2. P. 113-120.

301. Rose R., Perelson A.S. (1992). Immune Networks and Immune Responses. *Preprint* 92-12-058. Santa Fe Institute. *Lecture Notes of Biomathematics*, Ed. Levin S. 1992, V. 100.

302. Rosenstreich D.L. The macrophage, *Cellular function in immunity and inflammation*, Eds. J.J. Oppenheim, D.L. Rosenstreich. London, 1981. P. 127-159.

303. Roth M.G. Difference methods for stiff delay differential equations, *Dept. Comp. Sci., Rpt* UIUCDCS-R-8-1-12 Ph. D. Thesis: Univ. Illinois, 1980.

304. Sage A.R., Melsa J.L. *System Identification.* New York: Academic Press. 1971.

305. Segel L.A., Jaeger E. Reverse Engineering: A Model for T-Cell Activation, *Bull. Math. Biol.* 1984, 56. P. 687-722.

306. Segel L.A., Jaeger E., Elias D., Cohen I.R. A Quantitative Model of Autoimmune Disease and T Cell Vaccination: Why More Cells May Produce Less Effect, *Immunol. Today.* 1995.

307. Segel L.A., Perelson A.S. Exploiting the diversity of time scales in the immune system: a B cell antibody model. *Preprint of the Los Alamos National Laboratory.* LA-UR-91-196.

308. Segel L.A., Perelson A.S. Shape space analysis of immune networks, Cell to Cell Signalling. *From Experiments to Theoretical Models.* Academic Press Limited. 1989. P. 273-283.

309. Segel L.A., Perelson A.S. Shape-space: an approach to the evaluation of cross-reactivity effects, stability and controllability in the immune system, *Immunol. Lett.* 1989. 22. P. 91-134.

310. Semyonov B.F., Kaulen D.R., Balandin N.G. Cellular and Molecular Bases of Antiviral Immunity. Moscow, Medizina, 1982. 240 p. (In Russian).

311. Shevach E.M. Macrophages and other accessory cells, *Fundamental Immunology*, Ed. W.E. Paul. N. Y.: Raven Press, 1984. P. 71-107.

312. Sidorov I.A., Romanyukha A.A. Mathematical model of T cell proliferation *Mathematical Biosciences.* 1993, V. 115. N. 2. P. 187-232.

313. Sieburg H.B., McCutchan J.A., Clay O.K., Cabalerro L., Ostlund J. Simulation of HIV infection in artificial immune systems. *Physica D* 1990, 45. P. 208-227.

314. Skeel R. Analysis of fixed-stepsize methods, *SIAM J. on Numer. Anal.* 1976. V. 13. N. 5. P. 664-685.

315. Skeel R.D., Jackson L.W. Consistency of Nordsieck methods, *SIAM J. on Numer. Anal.* 1977. V. 14. N. 5. P. 910-924.

316. Skeel R.D., Jackson L.W. The stability of variable-stepsize Nordsieck methods, *SIAM J. on Numer. Anal.* 1983. V. 20. N. 4. P. 840-853.

317. Smirnova O.A. Mathematical model of immune reaction, *Vestn. Mosc. Univ. Ser. physica, astronomija*, 1975. N. 4. P. 485-486. (in Russian).

318. Smorodintsev A.A. *Influenza and its prophylaxis.* Leningrad: Medicine, 1984. 383 p. (in Russian).

319. Srinivas N.C., Pattabhi Ramacharyulu N.Ch. On the Model of a Virus Disease with Mutual Interference, *J. of Mathem. Anal. and Applicat.* 1992, 170. P. 276-290.

320. Stetter J. *Analysis of difference methods for ordinary differential equations.* Moscow: Mir, 1978. 464 p. (in Russian).

321. *Stiff Computation*, Ed. R.C. Aiken. Oxford University Press, 1985. 462 p.

322. Stewart J., Varela F.J. Dynamics of class of immune networks. II. Oscillatory activity of cellular and humoral components. *J. Theor. Biol.* 1990, 144. P. 103-115.

323. Stewart J., Varela F.J., Coutinho A. The relationship Between Connectivity and Tolerance as Revealed by Computer Simulation of the Immune Network: Some Lessons for an Understanding of Autoimmunity, *J. of Autoimmun.* 1989, 2 (Supplement). P. 15-23.

324. Swinton J., Schweitzer A.N., Anderson R.M. Two signal activation as an example of high zone tolerance: a mathematical exploration of the nature of the second signal. *J. Theor. Biol.* 1994, 169 P. 23-30.

325. Tabor E., Purcell R.H., London W.T., Gerety R.J. Use of and interpretation of results using inocula of hepatitis B virus with known infectivity titres, *J. Infect. Dis.* 1983. V. 147. N. 3. P. 531-534.

326. Takacs L., Osawa H., Toro I., Diamantstein T. Immunohistochemical localization of cells reacting with monoclonal antibodies directed against the IL-2 receptor of murine, rat and human origin, *Exp. Immunol.* 1985. V. 59. P. 37-44.

327. Tavernini L. The approximate solution of Volterra differential systems with state-dependent time lags, *SIAM J. on Numer. Anal.* 1978. V. 15. N. 5. P. 1039-1051.

328. *The Principles of Biology*, Ed. Libbert -Moscow: Mir, 1982. 437 p. (in Russian).

329. *The Respiratory System. Sec. 3*, Ed. Fishman. V. 3, Ed. P. T. Macklem.

330. *Theoretical and Experimental Insights into Immunology*, Eds. A.S. Perelson, G. Weisbuch. NATO ASI Series H. Heidelberg: Springer-Verlag. 1992. V. 66.

331. *Theoretical Immunology*, Eds. G. Bell, A. Perelson, G. Pimbley. Marcel Dekker, 1978. 660 p.

332. *Theoretical Immunology*, Ed. A. Perelson. Addison-Wesley, 1988. V. II, III.

333. Tikhonov A.I., Vasil'ev A.B., Sveschnikov A.G. *Differential equations.* Moscow: Nauka, 1986 (in Russian).

334. *Trends and Progress in System Identification*, Ed. P. Eykhoff. Oxford (U.K.): Pergamon Press, 1981.

335. Usmanov R.N., Zuev S.M. Parameterization in Mathematical Models of Diseases. *Rus. J. Numer. Anal. Math. Modelling.* 1993, V. 8. P. 275-284.

336. Varah J.M. A spline least squares method for numerical parameter estimation in differential equations, *SIAM J. Sci. Stat. Comput.* 1982. V. 3. N. 1. P. 28-46.

337. Varela F.J., and Stewart J. Dynamics of class of immune networks. I. Global behavior *J. Theor. Biol.* 1990, 144. P. 93-101.

338. Ventzel A.D., Freidlin O.O. *Fluctuations in dynamical system under the action of small random deviations.* Moscow: Nauka, 1979. 424 p. (in Russian).

339. Verbrugh H.A. et al. Staphylococcus aureus opsonization mediated via the classical and alternative pathways, *Immunology.* 1979. V. 36. N. 3. P. 391-397.

340. Vladimirov V.S. *Equations of mathematical physics.* Moscow: Nauka, 1988. 512 p (in Russian).

341. Waltman P. et al. A threshold model of antigen-antibody dynamics, *J. Theor. Biol.* 1977. V. 65. P. 499-512.

342. Watanabe D.S., Roth M.G. The stability of differential formulas for delay differential equations, *SIAM J. Numer. Anal.* 1985. V. 22. N. 1. P. 132-145.

343. Weisbuch G. A shape space approach to the dynamics of the immune system. *J. theor. Biol.* 1990, 143. 507-522.

344. Weisbuch G., De Boer R.J., Perelson A.S. Localized Memories in idiotypic networks. *J. Theor. Biol.* 1990, 146. P. 83-499.

345. Weisbuch G., Perelson A.S. Affinity Maturation and Learning in Immune Networks. In: *Molecular Evolution on Rugged Landscapes, SFI Studies in the Science of Complexity.* V. IX, Eds. A. Perelson, S. Kauffman. Addison-Wesley. 1991, P. 189-205.

346. Weissman I.L., Hood L.E., Wood W.V. *Essential Concepts in Immunology.* The Benjamin/Cummings Publ. 1978.

347. Wille D.R., Baker C.T.H. DELSOL - a numerical code for the solution of systems of delay differential equations. *Appl. Numer. Math.* 9, 1992. P. 223-234.

348. Zennaro M. Delay differential equations: theory and Numerics. In: Proceedings of the SERC/EPSRC Summer School (UK). 1994. 32 p.

349. Zhdanov V.M., Gaydamovich S.Ya. *Virology.* Moscow: Medicine, 1982, V. 1. 492 p. (in Russian).

350. Zhdanov V.M., Gaydamovich S.Ya. *Virology.* Moscow: Medicine, 1982, V. 2. 514 p. (in Russian).

351. Zinkernagel R.M. H-2 compatibility requirement for virus specific T-cell mediated cytolysis, *J. Exp. Med.* 1976. V. 143. P. 437-443.

352. Zinkernagel R.M. Immunity to viruses, *Fundamental immunology, Third edition*, Ed. W. Paul. N.Y.: Raven Press. 1993. P. 1211-1250.

353. Zinkernagel R.M., Hergartner H., Stitz L. On the role of viruses in the evolution of immune responses, *Brit. Med. Bull.* 1985. V. 41. P. 92-97.

354. Zinserling A.V. *Aetiology and pathological anatomy of acute respiratory infections.* Moscow: Medicina, 1977. 160 p. (in Russian).

355. Zuev S.M. Statistical estimation of immune response mathematical model coefficients, *Mathematical modeling in immunology and medicine. Proceedings of the IFIP TC 7 working conference on Mathematical modelling in immunology and Medicine in Moscow*, 1982. North-Holland Publishers Co. Amsterdam–New York–Oxford, 1982.

356. Zuev S.M. *Statistical estimation of parameters for mathematical models of diseases.* Moscow: Nauka, 1988. 192 p. (in Russian).

357. Zukhbaya T.M., Smirnova O.A. An experimental and mathematical analysis of lymphopoiesis dynamics under continuous irradiation, *Health Physics.* 1991, 61. P. 87-95.

358. Zverkina T.S. On the choice of integration method for equations with deviating argument, *Proc. of seminar: Theory of differential equations with deviating argument.* V. 11. 1969. P. 75-81. (in Russian).

Index

Other *Mathematics and Its Applications* titles of interest:

W.L. Miranker: *Numerical Methods for Stiff Equations and Singular Perturbation Problems*. 1980, 220 pp. ISBN 90-277-1107-0

K. Rektorys: *The Method of Discretization in Time and Partial Differential Equations*. 1982, 470 pp. ISBN 90-277-1342-1

L. Ixary: *Numerical Methods and Differential Equations and Applications*. 1984, 360 pp. ISBN 90-277-1597-1

B.S. Razumikhin: *Physical Models and Equilibrium Methods in Programming and Economics*. 1984, 368 pp. ISBN 90-277-1644-7

A. Marciniak: *Numerical Solutions of the N-Body Problem*. 1985, 256 pp. ISBN 90-277-2058-4

Y. Cherruault: *Mathematical Modelling in Biomedicine*. 1986, 276 pp. ISBN 90-277-2149-1

C. Cuvelier, A. Segal and A.A. van Steenhoven: *Finite Element Methods and Navier-Stokes Equations*. 1986, 500 pp. ISBN 90-277-2148-3

A. Cuyt (ed.): *Nonlinear Numerical Methods and Rational Approximation*. 1988, 480 pp. ISBN 90-277-2669-8

L. Keviczky, M. Hilger and J. Kolostori: *Mathematics and Control Engineering of Grinding Technology. Ball Mill Grinding*. 1989, 188 pp. ISBN 0-7923-0051-3

N. Bakhvalov and G. Panasenko: *Homogenisation: Averaging Processes in Periodic Media. Mathematical Problems in the Mechanics of Composite Materials*. 1989, 404 pp. ISBN 0-7923-0049-1

R. Spigler (ed.): *Applied and Industrial Mathematics*. Venice-1, 1989. 1991, 388 pp. ISBN 0-7923-0521-3

C.A. Marinov and P. Neittaanmaki: *Mathematical Models in Electrical Circuits. Theory and Applications*. 1991, 160 pp. ISBN 0-7923-1155-8

Z. Zlatev: *Iterative Improvement of Direct Solutions of Large and Sparse Problems*. 1991, 328 pp. ISBN 0-7923-1154-X

M. Vajtersic: *Algorithms for Elliptic Problems*. 1992, 352 pp. ISBN 0-7923-1918-4

V. Kolmanovskii and A. Myshkis: *Applied Theory of Functional Differential Equations*. 1992, 223 pp. ISBN 0-7923-2013-1

A.D. Egorov, P.I. Sobolevsky and L.A. Yanovich: *Functional Integrals: Approximate Evaluation and Applications*. 1993, 426 pp. ISBN 0-7923-2193-6

G.I. Marchuk: *Adjoint Equations and Analysis of Complex Systems*. 1994 ISBN 0-7923-3013-7

Other *Mathematics and Its Applications* titles of interest:

A. Bakushinsky and A. Goncharsky: *Ill-Posed Problems: Theory and Applications*. 1994, 256 pp. ISBN 0-7923-3073-0

G.N. Milstein: *Numerical Integration of Stochastic Differential Equations*. 1995, 170 pp. ISBN 0-7923-3213-X

V.V. Shaidurov: *Multigrid Methods for Finite Elements*. 1995, 331 pp. ISBN 0-7923-3290-3

A.N. Tikhonov, A.V. Goncharsky, V.V. Stepanov and A.G. Yagola: *Numerical Methods for the Solution of Ill-Posed Problems*. 1995, 264 pp. ISBN 0-7923-3583-X

H.W. Engl, M. Hanke and A. Neubauer: *Regularization of Inverse Problems*. 1996, 328 pp. ISBN 0-7923-4157-0

G.I. Marchuk: *Mathematical Modelling of Immune Response in Infectious Diseases*. 1997, 358 pp. ISBN 0-7923-4528-2

MIX
Papier aus verantwortungsvollen Quellen
Paper from responsible sources
FSC® C105338

If you have any concerns about our products,
you can contact us on
ProductSafety@springernature.com

In case Publisher is established outside the EU,
the EU authorized representative is:
Springer Nature Customer Service Center GmbH
Europaplatz 3, 69115 Heidelberg, Germany

Printed by Libri Plureos GmbH
in Hamburg, Germany